AF595551

Vegetable and Fruit Production

Vegetable and Fruit Production

Editors

Lord Abbey
Josephine Ampofo
Mason MacDonald

Basel • Beijing • Wuhan • Barcelona • Belgrade • Novi Sad • Cluj • Manchester

Editors
Lord Abbey
Dalhousie University
Truro, NS, Canada

Josephine Ampofo
McGill University
Montreal, QC, Canada

Mason MacDonald
Dalhousie University
Truro, NS, Canada

Editorial Office
MDPI
St. Alban-Anlage 66
4052 Basel, Switzerland

This is a reprint of articles from the Special Issue published online in the open access journal *Plants* (ISSN 2223-7747) (available at: https://www.mdpi.com/journal/plants/special_issues/Vegetable_Fruit).

For citation purposes, cite each article independently as indicated on the article page online and as indicated below:

Lastname, A.A.; Lastname, B.B. Article Title. *Journal Name* **Year**, *Volume Number*, Page Range.

ISBN 978-3-0365-9471-2 (Hbk)
ISBN 978-3-0365-9470-5 (PDF)
doi.org/10.3390/books978-3-0365-9470-5

Contents

About the Editors

Lord Abbey

Dr. Lord Abbey holds a BSc (Ghana), MSc (UK), and PhD (UK) in Horticulture and Crop Science, with a research focus on sustainable horticultural production systems for human health and well-being. Dr. Abbey has won various scholarships and awards, and is currently an Associate Professor (Amenity Horticulture) at Dalhousie University Faculty of Agriculture, where he teaches and supervises undergraduate and graduate students. He has received several industry, federal, and provincial government grants for his research work on clean technology and climate-smart agriculture. He has published over 96 research papers and book chapters (2023). He has worked extensively on soil nutrient management approaches for various horticultural crops and the development of biostimulants/biofertilizer derived from plant biomass and marine waste.

Josephine Ampofo

Dr. Josephine Ampofo has a background in Food Science and Technology. Her research focus is on non-thermal bioprocesses intended for the sustainable production of novel consumer-centric functional foods and ingredients. Dr. Josephine Ampofo completed her BSc (Hons) in Biological Sciences and MSc in Food Science and Technology at Kwame Nkrumah University of Science and Technology, Ghana, after which she pursued her PhD in Food Science and Agricultural Chemistry at McGill University, Canada. She is currently a Research Associate at McGill University, Department of Bioresource Engineering, where she performs research, writes grants, and mentors students. Her current research activities include the study of less harsh physical bioprocesses for the extraction and recovery of novel compounds from underutilized food crops as bio-functional food ingredients.

Mason MacDonald

Dr. Mason MacDonald is an Assistant Professor in Plant Ecophysiology at the Department of Plant, Food, and Environmental Sciences within the Faculty of Agriculture at Dalhousie University. Mason holds an MSc in Agriculture and a PhD in Plant Biology, both with an emphasis on environmental stress. As such, Mason heads a research lab on Environmental Stress and Plant Ecophysiological Response (ESPER). His research highlights include mapping out the physiological pathways of needle abscission in balsam fir, the development of integrated controlled environments to increase postharvest shelf-life, and the co-invention of foliar sprays to delay abscission. Mason's current work is focusing on the development of new seed preconditioning technologies and enhancement via nano-delivery mechanisms.

Editorial

Fruit and Vegetable Production

Lord Abbey [1,*], Mason MacDonald [1,*] and Josephine Ampofo [2]

[1] Department of Plant, Food, and Environmental Sciences, Faculty of Agriculture, Dalhousie University, Halifax, NS B2N 5E3, Canada
[2] Department of Food Science and Technology, University of California, Davis, CA 95616, USA; josephine.ampofo@mail.mcgill.ca
* Correspondence: labbey@dal.ca (L.A.); mason.macdonald@dal.ca (M.M.)

1. Background

Fruits and vegetables are generally known to contain important vitamins, fiber, essential minerals, and vital bioactive compounds that possess health benefits such as anti-inflammatory, antimicrobial, antioxidant, and anticancer properties [1]. These compounds can help protect against chronic diseases such as cancer, diabetes, and cardiovascular diseases. However, the composition of fruits and vegetables with regard to these compounds is dependent on plant genotypic, environmental, and management factors [2–4]. By nature, the production of fruits and vegetables requires high input and intensive management practices compared to other crops. As such, these crops are fragile and easily succumb to various environmental stressors during production and post-harvest processing, handling, and storage [5,6]. Consequently, collaboration between industry, researchers, and policymakers to develop or adopt novel technologies and techniques to improve fruit and vegetable crop growth, productivity, and harvest and edible qualities is paramount.

This Special Issue (Fruit and Vegetable Production) of the journal *Plants* focuses on the entire chain of fruit and vegetable production including post-harvest and marketing topics under field and greenhouse production systems. Therefore, it is not surprising that the information provided by this Special Issue will further strengthen the theory that effective collaboration between researchers and stakeholders across various food systems jurisdictions will continue to grow within these areas of research. This Special Issue comprises a collection of 12 peer-reviewed manuscripts covering basic and applied themes of crop response to organic amendments, biostimulants, irrigation, physiological stress, metabolic stimulation, and novel technologies for the efficient characterization of specialty crops, quality assessment, and cultivar evaluation.

Citation: Abbey, L.; MacDonald, M.; Ampofo, J. Fruit and Vegetable Production. *Plants* **2023**, *12*, 3125. https://doi.org/10.3390/plants12173125

Received: 4 August 2023
Revised: 22 August 2021
Accepted: 22 August 2023
Published: 30 August 2023

2. The Main Findings of This Special Issue

2.1. Spatial and Compositional Variations in the Fruit Characteristics of Papaya (Carica papaya cv. Tainung No. 2) during Ripening

Chung et al. (2023) investigated selected compositional changes of papaya during ripening, with particular emphasis on nutrition and metabolite distributions between the calyx and stem end of the fruit [7]. From their findings, total carbohydrates, total protein, nitrogen, and potassium showed maximum accumulation at the stem end, with fructose, glucose, magnesium, and manganese accumulating at the calyx end [7].

2.2. Magnetically Treated Water in Phaseolus vulgaris L.: An Alternative Method to Develop Organic Farming in Cuba

"Magnetic biostimulation" or "magnetopriming" involves exposing water used for priming or irrigation to a magnetic field, with this technological approach proven to be effective with several growth-stimulating effects during germination and plant development [8]. Boix et al. (2023) investigated the effect of this technology on the potential of common beans in organic farming, with their work reporting the ability of magnetically treated water to increase germination, stem length, vigor, leaf area, seed weight, and chlorophyll and carbohydrate content [9].

2.3. The Growth and Biochemical Composition of Microgreens Grown in Different Formulated Soilless Media

Saleh et al. (2022) evaluated the effectiveness of four different growing media on four different plant species: namely, kale (*Brassica oleracea* L. var. acephala), Swiss chard (*Beta vulgaris* var. cicla), arugula (*Eruca vesicaria* ssp. sativa), and pak choi (*Brassica rapa* var. chinensis) grown as microgreens. Two media were identified as having the most beneficial effect on microgreens: (1) 30% vermicast + 30% sawdust + 10% perlite + 30% PittMoss and (2) 30% vermicast + 20% sawdust + 20% perlite + 30% mushroom compost [10]. Each of the above growing media consistently increased the content of chlorophyll, carotenoids, phenolics, and flavonoids and increased antioxidant enzyme activity. The second of the above treatments also significantly increased yields [10].

2.4. Irrigation Effect on the Yield, Skin Blemishes, Phellem Formation, and Total Phenolics of Red Potatoes

Jiang et al. (2022) investigated the benefits of irrigation on Dark Red Norland potatoes known to be drought sensitive. The use of irrigation improved yields by 20.6% and reduced surface cracking by 48.5% [11]. Microscopy imaging analysis demonstrated that tubers from the rainfed trials formed higher numbers of suberized cell layers than those of the irrigated potatoes. In addition, surface cracking and silver patch skins had more suberized cell layers than the normal skins. The irrigated potatoes also had a significantly higher concentration of total polyphenols [11].

2.5. Changes in Soil Characteristics, Microbial Metabolic Pathways, TCA Cycle Metabolites, and Crop Productivity following the Frequent Application of Municipal Solid Waste Compost

The use of municipal solid waste compost is well established, but Abbey et al. (2022) investigated the effect of long-term changes in soil microbial composition and subsequent translation on plant metabolic pathways. Both annual and biennial applications of municipal compost increased yields, with annual application being the most effective technique [12]. Furthermore, the same effect was found with increased microbial function and tricarboxylic metabolites, thus suggesting the potential application of municipal compost as a novel long-term green solution for improving soil health [12].

2.6. Changes in the Biogenic Amines of Two Table Grapes (cv. Bronx Seedless and Italia) during Berry Development and Ripening

Incesu et al. (2022) studied the accumulation of biogenic amines in two varieties of grape berries during ripening. Although a difference in biogenic amine concentration between the varieties was observed, there was a consistent increase in concentration throughout the ripening process [13]. The concentration of most of the biogenic amines analyzed increased linearly from the beginning of berry touch to when the berries ripened during the harvesting stage. The most common biogenic amine in the grape varieties was putrescine, followed by histamine, agmatine, and tyramine.

2.7. The Field Application of a Vis/NIR Hyperspectral Imaging System for the Non-Destructive Evaluation of the Physicochemical Properties of 'Madoka' Peaches

One major challenge in assessing the physiological and biochemical characteristics of fruits and vegetables is the destructive approaches available for quality assessment. To help solve this problem, Jang et al. (2022) evaluated a vis/NIR hyperspectral imaging system as a non-destructive quality assessment method in peaches [14]. The coefficient of determination was higher than 0.8 for chromaticity and soluble solids content, although it was slightly lower (0.6 to 0.7) for firmness and titratable acidity. Their study demonstrates the potential of hyperspectral imaging as a method for the non-destructive estimation of specific quality characteristics, and as such, it would be interesting to investigate its use for other parameters.

2.8. Cultivar and Post-harvest Storage Duration Influence Fruit Quality and the Nutritional and Phytochemical Profiles of Soilless-Grown Cantaloupe and Honeydew Melons

Pulela et al. (2022) contrasted differences between honeydew melons and cantaloupes grown using Hygromix® growing media covered in a layer of vermiculite. The accumulated concentrations of vitamin C, β-carotene, zinc, phosphorous, potassium, magnesium, and calcium were all higher in the cantaloupes studied [15]. In general, total soluble solids, color intensity, pH, and flavonoids all increased during storage while vitamin C, flavonoids, and phenolics all decreased [15].

2.9. Maturation and the Post-Harvest Resting of Fruits Affect the Macronutrient and Protein Content of Sweet Pepper Seeds

Colombari et al. (2022) harvested sweet pepper fruits prior to physiological maturity in order to investigate the seed changes that occur during post-harvest resting (i.e., the time period between harvest and seed collection). The key findings of the study were that potassium, calcium, and magnesium decreased during fruit maturation while seed albumin, globulin, prolamin, and glutelin all increased [16].

2.10. The Effect of Pyroligneous Acid on the Productivity and Nutritional Quality of Greenhouse Tomato

Pyroligneous acid is a liquid obtained from the condensation of smoke from biodiesel or biochar formation and has been shown to be a potent biostimulant. Ofoe et al. (2022) applied concentrations between 0 and 2% as a root soak to tomato plants, which increased the number and mass of the tomato fruits [17]. The concentrations of several bioactive compounds such as total phenolics, flavonoids, and ascorbate were also improved by pyroligneous acid, as well as total dissolved solids [17]. It was shown through this study that pyroligneous acid could play a role in greenhouse tomato production.

2.11. Plant Growth-Promoting Rhizobacteria Improve the Growth and Fruit Quality of Cucumber under Greenhouse Conditions

Zapata-Sifuentes et al. (2022) studied the effect of three rhizobacteria species on cucumber plant growth in the hopes of reducing dependency on chemical fertilizer inputs. All three investigated species demonstrated improved cucumber growth parameters such as plant height, fruit length, and yield [18]. The researchers also observed significant accumulated levels of phenolics, flavonoids, and vitamin C and improved antioxidant capacity [18]. From the results of their study, it was found that rhizobacteria have the potential to improve both the quantity and quality of cucumbers.

2.12. Ascorbic Acid's Preconditioning Effect on Broccoli Seedling Growth and Photosynthesis under Drought Stress

MacDonald et al. (2022) explored the use of ascorbic acid as a seed priming agent in broccoli following previous success in other species. Broccoli was not grown to harvest in this experiment, but there were increases in critical parameters such as plant growth, photosynthesis, water use efficiency, leaf area, and membrane protection [19]. Perhaps more interesting is that these same critical parameters also increased during drought stress, pointing to ascorbic acid having a role in increased plant growth and drought stress mitigation.

Author Contributions: All authors contributed to the conception and design of this editorial for the Special Issue. M.M. and L.A. prepared the first draft of the article and J.A. contributed to the revision of the first draft. All authors contributed to the revision of the draft and approved the final version of the editorial. All authors have read and agreed to the published version of the manuscript.

Conflicts of Interest: The authors declare no conflict of interest.

References

1. Yahia, E.M.; Garcia-Solis, P.; Celis, M.E.M. Chapter 2: Contribution of fruits and vegetables to human nutrition and health. In *Postharvest Physiology and Chemistry of Fruits and Vegetables*, 1st ed.; Yahia, E.M., Ed.; Woodhead Publishing: Sawston, UK, 2017; pp. 19–45. [CrossRef]
2. Nezmer, B.; Al-Taher, F.; Abshiru, N. Phytochemical composition and nutritional value of different plant parts in two cultivated and wild purslane (*Portulaca oleracea* L.) genotypes. *Food Chem.* **2020**, *320*, 126621. [CrossRef]
3. Li, Y.; Kong, D.; Fu, F.; Sussman, M.R.; Wu, H. The effect of developmental and environmental factors on secondary metabolites in medicinal plants. *Plant Physiol. Biochem.* **2020**, *148*, 80–89. [CrossRef]
4. Koch, M.; Naumann, M.; Pawelzik, E.; Gransee, A.; Thiel, H. The importance of nutrient management for potato production Part I: Plant nutrition and yield. *Potato Res.* **2020**, *63*, 97–119. [CrossRef]
5. Cohen, I.; Zandalinas, S.I.; Huck, C.; Fritschi, F.B.; Mittler, R. Meta-analysis of drought and heat stress combination impact on crop yield and yield components. *Physiol. Plant.* **2021**, *171*, 66–76. [CrossRef] [PubMed]
6. Elik, A.; Yanik, D.K.; Istanbullu, Y.; Guzelsoy, N.A.; Yavuz, A.; Gogus, F. Strategies to reduce post-harvest losses for fruits and vegetables. *Int. J. Sci. Technol. Res.* **2019**, *5*, 29–39. [CrossRef]
7. Chung, S.W.; Jang, Y.J.; Kim, S.; Kim, S.C. Spatial and compositional variations in fruit charcateristics of papaya (*Carica papaya* cv. Tainung No. 2) during ripening. *Plants* **2023**, *12*, 1465. [CrossRef] [PubMed]
8. Sarraf, M.; Kataria, S.; Taimourya, H.; Santos, L.O.; Menegatti, R.D.; Jain, M.; Ihtisham, M.; Liu, S. Magnetic field (MF) applications in plants: An overview. *Plants* **2020**, *9*, 1139. [CrossRef] [PubMed]
9. Boix, Y.F.; Dubois, A.F.; Quintero, Y.P.; Aleman, E.I.; Victorio, C.P.; Aguilera, J.G.; Betancourt, M.N.; Morales-Aranibar, L. Magnetically treated water in *Phaseolus vulgaris* L.: An alternative to develop organic farming in Cuba. *Plants* **2023**, *12*, 340. [CrossRef] [PubMed]
10. Saleh, R.; Gunupuru, L.R.; Lada, R.; Nams, V.; Thomas, R.H.; Abbey, L. Growth and biochemical composition of microgreens grown in different formulated soilless media. *Plants* **2022**, *11*, 3546. [CrossRef] [PubMed]
11. Jiang, M.; Shinners-Carnelley, T.; Gibson, D.; Jones, D.; Joshi, J.; Wang-Pruski, G. Irrigation effect on yield, skin blemishes, phellem formation, and total phenolics of red potatoes. *Plants* **2022**, *11*, 3523. [CrossRef] [PubMed]
12. Abbey, L.; Yurgel, S.N.; Asunni, O.A.; Ofoe, R.; Ampofo, J.; Gunupuru, L.R.; Ajeethan, N. Changes in soil characteristics, microbial metabolic pathways, TCA cycle metabolites and crop productivity following frequent application of municipal solid waste compost. *Plants* **2022**, *11*, 3153. [CrossRef] [PubMed]
13. Incesu, M.; Karakus, S.; Hajizadeh, H.S.; Ates, F.; Turan, M.; Skalicky, M.; Kaya, O. Changes in biogenic amines of two table grapes (cv. Bronx seedless and Italia) during berry development and ripening. *Plants* **2022**, *11*, 2845. [CrossRef] [PubMed]
14. Jang, K.E.; Kim, G.; Shin, M.H.; Cho, J.G.; Jeong, J.H.; Lee, S.K.; Kang, D.; Kim, J.G. Field application of a vis/NIR hyperspectral imaging system for nondestructive evaluation of physiochemical properties in 'Madoka' peaches. *Plants* **2022**, *11*, 2327. [CrossRef] [PubMed]
15. Pulela, B.L.; Maboko, M.M.; Soundy, P.; Amoo, S.O. Cultivar and postharvest storage duration influence fruit quality, nutritional and phytochemical profiles of soilless-grown cantaloupe and honeydew melons. *Plants* **2022**, *11*, 2136. [CrossRef] [PubMed]
16. Colombari, L.F.; Chamma, L.; da Silva, G.F.; Zanetti, W.A.L.; Putti, F.F.; Cardoso, A.I.I. Maturation and post-harvest resting of fruits affects the macronutrients and protein content in sweet pepper seeds. *Plants* **2022**, *11*, 2084. [CrossRef] [PubMed]
17. Ofoe, R.; Qin, D.; Gunupuru, L.R.; Thomas, R.H.; Abbey, L. Effect of pyroligneous acid on the productivity and nutritional quality of greenhouse tomato. *Plants* **2022**, *11*, 1650. [CrossRef] [PubMed]
18. Zapata-Sifuentes, G.; Hernandez-Montiel, L.G.; Saenz-Mata, J.; Fortis-Hernandez, M.; Blanco-Contreras, E.; Chiquito-Contreras, R.G.; Preciado-Rangel, P. Plant growth-promoting rhizobacteria improve growth and fruit quality of cucumber under greenhouse conditions. *Plants* **2022**, *11*, 1612. [CrossRef] [PubMed]
19. MacDonald, M.T.; Kannan, R.; Jayaseelan, R. Ascorbic acid preconditioning effect on broccoli seedling growth and photosynthesis under drought stress. *Plants* **2022**, *11*, 1324. [CrossRef] [PubMed]

Article

Spatial and Compositional Variations in Fruit Characteristics of Papaya (*Carica papaya* cv. Tainung No. 2) during Ripening

Sun Woo Chung, Yeon Jin Jang, Seolah Kim and Seong Cheol Kim *

Research Institute of Climate Change and Agriculture, National Institute of Horticultural and Herbal Science, Jeju 63240, Republic of Korea; jsw599@korea.kr (S.W.C.)
* Correspondence: kimsec@korea.kr; Tel.: +82-647412560

Abstract: Papaya fruit (*Carica papaya*) has different degrees of ripening within each fruit, affecting its commercial market value. The fruit characteristics of "Tainung No. 2" Red papaya were investigated at the stem-end, middle, and calyx-end across 3 ripening stages and categorized based on fruit skin coloration: unripe at 16 weeks after anthesis (WAA), half-ripe at 18 WAA, and full-ripe at 20 WAA. The fruits maintained an elliptical shape during ripening with a ratio of 2.36 of the length to the width. The peel and pulp color changed from green to white to yellow during ripening, regardless of the three parts. In the pulp, soluble solid contents increased, and firmness decreased during ripening but did not differ among the three parts. Individual nutrient contents, including metabolites and minerals, changed dynamically between the ripening stages and fruit parts. Total carbohydrates and proteins, N, and K, were accumulated more at the stem-end during ripening; meanwhile, fructose, glucose, Mg, and Mn were accumulated more at the calyx-end. In the principal component analysis, ripening stages and fruit parts were distinctly determined by the first and second principal components, respectively. Understanding the nutrient and metabolite dynamics during ripening and their distribution within the fruit can help optimize cultivation practices, enhance fruit quality, and ultimately benefit both growers and consumers.

Keywords: *Carica papaya*; cultivation practice; fruit development; fruit tree; greenhouse; mineral distribution; papaya fruit; skin coloration; nutrient; ripening stages; tropical

Citation: Chung, S.W.; Jang, Y.J.; Kim, S.; Kim, S.C. Spatial and Compositional Variations in Fruit Characteristics of Papaya (*Carica papaya* cv. Tainung No. 2) during Ripening. *Plants* **2023**, *12*, 1465. https://doi.org/10.3390/plants12071465

Academic Editors: Lord Abbey, Josephine Ampofo and Mason MacDonald

Received: 2 March 2023
Revised: 24 March 2023
Accepted: 26 March 2023
Published: 27 March 2023

1. Introduction

Papaya (*Carica papaya*) is a popular fruit crop known for its sweet and distinct flavor. The global production of papaya has continued to grow over the years, with an estimated production of 14.10 million metric tons in 2021, making it the fourth most widely cultivated tropical fruit behind banana, mango, and pineapple [1]. Papaya originated from the lowland of Mexico to Panama and is found primarily in tropical regions between 23° N and S latitudes because the cultivation requires high temperatures year-round [2]. Subtropical regions, including Mediterranean countries, can also cultivate papaya; some of these regions have to use protected facilities, including a greenhouse, because low temperatures during winter affect fruit set, growth, and production [3–5]. Recently, due to the increasing demand for papaya and the changing climate, the cultivation has expanded to temperate regions such as Far East Asia [6–8].

Papaya fruits have nutraceutical values, which render them important in human diets. These papaya parts are rich in macro- and micronutrients to varying degrees [9]. Papaya fruit is abundant in bioactive compounds, including phenolics, carotenoids, saponins/triterpenoids, and ascorbic acid, which possess various medicinal properties [2]. Papaya fruits exhibit cell protection against oxidative stress, which causes or progresses several diseases, such as cancer, metabolic disorders, and cardiovascular diseases [9]. In addition, papaya fruit displays anti-inflammatory properties by regulating signaling pathways such as NF-κB and MAPKs. NF-κB, a transcription factor, plays a role in the production of various pro-inflammatory mediators such as iNOS, COX-2, and cytokines [7]. The buildup of excessive

NO and increased COX-2 expression are linked to human diseases such as cancer and inflammation. Research on the bioactive components of papaya fruit is needed to fully understand the extent of their benefits for human health.

Papaya fruit quality is determined by various characteristics that are required from the market. Papaya fruits have different shapes as sex forms, including males, females, and hermaphrodites. Markets prefer the hermaphrodite fruit shape, which tends to be pear-shaped or elongated [2]. The nutritional value of papaya fruit depends on the fruit ripening stages. Papaya fruit ripening is accompanied by physiochemical changes, including sugar metabolism, peel color changes, and pulp softening [10]. The more ripened fruit shows different physicochemical characteristics, such as moisture, titratable acidity, and soluble solid contents, as well as higher antioxidant activity and higher total phenolic and flavonoid contents [7,11]. Minerals, according to the fruit ripening progress, have different distributions depending on the fruit ripening stage [12]. The fruit is consumed regardless of ripeness and can be eaten either raw or in processed forms [9], with unripe fruits being used as a vegetable and ripe fruits being used as fruit [4].

Ripening initiation is a critical stage that varies significantly across different fruit species, resulting in uneven ripening [13,14]. It has been reported that the initiation of ripening occurs in specific fruit regions, causing distinct color changes [14,15]. For instance, tomatoes start ripening near the calyx [14], bananas start from their distal ends [15], and apples begin ripening from the stem-end [13]. Variations in fruit qualities have also been observed in apples [13,16], bananas [15], and grapes [17]. These differences could be attributed to genotypes and environmental factors, such as sun exposure and fruit skin temperatures [13]. The different ripening degrees of fruit parts may lead to a loss of consumer confidence [13,18,19]. Uneven fruit skin color has been reported in papaya [18], but the other characteristics have yet to be well studied in different parts.

In this study, we investigated morphological, physicochemical, and nutritional characteristics in different parts, ranging from the stem-end to the calyx-end, of "Tainung No.2" papaya fruits during ripening. We also investigated the relationship between the fruit characteristics, parts, and ripening stages. The fruit characteristics and their relationships would be contributed to a commercially valuable database associated with papaya fruit physiology.

2. Results

The fruit length, the width, the ratio of the length to the width, and the weight did not change during ripening (Table 1). The length of the papaya fruit was an average of 23.6 mm, ranging from 19.2 to 25.8 mm; the width was, on average, 10.0 mm, ranging from 9.6 to 10.3 mm. The ratio of the length to width ranged from 1.9 to 2.7, indicating the elliptical shape of the papaya fruit. The fresh weight of the fruit ranged from 699.6 to 997.9 g, with an average of 854.5 g.

Table 1. Fruit length (mm), width (mm), the ratio of the length to the width, and weight (g FW) of "Tainung No. 2" Red papaya (*Carica papaya*) at 3 parts across ripening stages: unripe at 16 weeks after anthesis (WAA), half-ripe at 18 WAA, and full-ripe at 20 WAA.

Stage	Length	Width	Ratio of Length to Width	Weight
	mm			g
Unripe	25.9 ± 1.40 [a†]	9.6 ± 0.54 [a]	2.7 ± 0.27 [a]	865.9 ± 123.92 [a]
Half-ripe	25.8 ± 0.80 [a]	10.3 ± 1.11 [a]	2.5 ± 0.32 [a]	997.9 ± 110.92 [a]
Full-ripe	19.2 ± 5.08 [a]	10.0 ± 0.73 [a]	1.9 ± 0.67 [a]	699.6 ± 200.26 [a]

[†] Different letters within columns indicate significant differences by Duncan's multiple range test at $p < 0.05$.

The peel color of the papaya fruit changed during ripening (Figure 1 and Table 2). The peel lightness decreased at the half-ripe and full-ripe stages compared with the unripe stage (Table 2). The half-ripe and full-ripe stages became redder during ripening than in the unripe stage. The b value increased during ripening, indicating that the pulp became yellow. The h° between the parts did not change at each ripening stage. The h° decreased in the half-ripe and full-ripe stages compared to that in the unripe stage; the values were, on average, 129.9, 75.5, and 58.0 at the unripe, half-ripe, and full-ripe stages, respectively. The h° of the unripe stage was significantly higher than those of the other two stages. The h° tended to decrease during fruit ripening continuously. The trend was obviously shown in the calyx-end parts of each stage, in which the h° significantly decreased. The C* increased in the half-ripe and full-ripe stages (51.9 and 60.0, respectively) compared to the unripe stage (13.2); there was no significant difference between the half-ripe and full-ripe stages.

Figure 1. Inner and outer flesh of papaya (*Carica papaya*) fruit across different ripening stages: unripe (**left**), half-ripe (**middle**), full-ripe (**right**). Bar = 10 mm.

Table 2. Peel chromaticity of "Tainung No. 2" Red papaya (*Carica papaya*) at 3 parts across ripening stages: unripe at 16 weeks after anthesis (WAA), half-ripe at 18 WAA, and full-ripe at 20 WAA.

Stage	Position	Peel Chromaticity				
		L*	a*	b*	h°	C*
Unripe	Stem-end	35.5 ± 1.66 [b†]	-8.1 ± 0.45 [cd]	9.3 ± 0.90 [b]	131.2 ± 1.60 [a]	12.4 ± 0.95 [b]
	Middle	34.3 ± 0.58 [b]	-8.7 ± 0.43 [cd]	9.0 ± 0.20 [b]	132.2 ± 1.13 [a]	12.2 ± 0.41 [b]
	Calyx-end	34.9 ± 1.01 [b]	-8.8 ± 1.35 [d]	12.2 ± 3.36 [b]	126.4 ± 3.71 [a]	15.0 ± 3.48 [b]
Half-ripe	Stem-end	58.0 ± 4.55 [a]	18.2 ± 12.01 [ab]	51.0 ± 6.20 [a]	71.4 ± 1.12 [bc]	54.8 ± 8.55 [a]
	Middle	55.6 ± 2.21 [a]	13.1 ± 14.08 [abc]	48.5 ± 2.16 [a]	75.9 ± 11.60 [bc]	51.3 ± 6.48 [a]
	Calyx-end	55.7 ± 2.61 [a]	10.1 ± 12.46 [bcd]	47.6 ± 3.48 [a]	79.1 ± 13.85 [b]	49.5 ± 6.50 [a]
Full-ripe	Stem-end	56.8 ± 1.85 [a]	32.5 ± 2.25 [a]	53.9 ± 3.29 [a]	58.9 ± 12.64 [bc]	62.9 ± 3.74 [a]
	Middle	54.5 ± 1.96 [a]	31.7 ± 1.42 [a]	51.5 ± 2.62 [a]	58.4 ± 1.60 [bc]	60.5 ± 2.44 [a]
	Calyx-end	56.0 ± 1.81 [a]	33.5 ± 0.95 [a]	51.0 ± 3.79 [a]	56.7 ± 1.24 [c]	61.1 ± 3.68 [a]

† Different letters within columns indicate significant differences by Duncan's multiple range test at $p < 0.05$. L*—the lightness of colors, ranging from 0 to 100 (0, black; 100, white); a*—negative for green and positive for red; b*—negative for blue and positive for yellow; h°—hue angle calculated as $\tan^{-1}$ (b*/a*); C*—chroma calculated as $(a^{*2} + b^{*2}) \times 1/2$.

Pulp characteristics were investigated during fruit parts and ripening (Table 3). The pulp color changed from unripe to half-ripe stages (Table 3) but not afterward. The lightness became darker in the half-ripe and full-ripe stages (55.8) than in the unripe stages (75.1). The a* and b* values of the unripe stage were lower than those of the half-ripe and full-ripe stages, indicating a greenish and a yellowish color, respectively. The significance of the h° and C* during ripening was consistent with the L, a*, and b* values. Soluble solid contents did not change between the fruit parts at each ripening stage but changed significantly during ripening, from the unripe stage (3.9 °Brix) to the half-ripe and full-ripe stages

(10.8 °Brix). Pulp firmness did not change at the fruit parts in each ripening stage. The firmness did not change between the unripe and half-ripe stages and decreased only at the full-ripe stage up to 96% (from 9.75 N to 0.37 N).

Table 3. Pulp chromaticity, soluble solid contents, and firmness (N) of "Tainung No. 2" Red papaya (*Carica papaya*) at 3 parts across ripening stages unripe at 16 weeks after anthesis (WAA), half-ripe at 18 WAA, and full-ripe at 20 WAA.

Stage	Position	Chromaticity					SSC	Firmness
		L*	a*	b*	h°	C*	°Brix	N
Unripe	Stem-end	75.1 ± 1.12 a†	−3.7 ± 0.98 b	23.7 ± 3.40 c	98.8 ± 1.10 a	24.0 ± 3.51 c	4.6 ± 0.31 b	11.1 ± 0.65 a
	Middle	74.4 ± 3.10 a	−4.2 ± 0.74 b	24.3 ± 3.16 c	99.8 ± 1.13 a	24.7 ± 3.21 c	3.4 ± 0.92 b	11.7 ± 0.91 a
	Calyx-end	75.7 ± 3.01 a	−3.3 ± 0.87 b	22.2 ± 3.35 c	98.5 ± 0.92 a	22.4 ± 3.44 c	3.7 ± 0.21 b	11.5 ± 0.71 a
Half-ripe	Stem-end	57.8 ± 0.46 b	28.6 ± 1.23 a	38.5 ± 1.12 b	53.4 ± 1.98 b	48.0 ± 0.20 ab	11.3 ± 0.46 a	8.2 ± 1.86 a
	Middle	57.8 ± 1.14 b	30.1 ± 2.81 a	40.2 ± 3.77 ab	53.2 ± 2.17 b	50.3 ± 4.32 ab	9.7 ± 0.50 a	7.5 ± 4.14 a
	Calyx-end	58.5 ± 2.51 b	28.6 ± 2.17 a	38.1 ± 0.26 b	53.1 ± 1.89 b	47.6 ± 1.51 b	9.4 ± 0.93 a	8.5 ± 2.41 a
Full-ripe	Stem-end	54.1 ± 4.10 b	30.6 ± 4.04 a	42.6 ± 2.45 ab	54.5 ± 2.04 b	52.5 ± 4.34 ab	11.3 ± 1.80 a	0.4 ± 0.29 b
	Middle	55.2 ± 1.37 b	31.9 ± 2.67 a	46.7 ± 2.64 a	55.7 ± 2.99 b	56.6 ± 2.28 a	11.7 ± 0.40 a	0.3 ± 0.11 b
	Calyx-end	51.3 ± 4.25 b	33.7 ± 1.59 a	42.3 ± 3.22 ab	51.4 ± 2.44 b	54.1 ± 2.77 ab	11.4 ± 0.29 a	0.4 ± 0.09 b

† Different letters within columns indicate significant differences by Duncan's multiple range test at $p < 0.05$. L*—the lightness of colors, ranging from 0 to 100 (0, black; 100, white); a*—negative for green and positive for red; b*—negative for blue and positive for yellow; h°—hue angle calculated as $\tan^{-1}$ (b*/a*); C*—chroma calculated as $(a^{*2} + b^{*2}) \times 1/2$. SSC—soluble solid contents.

The water contents continuously decreased during fruit ripening (Table 4). The water contents of the unripe, half-ripe, and full-ripe stages were 93.9, 92.5, and 91.2%, respectively. The water content was not different among the parts of the unripe and half-ripe stages. In the full-ripe stage, the content was the lowest in the stem-end (90.6%) over the other parts (91.5%). The total carbohydrates tended to increase during fruit ripening, showing the differences among the parts (Table 4). The lowest amount was 4.5 g/100 g at the stem-end and the calyx-end of the unripe stage; conversely, the highest was 7.4 g/100 g at the stem-end of the full-ripe stage. The accumulation rate of carbohydrates at the stem-end increased from 120% to 137% during ripening, while the rate at the middle and calyx-end ranged from 120% to 110% and from 131% to 117%, respectively. Protein contents did not show distinct tendencies during ripening in part of the fruit, except that the accumulation from the stem-end of each stage was higher than that in the other parts. In addition, the lipid contents were not detected in any fruit part during the ripening.

Table 4. Water, carbohydrate, protein, and lipid contents of "Tainung No. 2" Red papaya (*Carica papaya*) at 3 parts across ripening stages: unripe at 16 weeks after anthesis (WAA), half-ripe at 18 WAA, and full-ripe at 20 WAA.

Stage	Position	Water	Carbohydrate	Protein	Lipid
		%		g/100 g FW	
Unripe	Stem-end	94.0 ± 0.03 a†	4.5 ± 0.02 e	1.06 ± 0.03 cd	nd
	Middle	93.7 ± 0.15 a	5.1 ± 0.15 d	0.91 ± 0.01 de	nd
	Calyx-end	93.9 ± 0.02 a	4.5 ± 0.05 e	1.15 ± 0.05 bc	nd
Half-ripe	Stem-end	92.6 ± 0.02 b	5.4 ± 0.05 d	1.29 ± 0.01 ab	nd
	Middle	92.5 ± 0.04 b	6.1 ± 0.02 c	0.93 ± 0.03 de	nd
	Calyx-end	92.3 ± 0.03 b	5.9 ± 0.03 c	1.11 ± 0.01 c	nd
Full-ripe	Stem-end	90.6 ± 0.08 d	7.4 ± 0.08 a	1.39 ± 0.02 a	nd
	Middle	91.4 ± 0.04 c	6.7 ± 0.03 b	1.20 ± 0.02 bc	nd
	Calyx-end	91.6 ± 0.05 c	6.9 ± 0.04 b	0.88 ± 0.01 e	nd

† Different letters within columns indicate significant differences by Duncan's multiple range test at $p < 0.05$. Nd—not detected.

Three soluble sugars were determined, and two were detected during ripening (Table 5). Sugar contents increased during ripening and were different from that of the fruits. Fructose contents were not different among the parts at the unripe stage: the fructose contents in the calyx-end of the two other stages were approximately 110% higher than that in the middle and stem-end. The glucose content in each part of the fruit significantly increased during ripening. However, these contents were not different at the three parts within each ripening stage, except for the full-ripe stage, in which the glucose content of the calyx-end was higher than that of the stem-end.

Table 5. Changes in the content of soluble sugars of "Tainung No. 2" Red papaya (*Carica papaya*) pulp at 3 parts across ripening stages: unripe at 16 weeks after anthesis (WAA), half-ripe at 18 WAA, and full-ripe at 20 WAA.

Stage	Position	Fructose	Glucose	Sucrose
			$g{\cdot}100\ g^{-1}$ FW	
Unripe	Stem-end	1.1 ± 0.01 [e†]	1.3 ± 0.00 [d]	nd
	Middle	1.1 ± 0.08 [e]	1.3 ± 0.03 [d]	nd
	Calyx-end	1.4 ± 0.07 [e]	1.6 ± 0.03 [d]	nd
Half-ripe	Stem-end	1.8 ± 0.17 [d]	2.0 ± 0.07 [c]	nd
	Middle	1.9 ± 0.23 [d]	2.1 ± 0.09 [c]	nd
	Calyx-end	2.1 ± 0.12 [c]	2.4 ± 0.04 [c]	nd
Full-ripe	Stem-end	2.5 ± 0.01 [bc]	2.9 ± 0.01 [b]	nd
	Middle	2.7 ± 0.14 [ab]	3.1 ± 0.17 [ab]	nd
	Calyx-end	2.9 ± 0.05 [a]	3.4 ± 0.05 [a]	nd

† Different letters within columns indicate significant differences by Duncan's multiple range test. Nd—not detected.

In addition to these structural compounds, bioactive compounds and minerals are essential components of fruits. In the papaya fruit, ascorbic acid contents increased up to approximately 300% from unripe to half-ripe and approximately 140% from half-ripe to full-ripe in all 3 parts during ripening (Table 6). However, there were no differences between the ripening stages. The proportion of mineral compounds varied between fruit parts and ripening stages (Tables 7 and 8). The K proportion was about 2.85% during ripening, followed by N (0.80%), P (0.37%), Ca (0.24%), Mg (0.13%), Fe (26.05 ppm), Cu (2.73 ppm), Zn (11.11 ppm), and Mn (1.13 ppm). The proportion of K in each part was maintained during ripening, whereas those of other minerals fluctuated during ripening. The change patterns of the N proportion were distinctly different for each part: increase at the stem-end in the full-ripe stage, decrease at the middle in the half-ripe stage, decrease and then increase at the calyx-end in the half-ripe and full-ripe stage, respectively. The P proportion changed only at the stem-end between the unripe and full-ripe stages. The Ca proportion steadily decreased at the stem-end and middle parts during ripening; that of the calyx-end decreased in the half-ripe stage and was maintained up to the full-ripe stage. The Mg proportion of all fruit parts decreased from the unripe to half-ripe stages and then remained in the full-ripe stage. The Fe proportion of all the parts decreased from the unripe to half-ripe stages. Subsequently, the Fe proportion of the stem-end did not change, and those of the middle and calyx-end increased at the full-ripe stage. The Cu proportion decreased only in the calyx-end from the unripe to half-ripe stages; all the other proportions did not differ between the parts and ripening stages. The Zn proportion decreased from the unripe to the half-ripe stage and was maintained between the parts and ripening stages. The Mn proportion at the stem-end did not change during ripening, that at the middle increased only in the full-ripe stage, and that at the calyx-end decreased and then increased during ripening.

The fruit characteristics of papaya were analyzed by using principal component analysis (PCA) to identify patterns and relationships among variables. In this study, the PCA results showed that the fruit characteristics were distributed across each PC, with the PCs arranged according to the size of their variance, as shown in Figure 2. PC1 and PC2 accounted for the majority of the variance at 85.36%. Specifically, PC1 explained 71.78% of the variance and was associated with chromaticity, soluble solid contents, and ascorbic acid. These fruit characteristics had a high correlation value of approximately 0.98 on an absolute average, contributing to 59.78% of the variance in PC1. Notably, peel h°, pulp L, and pulp h° had negative correlations with PC1, while the other fruit characteristics had positive correlations. Overall, PC1 was indicative of papaya fruit ripening. In contrast, PC2 accounted for 13.58% of the variance and reflected the different parts of the fruit.

Table 6. Changes in the content of ascorbic acid at each part of "Tainung No. 2" Red papaya (*Carica papaya*) pulp at 3 parts across ripening stages: unripe at 16 weeks after anthesis (WAA), half-ripe at 18 WAA, and full-ripe at 20 WAA.

Stage	Position	Ascorbic Acid
		$mg{\cdot}100\ g^{-1}$ FW
Unripe	Stem-end	8.7 ± 0.43 ^{c†}
	Middle	6.8 ± 0.20 ^{c}
	Calyx-end	6.9 ± 0.37 ^{c}
Half-ripe	Stem-end	21.9 ± 0.23 ^{b}
	Middle	23.2 ± 0.53 ^{b}
	Calyx-end	22.1 ± 1.10 ^{b}
Full-ripe	Stem-end	32.7 ± 0.97 ^{a}
	Middle	31.4 ± 0.97 ^{a}
	Calyx-end	30.7 ± 1.13 ^{a}

† Different letters within columns indicate significant differences by Duncan's multiple range test.

Table 7. Changes in macronutrients of "Tainung No. 2" Red papaya (*Carica papaya*) pulp at 3 parts across ripening stages: unripe at 16 weeks after anthesis (WAA), half-ripe at 18 WAA, and full-ripe at 20 WAA.

Stage	Position	N	P	K	Ca	Mg
				$mg{\cdot}g^{-1}$ FW		
Unripe	Stem-end	0.7 ± 0.01 ^{def†}	0.5 ± 0.02 ^{a}	3.3 ± 0.13 ^{a}	0.4 ± 0.03 ^{a}	0.15 ± 0.010 ^{b}
	Middle	0.8 ± 0.01 ^{bcd}	0.4 ± 0.02 ^{bc}	2.8 ± 0.10 ^{b}	0.4 ± 0.04 ^{a}	0.20 ± 0.010 ^{a}
	Calyx-end	1.2 ± 0.07 ^{a}	0.4 ± 0.03 ^{b}	2.6 ± 0.13 ^{b}	0.3 ± 0.02 ^{b}	0.16 ± 0.010 ^{b}
Half-ripe	Stem-end	0.7 ± 0.08 ^{ef}	0.4 ± 0.01 ^{ab}	3.5 ± 0.19 ^{a}	0.3 ± 0.02 ^{bc}	0.09 ± 0.000 ^{e}
	Middle	0.6 ± 0.02 ^{f}	0.3 ± 0.01 ^{c}	2.7 ± 0.14 ^{b}	0.3 ± 0.02 ^{bc}	0.12 ± 0.000 ^{cd}
	Calyx-end	0.8 ± 0.04 ^{cde}	0.4 ± 0.02 ^{b}	2.6 ± 0.10 ^{b}	0.2 ± 0.01 ^{cd}	0.13 ± 0.000 ^{c}
Full-ripe	Stem-end	0.9 ± 0.07 ^{bc}	0.4 ± 0.04 ^{bc}	3.3 ± 0.25 ^{a}	0.2 ± 0.01 ^{d}	0.09 ± 0.000 ^{e}
	Middle	0.7 ± 0.01 ^{def}	0.3 ± 0.03 ^{c}	2.5 ± 0.19 ^{b}	0.1 ± 0.00 ^{d}	0.11 ± 0.000 ^{de}
	Calyx-end	0.9 ± 0.07 ^{b}	0.4 ± 0.02 ^{ab}	2.5 ± 0.14 ^{b}	0.1 ± 0.00 ^{d}	0.13 ± 0.000 ^{c}

† Different letters within columns indicate significant differences by Duncan's multiple range test.

Table 8. Changes in micronutrients of "Tainung No. 2" Red papaya (*Carica papaya*) pulp at 3 parts across ripening stages (unripe, half-ripe, and full-ripe).

Stage	Position	Cu	Fe	Mn	Zn
			$\mu g{\cdot}g^{-1}$ FW		
Unripe	Stem-end	0.28 ± 0.033 ^{b†}	2.68 ± 0.074 ^{bcd}	0.09 ± 0.005 ^{cd}	1.36 ± 0.095 ^{b}
	Middle	0.30 ±0.027 ^{b}	2.94 ± 0.134 ^{ab}	0.10 ± 0.010 ^{cd}	1.78 ± 0.164 ^{a}
	Calyx-end	0.42 ± 0.029 ^{a}	3.12 ± 0.159 ^{a}	0.27 ± 0.018 ^{a}	1.73 ± 0.154 ^{a}
Half-ripe	Stem-end	0.23 ± 0.024 ^{b}	2.19 ± 0.070 ^{ef}	0.05 ± 0.004 ^{d}	0.80 ± 0.042 ^{c}
	Middle	0.22 ± 0.028 ^{b}	2.14 ± 0.063 ^{f}	0.05 ± 0.003 ^{d}	0.77 ± 0.063 ^{c}
	Calyx-end	0.30 ± 0.042 ^{b}	2.41 ± 0.083 ^{def}	0.07 ± 0.006 ^{cd}	0.84 ± 0.068 ^{c}
Full-ripe	Stem-end	0.24 ± 0.023 ^{b}	2.51 ± 0.149 ^{cde}	0.09 ± 0.016 ^{cd}	0.81 ± 0.070 ^{c}
	Middle	0.24 ± 0.010 ^{b}	2.62 ± 0.060 ^{bcd}	0.11 ± 0.019 ^{c}	0.86 ± 0.038 ^{c}
	Calyx-end	0.24 ± 0.010 ^{b}	2.83 ± 0.075 ^{abc}	0.19 ± 0.040 ^{b}	1.06 ± 0.085 ^{c}

† Different letters within columns indicate significant differences by Duncan's multiple range test.

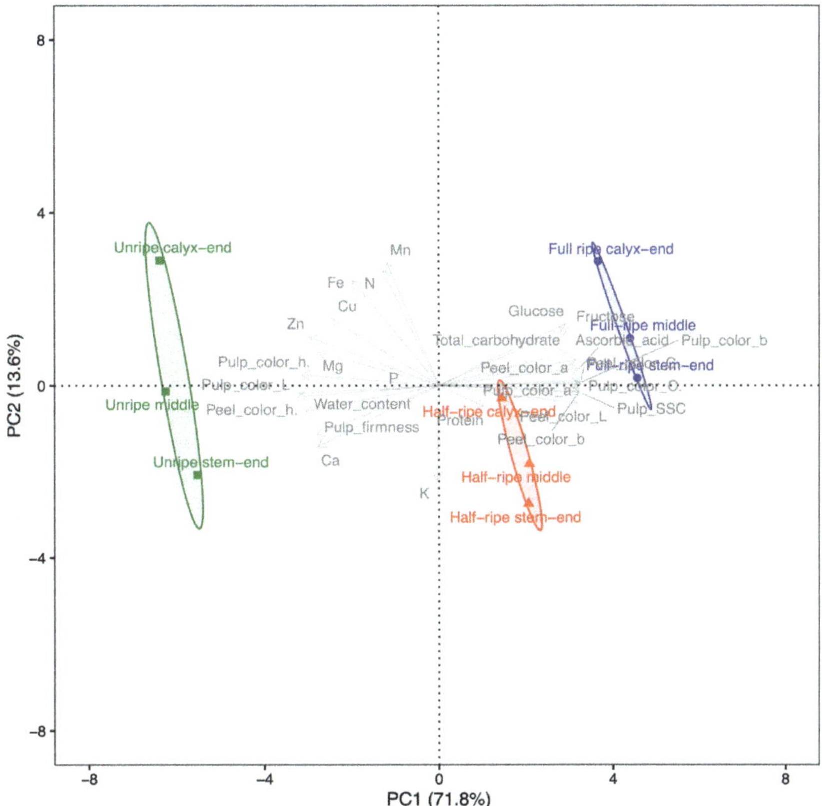

Figure 2. Principal component (PC) analysis scores of fruit qualities with PC scores of means on the first 2 PC axes for "Tainung No. 2" Red papaya (*Carica papaya*) at 3 parts across ripening stages with 95% confidence of ellipses for the mean of ripening stages. Percentages in parenthesis of each axis represent variances of each PC. Green—unripe at 16 weeks after anthesis (WAA); red—half-ripe at 18 WAA; blue—full-ripe at 20 WAA.

3. Discussion

In our result, the ratio of the length to the width of "Tainung No.2" was an average of 2.4 during the entire ripening stage, which is consistent with the general shape of hermaphroditic fruits [20]. In addition to this ratio, the characteristics associated with the fruit volume, including length and width, did not change during ripening. The growth of papaya fruit follows a sigmoidal pattern, with size development being completed before the onset of ripening [3]. This is due to the cessation of cell expansion and enlargement, with carbon sources being redirected toward metabolic changes such as the accumulation of pigments and soluble sugars [13,21]. This redistribution of carbon sources is critical for the development of desirable fruit qualities such as flavor, aroma, and color during ripening [13].

The degree of fruit ripeness is a consumer-driven trait and has significant importance during picking, packing, and transportation [18]. Fruit color is an essential indicator of ripeness in fresh fruits, including papaya [18] and mangoes [22], providing visual cues for marketable values. In this study, "Tainung No.2" papaya fruit showed distinct changes in both peel and pulp colors during ripening, with the colors evenly distributed throughout the fruit. Along with changes in color, the soluble solid contents increased while the firmness decreased during ripening, regardless of the fruit part. These changes are commonly observed during the ripening of various fruits, including papaya [18,23] and mangoes [22]. The softening of papaya fruit is highly correlated with the sweetening process that is associated with soluble solid contents, possibly due to the easier release of cellular contents in fully ripened tissue [23]. The results suggest that color changes and changes in soluble solid contents and firmness can be used as reliable indicators of papaya fruit ripeness.

The present study demonstrated that individual metabolites in papaya fruit exhibited unique accumulation patterns in different fruit parts during ripening. Although research on differences among tissue zones has been conducted on apples [13,19,24,25], there is limited information to determine the characteristics of each part of papaya fruit, with no comparative data in previous studies on papaya fruits. Doerflinger, Miller, Nock, and Watkins [13] compared the carbohydrate concentration of the stem-end, middle, and calyx-end in three apple cultivars ("Empire", "Honeycrisp", and "Gala") during ripening. In the "Empire" cultivar, the concentration of starch was the highest at the calyx-end and the lowest at the stem-end. "Honeycrisp" and "Empire" had the highest concentration of sorbitol in the calyx-end, whereas the concentration was highest in the stem-end in "Gala". The distribution differences of glucose, fructose, and sucrose were similar in all three cultivars: higher fructose and glucose concentrations in the stem-end and higher sucrose concentrations in the calyx-end of the fruit. In papaya fruit, soluble sugars are accumulated mainly when the fruit remains attached to the plant [23]. At the unripe stage, glucose is prevalent among the soluble sugars, and during ripening, sucrose becomes the predominant sugar with the modification of the soluble sugar profiles [26]. Chan Jr and Kwok [27] reported that sucrose levels varied from 1.8% to 8.0% during ripening. In this study, "Tainung No. 2" Red papaya fruit accumulated more total carbohydrates in the stem-end than in the other parts. The main soluble sugars in papaya are glucose, fructose, and glucose [28]; however, their compositions vary among cultivars [28]. In "Tainung No. 2" papaya fruit, only fructose and glucose were identified during ripening and were accumulated more in the calyx-end than in other parts. In addition to carbohydrates, more protein was accumulated at the stem-end. These differences could be associated with different development rates in different tissue zones. An increase in total primary amounts is associated with carbon allocation from the photosynthetic organs before ripening. Meanwhile, the accumulation of soluble sugars is one of the ripening processes. In papaya fruit, the accumulation rate rapidly increases during ripening [29]. Therefore, changes in primary metabolites are more pronounced and active during the ripening process of "Tainung No. 2" papaya fruit. The differences in the structure and size of cells could also lead to developmental differences within the fruit. However, such differences vary by each species [13] and its cultivars [30,31].

Establishing nutrient absorption and accumulation patterns is critical for planning optimum nutrient supply and improving the influence on fruit quality [32]. Macro- and micro-element accumulation shows dynamic variance during ripening. In this study, all elements except N at the stem-end decreased or were maintained at the full-ripe stage, despite different patterns at the half-ripe stage among fruit parts. The decreases in the elements have also been shown in various fruits, including papaya [12] and apple fruit [33,34]. In apple pulp, when fruits are compared approximately 60 and 120 days after full bloom for the elements N, P, K, Mg, and S, there is a slight decrease, but Ca, Fe, Cu, Mn, and B contents had no significant differences [34]. In papaya pulp, Ca, K, P, and Mg decreased to 75.56%, 38.67%, 66.46%, and 50.00%, respectively, at the full-ripe stage, compared to the unripe stage. [12].

All papaya fruit samples were distinctly divided according to fruit characteristics. PCA revealed that the ripening stage was accounted for by PC1, while different fruit parts were accounted for by PC2. PC1 was associated with chromaticity, soluble solid contents, and ascorbic acid, making it a suitable indicator for selecting the most appropriate stage of ripening based on these characteristics. However, it is crucial to consider both fruit characteristics and fruit parts when evaluating the quality and nutritional value of papaya fruit. These findings suggest that the ripening process and nutrient accumulation in papaya fruit are complex and occur differently in different parts of the fruit, highlighting the importance of considering both fruit characteristics and fruit parts when evaluating the quality and nutritional value of papaya fruit.

4. Materials and Methods

4.1. Plant Materials

"Tainung No. 2" papaya seeds were collected from the Research Institute of Climate Change and Agriculture, the National Institute of Horticultural and Herbal Science, Jeju, Republic of Korea (33° 28 N′, 126° 31′ E). The seeds were sown on 10 November 2015 in black plastic pots (200 mm in diameter, 300 mm in length, and 30 L in volume) in a medium containing 90% commercial grow media (4% zeolite, 7% perlite, 6% vermiculite, 68% coco peat, and 14% peat moss) (Baroker, Seoul Bio Co., Ltd., Gyeongju, Republic of Korea) and 10% coarse sand (by volume), according to Joa et al. [35]. Plantlets were grown in a greenhouse of an experimental orchard in the Institute of Climate Change and Agriculture and then transplanted on 8 June 2016 into the ground of the same greenhouse. These trees were cultivated according to the standard guidelines for papaya cultivation [35]: the minimum temperature during winter was maintained at 15 °C in the greenhouse by using hot air blowers to prevent papaya trees from chilling injury.

The fruit was harvested in July 2018, owing to a short juvenile phase [5], and was categorized into 3 ripening stages based on weeks after anthesis (WAA), according to Zuhair, Aminah, Sahilah, and Eqbal [11]: (1) unripe at 16 WAA, (2) half-ripe at 18 WAA, and (3) full-ripe at 20 WAA. At each stage, a total of nine fruits were harvested in three biological replicates. After the length (mm), width (mm), and weight (g) of each fruit were measured, each fruit was dissected into three parts of equal lengths: stem-end, middle, and calyx-end for further analyses.

4.2. Determination of Colors, Soluble Solid Contents, and Firmness

The peel and pulp colors of papaya fruits were measured at the stem-end, middle, and calyx-end at the three ripening stages by using a spectrophotometer (CM700d, Minolta Co., Osaka, Japan) and described by the CIE L*, a*, and b* color space coordinates [36]. The L* value represents the lightness of colors, ranging from 0 to 100 (0, black; 100, white). The a* value was negative for green and positive for red. The b* value was negative for blue and positive for yellow. The values were measured at three midpoint regions of each fruit part. In addition to the CIE coordinates, the hue angle (h°) was calculated as tan−1 (b*/a*), which implies visual color appearance; 0° was represented by red-purple; 90° was represented by yellow; 180° was represented by bluish-green; 270° was represented by blue. The chroma (C*) was computed as $(a^{*2} + b^{*2}) \times 1/2$, which describes the quality of the color intensity or saturation.

The soluble solid contents (°Brix) were measured by using a digital refractometer (HI98801, Hanna Instruments Inc., Woonsocket, RI, USA). Peel firmness was measured by using a texture analyzer (TA-XT express, Stable Micro System, Godalming, UK). A puncture test was conducted by using a 2.0 mm diameter cylindrical probe (Stable Micro System) at 3 different points along the fruit equator with a cross-head speed of 2.0 mm s^{-1} and a penetration depth of 5.0 mm.

4.3. Determination of Water Proportion and Carbohydrate, Protein, and Lipid Contents

The fruit was dehydrated at 105 °C in a drying oven up to a constant weight to determine the proportion of water content compared with the total fresh weight of each sample, which was expressed as a percentage (%). To determine carbohydrate, protein, and lipid contents, all the samples were lyophilized by using a freeze dryer and finely ground by using a mortar and pestle. Total carbohydrate content was determined according to the procedure of Jermyn [37]. Protein concentration was determined by using Micro-Kjeldahl's apparatus [38]. Lipid concentration was determined by using the Soxhlet extraction method [39]. All analyses were performed according to the official analysis methods of the Association of Official Analytical Chemists (AOAC) [40].

4.4. Determination of Free Sugars

Free sugars were extracted according to the method described by Oh et al. [41] with some modifications. In total, 3 g of ground fruits was added to 50 mL of 50% acetonitrile. The samples were extracted thrice via the ultrasonic extraction method for 6 h. The extract was diluted and filtered through Sep-Pak C18 cartridge column (Waters, Milford, MA, USA) and a 0.45 μm pore size micro-filter (Woongki Science Co., Ltd., Seoul, Republic of Korea). After the sample preparation, the free sugars were separated in a Prevail TM Carbohydrate ES column (4.6 × 250 mm, 5 μm, Grace, Deerfield, IL, USA) that was equipped with an HPLC system (Waters 2695, Waters Associate Inc., Milford, MA, USA). The eluents were passed through the column at a flow rate of 0.8 mL/min by using acetonitrile: distilled water (70:30, *v*/*v*). The chromatographic peak corresponding to free sugar was identified by comparing the retention times with those of glucose, fructose, and sucrose (Sigma, St. Louis, MO, USA) as standards. Concentrations were calculated by using the calibration curve that was generated from the standard solutions.

4.5. Determination of Ascorbic Acids

Ascorbic acid was extracted by using the procedure described by Rizzolo et al. [42], with some modifications. Thirty grams of samples were added to 25 mL of 6% metaphosphoric acid. The sample was homogenized and then centrifuged at 10,000× *g* at 4 °C for 10 min. The supernatant was filtrated through filter paper (Whatman No. 4, Whatman plc, Kent, UK), a membrane filter (MF-milliporeTM 0.22 μm, Merck KGaA, Darmstadt, Germany), and a Waters Sep-Pak C18 column (Waters, Milford, CT, USA). Ascorbic acids in the prepared samples were separated by using a Waters Bonda Pak NH2 (3.9 × 300 mm) column (Waters) that was equipped with an HPLC system (Waters TM 600 controller, Waters TM 616 Pump, Waters 717 plus auto Sampler, Waters TM 486 tunable absorbance detector). The eluent was passed through the column by using 5 mM KH_2PO_4 (pH 4.6): acetonitrile (30:70, *v*/*v*) (A) and acetonitrile (B) by using a linear gradient from 100% A to 40% A in 50 min at a flow rate of 1 mL/min. The chromatographic peak corresponding to ascorbic acid in the samples was identified by comparing its retention times with those of the standard (Sigma). Concentrations were calculated by using the calibration curve that was generated from standard solutions prepared for the standard. The limit of detection (LOD) and limit of quantification (LOQ) for ascorbic acid was determined to ascertain the linearity of the method, as previously described by [43,44]; LOD of 0.003 $\mu g \cdot \mu L^{-1}$ and LOQ of 0.013 $\mu g \cdot \mu L^{-1}$.

4.6. Determination of Macro- and Micro-Elements

The proportions of nitrogen and phosphorus contents were determined by using the Kjeldahl method [45] and the molybdenum blue method [46], respectively. In total, 3 g of fruit was weighed and heated in a furnace for 4 h at 550 °C. The crucible was cooled down in a desiccator and dissolved in 2.5 mL of a decomposition solution (HNO_3:H_2SO_4:$HClO_4$ (10:1:4, *v*/*v*/*v*)). The solution was filtered and diluted up to 100 mL by using distilled water. Exchangeable cations (K, Ca, and Mg) and minerals (Fe, Mn, Zn, and Cu) were determined by inductively coupled plasma spectrophotometry (ICP-Integra XL, GBC Scientific Equipment Pty Ltd., Braeside, Australia) following the procedure of Zarcinas et al. [47]. The results were obtained by using a working standard of 1000 ppm for each sample [41].

4.7. Statistical Analyses

To determine statistical significance, data were analyzed by using the R 4.2.2 [48]. Statistical significances were determined by using a two-way analysis of variance with the agricolae v1.3-5 package [49]. The means of fruit characteristics were compared by using Duncan's multiple range tests at $p < 0.05$.

PCA was performed on fruit characteristics by using the mean values of replicate samples. The Pearson correlation coefficient was used to calculate the correlation matrix among the fruit characteristics, with a significance level of $p < 0.05$. The PCA was conducted with raw data by using the factoextra v1.0.7 package in R [50]. The variables were standardized by subtracting the mean and then dividing the result by the standard deviation of each original variable to assign each weight in the analysis, according to Abdi and Williams [51].

5. Conclusions

Changes in color, soluble solid contents, and firmness can be reliable indicators of papaya fruit ripeness. Moreover, individual metabolites showed unique accumulation patterns in different fruit parts during ripening, with the stem-end accumulating more total carbohydrates and protein. The nutrient accumulation patterns exhibited dynamic variance during ripening, emphasizing the need to consider both fruit characteristics and parts when evaluating the quality and nutritional value of papaya fruit. Overall, our findings provide valuable insights into the ripening process and nutrient accumulation in papaya fruit, which can help optimize nutrient supply and improving fruit quality.

Author Contributions: Conceptualization, S.W.C. and S.C.K.; methodology, S.W.C. and S.K.; software, S.W.C. and Y.J.J.; validation, S.W.C., Y.J.J. and S.C.K.; formal analysis, S.W.C. and S.K.; investigation, S.W.C. and S.K.; resources, Y.J.J. and S.C.K.; data curation, S.W.C.; writing—original draft preparation, S.W.C.; writing—review and editing, S.W.C. and S.C.K.; visualization, S.W.C.; supervision, S.C.K.; project administration, S.C.K.; funding acquisition, S.C.K. All authors have read and agreed to the published version of the manuscript.

Funding: This study was carried out with the support of the "Cooperative Research Program for Agriculture Science and Technology Development (Project No. PJ01258401)" Rural Development Administration, Republic of Korea.

Data Availability Statement: The data presented in the study are included in this article. Further inquiries can be directed to the corresponding author.

Conflicts of Interest: The authors declare no conflict of interest. The funders had no role in the design of the study; in the collection, analyses, or interpretation of data; in the writing of the manuscript; or in the decision to publish the results.

References

1. FAO. FAOSTAT Statistical Database. Available online: https://www.fao.org/faostat/en/ (accessed on 1 March 2023).
2. Daagema, A.A.; Orafa, P.N.; Igbua, F.Z. Nutritional Potentials and Uses of Pawpaw (*Carica papaya*): A Review. *Eur. J. Nutr. Food Saf.* **2020**, *12*, 52–66. [CrossRef]
3. Salinas, I.; Hueso, J.J.; Cuevas, J. Fruit growth model, thermal requirements, and fruit size determinants in papaya cultivars grown under subtropical conditions. *Sci. Hortic.* **2018**, *246*, 1022–1027. [CrossRef]
4. Gunes, E.; Gübbük, H. Growth, yield and fruit quality of three papaya cultivars grown under protected cultivation. *Fruits* **2012**, *67*, 23–29. [CrossRef]
5. Salinas, I.; Hueso, J.J.; Cuevas, J. Determination of the Best Planting Season for the Protected Cultivation of Papaya. *Horticulturae* **2022**, *8*, 738. [CrossRef]
6. Ogata, T.; Yamanaka, S.; Shoda, M.; Urasaki, N.; Yamamoto, T. Current status of tropical fruit breeding and genetics for three tropical fruit species cultivated in Japan: Pineapple, mango, and papaya. *Breed. Sci.* **2016**, *66*, 69–81. [CrossRef]
7. Jeon, Y.A.; Chung, S.W.; Kim, S.C.; Lee, Y.J. Comprehensive assessment of antioxidant and anti-inflammatory properties of papaya extracts. *Foods* **2022**, *11*, 3211. [CrossRef]
8. Jeong, U.S.; Kim, S.; Chae, Y.-W. Analysis on the cultivation trends and main producing areas of subtropical crops in Korea. *J. Korea Acad.-Ind. Coop. Soc.* **2020**, *21*, 524–535. [CrossRef]
9. Dotto, J.M.; Abihudi, S.A. Nutraceutical value of *Carica papaya*: A review. *Sci. Afr.* **2021**, *13*, e00933. [CrossRef]
10. Farina, V.; Tinebra, I.; Perrone, A.; Sortino, G.; Palazzolo, E.; Mannino, G.; Gentile, C. Physicochemical, nutraceutical, and sensory traits of six papaya (*Carica papaya* L.) Cultivars Grown in Greenhouse Conditions in the Mediterranean Climate. *Agronomy* **2020**, *10*, 501. [CrossRef]

11. Zuhair, R.; Aminah, A.; Sahilah, A.; Eqbal, D. Antioxidant activity and physicochemical properties changes of papaya (*Carica papaya* L. cv. Hongkong) during different ripening stage. *Int. Food Res. J.* **2013**, *20*, 1653–1659.
12. Chukwuka, K.S.; Iwuagwu, M.; Uka, U.N. Evaluation of nutritional components of *Carica papaya* L. at different stages of ripening. *IOSR J. Pharm.* **2013**, *6*, 13–16. [CrossRef]
13. Doerflinger, F.C.; Miller, W.B.; Nock, J.F.; Watkins, C.B. Variations in zonal fruit starch concentrations of apples—A developmental phenomenon or an indication of ripening? *Hortic. Res.* **2015**, *2*, 15047. [CrossRef] [PubMed]
14. Nguyen, C.V.; Vrebalov, J.T.; Gapper, N.E.; Zheng, Y.; Zhong, S.; Fei, Z.; Giovannoni, J.J. Tomato *GOLDEN2-LIKE* transcription factors reveal molecular gradients that function during fruit development and ripening. *Plant Cell* **2014**, *26*, 585–601. [CrossRef]
15. Mustaffa, R.; Osman, A.; Yusof, S.; Mohamed, S. Physico-chemical changes in Cavendish banana (*Musa cavendishii* L. var Montel) at different positions within a bunch during development and maturation. *J. Sci. Food Agric.* **1998**, *78*, 201–207. [CrossRef]
16. Singha, S.; Baugher, T.A.; Townsend, E.C.; D'Souza, M.C. Anthocyanin distribution in "Delicious" apples and the relationship between anthocyanin concentration and chromaticity values. *J. Am. Soc. Hortic. Sci.* **1991**, *116*, 497–499. [CrossRef]
17. Ryu, S.; Han, J.H.; Cho, J.G.; Jeong, J.H.; Lee, S.K.; Lee, H.J. High temperature at veraison inhibits anthocyanin biosynthesis in berry skins during ripening in "Kyoho" grapevines. *Plant Physiol. Biochem.* **2020**, *157*, 219–228. [CrossRef]
18. Barragán-Iglesias, J.; Méndez-Lagunas, L.; Rodríguez-Ramírez, J. Ripeness indexes and physicochemical changes of papaya (*Carica papaya* L. cv. Maradol) during ripening on-tree. *Sci. Hortic.* **2018**, *236*, 272–278. [CrossRef]
19. Lee, J.; Cheng, L.; Rudell, D.R.; Nock, J.F.; Watkins, C.B. Antioxidant metabolism in stem and calyx end tissues in relation to flesh browning development during storage of 1-methylcyclopropene treated "Empire" apples. *Postharvest Biol. Technol.* **2018**, *149*, 66–73. [CrossRef]
20. Kung, Y.-J.; Yu, T.-A.; Huang, C.-H.; Wang, H.-C.; Wang, S.-L.; Yeh, S.-D. Generation of hermaphrodite transgenic papaya lines with virus resistance via transformation of somatic embryos derived from adventitious roots of in vitro shoots. *Transgenic Res.* **2009**, *19*, 621–635. [CrossRef]
21. Seymour, G.B.; Chapman, N.H.; Chew, B.L.; Rose, J.K.C. Regulation of ripening and opportunities for control in tomato and other fruits. *Plant Biotechnol. J.* **2013**, *11*, 269–278. [CrossRef]
22. Chung, S.W.; Oh, H.; Lim, C.K.; Jeon, M.K.; An, H.J. Fruit characteristics of ten greenhouse-grown mango varieties during postharvest ripening at ambient temperature and relative humidity. *Int. J. Fruit Sci.* **2021**, *21*, 1073–1085. [CrossRef]
23. Gomez, M.; Lajolo, F.; Cordenunsi, B. Evolution of soluble sugars during ripening of papaya fruit and its relation to sweet taste. *J. Food Sci.* **2002**, *67*, 442–447. [CrossRef]
24. Lewis, T.L. The rate of uptake and longitudinal distribution of potassium, calcium, and magnesium in the flesh of developing apple fruits of nine cultivars. *J. Hortic. Sci.* **1980**, *55*, 57–63. [CrossRef]
25. Angmo, T.; Rehman, M.U.; Mir, M.M.; Bhat, B.H.; Bhat, S.A.; Kosser, S.; Ahad, S.; Sharma, A. Abscisic acid application regulates vascular integrity and calcium allocation within apple fruits. *Can. J. Plant Sci.* **2022**, *102*, 964–972. [CrossRef]
26. Paull, R.E. Pineapple and papaya. In *Biochemistry of Fruit Ripening*; Seymour, G.B., Taylor, J.E., Tucker, G.A., Eds.; Chapman & Hall: London, UK, 1996; pp. 302–315.
27. Chan, H.T.; Kwok, S.C.M. Importance of enzyme inactivation prior to extraction of sugars from papaya. *J. Food Sci.* **1975**, *40*, 770–771. [CrossRef]
28. Kelebek, H.; Selli, S.; Gubbuk, H.; Gunes, E. Comparative evaluation of volatiles, phenolics, sugars, organic acids, and antioxidant properties of Sel-42 and Tainung papaya varieties. *Food Chem.* **2015**, *173*, 912–919. [CrossRef] [PubMed]
29. Selvaraj, Y.; Pal, D.; Subramanyam, M.; Iyer, C. Changes in the chemical composition of four cultivars of Papaya (*Carica papaya* L.) during growth and development. *J. Hortic. Sci.* **1982**, *57*, 135–143. [CrossRef]
30. Bain, J.M.; Robertson, R. The physiology of growth in apple fruits I: Cell size, cell number, and fruit development. *Aust. J. Biol. Sci.* **1951**, *4*, 75–107. [CrossRef]
31. Leshem, Y.Y.; Ferguson, I.B.; Grossman, S. On Ethylene, calcium, and oxidative mediation of whole apple fruit senescence by core control. In Proceedings of the Ethylene: Biochemical, Physiological and Applied Aspects, An International Symposium, Oiryat Anavim, Israel, 9–12 January 1984; Fuchs, Y., Chalutz, E., Eds.; Springer: Dordrecht, The Netherlands, 1984; pp. 111–120. [CrossRef]
32. Casero, T.; Torres, E.; Alegre, S.; Recasens, I. Macronutrient accumulation dynamics in apple fruits. *J. Plant Nutr.* **2017**, *40*, 2468–2476. [CrossRef]
33. Nachtigall, G.R.; Dechen, A.R. Seasonality of nutrients in leaves and fruits of apple trees. *Sci. Agric.* **2006**, *63*, 493–501. [CrossRef]
34. Mota, M.; Martins, M.J.; Policarpo, G.; Sprey, L.; Pastaneira, M.; Almeida, P.; Maurício, A.; Rosa, C.; Faria, J.; Martins, M.B.; et al. Nutrient content with different fertilizer management and influence on yield and fruit quality in apple cv. Gala. *Horticulturae* **2022**, *8*, 713. [CrossRef]
35. Joa, J.-H.; Chun, S.-J.; Lim, C.K.; Choi, K.-S.; Kim, S.-C. *Papaya: Manual for Cultivation and Integreted Pest Management*; Rural Development Administration: Jeju, Republic of Korea, 2012.
36. McGuire, R.G. Reporting of objective color measurements. *Hortscience* **1992**, *27*, 1254–1255. [CrossRef]
37. Jermyn, M. Increasing the sensitivity of the anthrone method for carbohydrate. *Anal. Biochem.* **1975**, *68*, 332–335. [CrossRef] [PubMed]

38. Chromý, V.; Vinklárková, B.; Šprongl, L.; Bittová, M. The Kjeldahl method as a primary reference procedure for total protein in certified reference materials used in clinical chemistry I: A review of Kjeldahl methods adopted by laboratory medicine. *Crit. Rev. Anal. Chem.* **2015**, *45*, 106–111. [CrossRef] [PubMed]
39. de Castro, M.L.; Priego-Capote, F. Soxhlet extraction: Past and present panacea. *J. Chromatogr. A* **2010**, *1217*, 2383–2389. [CrossRef] [PubMed]
40. Cunniff, P.; Washington, D. Official methods of analysis of AOAC international. *J. AOAC Int.* **1997**, *80*, 127A.
41. Oh, H.-J.; Jeon, S.-B.; Kang, H.-Y.; Yang, Y.-J.; Kim, S.-C.; Lim, S.-B. Chemical composition and antioxidative activity of kiwifruit in different cultivars and maturity. *J. Korean Soc. Food Sci. Nutr.* **2011**, *40*, 343–349. [CrossRef]
42. Rizzolo, A.; Forni, E.; Polesello, A. HPLC assay of ascorbic acid in fresh and processed fruit and vegetables. *Food Chem.* **1984**, *14*, 189–199. [CrossRef]
43. Mannino, G.; Occhipinti, A.; Maffei, M.E. Quantitative determination of 3-*O*-acetyl-11-Keto-β-Boswellic acid (AKBA) and other Boswellic acids in *Boswellia sacra* Flueck (syn. B. carteri Birdw) and *Boswellia serrata* Roxb. *Molecules* **2016**, *21*, 1329. [CrossRef]
44. Vigliante, I.; Mannino, G.; Maffei, M.E. Chemical characterization and DNA fingerprinting of *Griffonia simplicifolia* Baill. *Molecules* **2019**, *24*, 1032. [CrossRef]
45. Bradstreet, R.B. Kjeldahl method for organic nitrogen. *Anal. Chem.* **1954**, *26*, 185–187. [CrossRef]
46. Crouch, S.R.; Malmstadt, H.V. Mechanistic investigation of molybdenum blue method for determination of phosphate. *Anal. Chem.* **1967**, *39*, 1084–1089. [CrossRef]
47. Zarcinas, B.A.; Cartwright, B.; Spouncer, L.R. Nitric acid digestion and multi-element analysis of plant material by inductively coupled plasma spectrometry. *Commun. Soil Sci. Plant Anal.* **1987**, *18*, 131–146. [CrossRef]
48. R Core Team. *R: A Language and Environment for Statistical Computing*; R Core Team: Vienna, Austria, 2022.
49. de Mendiburu, F. Agricolae: Statistical Procedures for Agricultural Research. Available online: https://github.com/myaseen208/agricolae/ (accessed on 1 January 2023).
50. Kassambara, A.; Mundt, F. Factoextra: Extract and Visualize the Results of Multivariate Data Analyses. Available online: https://github.com/kassambara/factoextra/ (accessed on 1 January 2023).
51. Abdi, H.; Williams, L.J. Principal component analysis. *Wiley Interdiscip. Rev. Comput. Stat.* **2010**, *2*, 433–459. [CrossRef]

Article

Magnetically Treated Water in *Phaseolus vulgaris* L.: An Alternative to Develop Organic Farming in Cuba

Yilan Fung Boix [1], Albys Ferrer Dubois [1], Yanaisy Perez Quintero [1], Elizabeth Isaac Alemán [1], Cristiane Pimentel Victório [2], Jorge González Aguilera [3,*], Malgreter Noguera Betancourt [3] and Luis Morales-Aranibar [4]

1 National Center for Applied Electromagnetism, Santiago de Cuba 90600, Cuba
2 Faculdade de Ciências Biológicas e da Saúde, Universidade do Estado do Rio de Janeiro, Campus Zona Oeste—UERJ-ZO, Rio de Janeiro 23070-200, Brazil
3 Pantanal Editora, Rua Abaete, 83, Sala B, Centro, Nova Xavantina 78690-000, Brazil
4 The Office of Innovation, Technology Transfer and Intellectual Property at the National Intercultural University of Quillabamba, Universidad Nacional Intercultural de Quillabamba, Cusco 08741, Peru
* Correspondence: j51173@yahoo.com

Abstract: *Phaseolus vulgaris* L. (common bean) significantly contributes to the human diet due to its protein, vitamin and mineral contents, making it one of the major edible plant species worldwide. Currently, the genetic resources conserved in germplasm banks in Cuba have experienced a loss of viability, which makes their propagation difficult. Magnetically treated water has been used to improve the response of seeds and plants of different species. However, there is little experimental evidence on the cultivation of the common bean irrigated with magnetically treated water or its positive effects on seed germination recovery and its effects on physiological, anatomical and morphological characteristics. This study aims to evaluate the growth and development of common bean with magnetically treated water as an alternative to rejuvenate the seeds for organic agriculture. A two-group experimental design was used: a group of plants irrigated with water without a magnetic field and a group of plants irrigated with water treated with a magnetic field at induction in the range of 100 to 150 mT. There was an increase of 25% in the percentage of germination; the stomatal anatomical structures behaved normally; and the stem length, vigor index, leaf area and seed weight increased by 35, 100, 109 and 16%, respectively. The concentrations of chlorophyll a, chlorophyll b pigments and carbohydrates in the plants grown with magnetically treated water were also stimulated in relation to control plants with increments of 13, 21 and 26%, respectively. The technology employed in this study did not have negative effects on the plant nor did it affect the presence of structures or the net content of the assessed compounds. Its use in the cultivation of *Phaseolus vulgaris* L. might represent a viable alternative for the improvement of the plant in organic farming production.

Keywords: common bean; magnetic field; photosynthesis; physiological parameters; seeds

Citation: Boix, Y.F.; Dubois, A.F.; Quintero, Y.P.; Alemán, E.I.; Victório, C.P.; Aguilera, J.G.; Betancourt, M.N.; Morales-Aranibar, L. Magnetically Treated Water in *Phaseolus vulgaris* L.: An Alternative to Develop Organic Farming in Cuba. *Plants* **2023**, *12*, 340. https://doi.org/10.3390/plants12020340

Academic Editors: Lord Abbey, Josephine Ampofo, Mason MacDonald and Othmane Merah

Received: 7 December 2022
Revised: 20 December 2022
Accepted: 4 January 2023
Published: 11 January 2023

1. Introduction

Phaseolus vulgaris L. (common bean) belongs to the group of leguminous plants with edible seeds containing a high nutritional value. Beans are grown on all five continents and are an essential staple food, especially in Central and South America [1]. The common bean belongs to the Leguminosae family, which comprises 727 genera and approximately 19,000 species. Of these species, only five belonging to the genus Phaseolus are cultivated worldwide to form part of human food (*P. vulgaris* L., *P. coccineus* L., *P. acutifolius* A. Gray, *P. lunatus* L. and *P. polyanthus* Greenman) [2,3].

Phaseolus vulgaris L. is preferred within the bean species for economic and scientific purposes [4]. The bean is native to America, and its domestication has taken place in northern Andean and Mesoamerican populations as two gene pools [2,5]. The common bean was brought to Europe as an ornamental plant, and over time, it began to be cultivated in almost all parts of the world [2,6]. Common bean is an early maturing crop with a reasonable degree of drought tolerance, which has led it to play an essential role in farmers' strategies in drought-prone lowland regions [7].

In Cuba, the consumption of common bean is very popular due to its high caloric content, which includes, among other elements, phosphorus, vitamins and iron. There are more than 20 improved bean varieties and cultivars in the country [8,9]. Crop yields are 9723 and 110,765 kg ha^{-1} for state and nonstate agricultural sectors, respectively, which cannot meet demand [10]. Over the past decade, the nonstate agricultural sector, mainly consisting of farms and small plots with very diverse conditions and low availability of agrochemical and energetic inputs, was mostly in charge of bean production in our country [11,12]. Many of these productions are mainly based on organic agriculture.

The organic agricultural production system is promoted as a low-cost ecological production system that improves human and soil health, improves the environment and increases agricultural sustainability [13]. Organic agriculture includes cultivation by small, low-income farmers. These low-input productions, if properly managed, can avoid susceptibility to water stress and low soil fertility and avoid storage pests, which are the main constraints to the yield of most crops [14]. In Cuba, the country's conditions have forced the more extensive use of organic agriculture as a way of achieving higher quality production with lower inputs, and beans have been one of those crops produced on this ecological basis.

For better crop yields of *Phaseolus vulgaris* L. and to meet the Cuban population's demand, new technologies that would allow large-scale bean production with the use of environmentally friendly techniques and better seed quality are needed. The success of agricultural production initially depends on the quality of the seeds planted. The term "seed quality" refers to the ability of the seed to germinate and have the vigor to develop in the new growing environment [15]. It is well-known that seed germination and viability can vary greatly from year to year and from one production site to another, as well as depending on how they are conserved [15–17]. Currently, advances in seed technology that stimulate growth and high vigor in the seedling phase have been obtained using films with silver nanoparticles on watermelon seeds (*Citrullus lanatus* L.) [18]; bioplastics consisting of modified starch and chitin mixed with *Bacillus subtilis*, used as a coating for corn seeds [19]; and pelleting with cellulose, diatomaceous earth and soy protein in broccoli seeds (*Brassica oleracea* L.) [20], among other techniques, such as the use of magnetically treated water [21].

One feasible technology is the application of magnetic fields to irrigation water in agricultural, industrial and household irrigation systems with the use of high-performance permanent magnet devices. In Cuba, specifically in Santiago de Cuba province, GREMAG®® technology was employed for the magnetic treatment of irrigation water [21]. This technology has been used in different plant species, such as medicinal plants [22], ornamental plants [23] and vegetable plants [24], with satisfactory results.

This technology improves plant growth variables and nutrient uptake [24–28]. This type of treatment has been used to promote the best performance of several crops, such as *Solanum lycopersicum* L. [24,29–32], *Capsicum annuum* L. [33,34], *Cucumis sativus* L. [35], *Vicia faba* L. [36], *Beta vulgaris* [37], *Raphanus sativus* [38], *Cicer arietinum* L. [39] and *Phaseolus vulgaris* L. [40–42]. However, there is little experimental evidence on the cultivation of *Phaseolus vulgaris* L. irrigated with magnetically treated water or its positive effects on seed germination recovery and its effects on physiological, anatomical and morphological characteristics.

The objective of this paper is to assess the growth and development of *Phaseolus vulgaris* L. sown with magnetically treated water, which may improve the development and production of high-quality beans for human consumption.

2. Results

2.1. Determination of the Effect of Irrigation with Magnetically Treated Water on the Germination of Phaseolus vulgaris L.

The germination percentage showed statistically marked differences between the mean values of the two experimental groups with a confidence level of 95.0%. Plants grown with magnetically treated water showed 75% germination compared to 50% germination in control plants (Figure 1). The difference between treatments represents a gain of 25% in favor of the magnetic treatment of irrigation water (Figure 1).

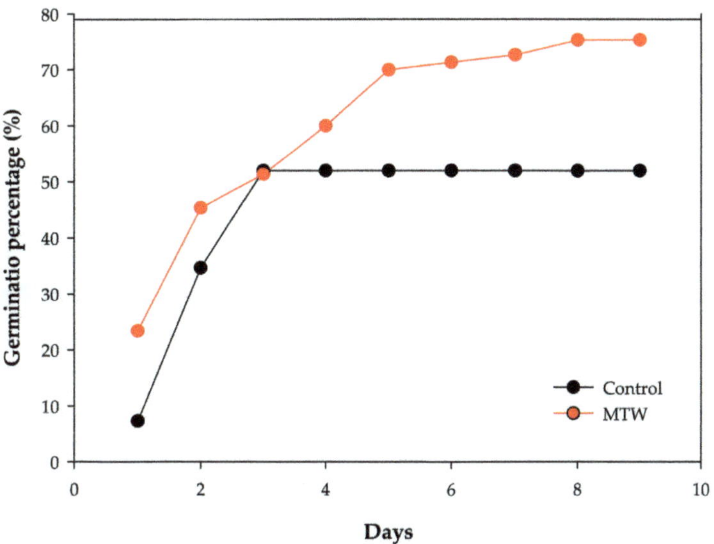

Figure 1. Germination percentage of *Phaseolus vulgaris* L. grown with magnetically treated water ($p < 0.05$). Control: plants grown without magnetically treated water, MTW: plants grown with magnetically treated water under a static magnetic field between 100 and 160 mT.

Similar results were reported for seed germination rates for both experimental groups from day 1 to 2. From day 3 to 8, the germination percentage increased in the plant group treated with magnetically treated water, while it stayed constant in the control group. This suggests that magnetic field application had a positive impact on seed germination and that it can be effective from the onset of the sowing of *Phaseolus vulgaris* L., regardless of the species variety.

Mean values of the hypocotyl length and vigor index in *Phaseolus vulgaris* L. plants treated with magnetized water were markedly different compared to the control plant group, with a 95% probability level (Table 1). The vigor index values suggest statistically significant differences ($p < 0.05$) between the two experimental groups. The length of the hypocotyl also increased within the same range; these data are of great importance for the assessment of the vigor index.

Table 1. Mean values of hypocotyl length and vigor index of *Phaseolus vulgaris* L. grown with magnetically treated water.

Experimental Group [1]	Hypocotyl Length (cm)	Vigor Index (%)
Control	1.28 ± 0.36	66.77 ± 18.44
MTW	1.78 ± 0.58 *	133.79 ± 43.70 *

[1] Control: plants grown without magnetically treated water, MTW: plants grown with magnetically treated water under a static magnetic field between 100 and 160 mT. * indicates statistically significant differences ($p < 0.05$).

2.2. Determination of the Effect of Irrigation with Magnetically Treated Water on the Anatomical and Morphological Characters of *Phaseolus vulgaris* L.

The stomatal area estimates showed statistically significant values of 35.32 ± 3.11 μm^2 for plants treated with MTW compared to control plants (25.54 ± 2.27 μm^2) (Figure 2A). Despite the fact that stomatal density was not significantly different between the two experimental groups (Figure 3), from a biological point of view, it was greater for plants grown with MTW ($4.9 \times 10^{-4} \pm 1.31$ stomata μm^2) compared with the control ($3.93 \times 10^{-4} \pm 0.82$ stomata μm^2) (Figure 2B).

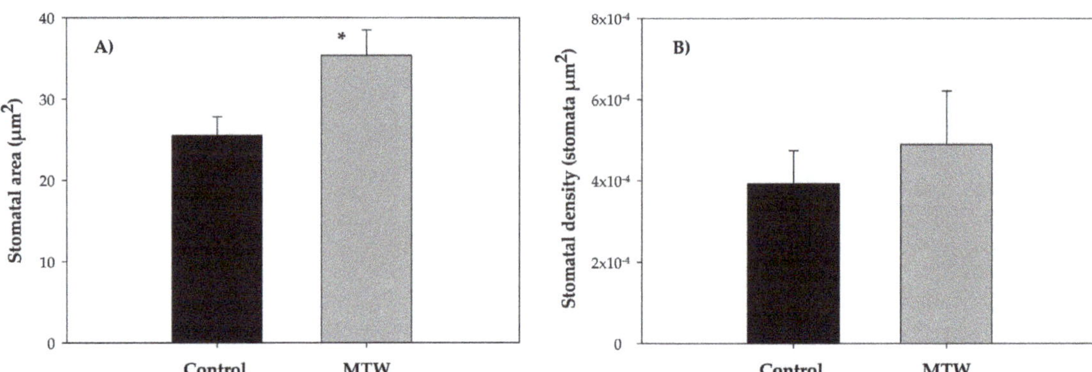

Figure 2. Mean values of stomatal area (**A**) and stomatal density (**B**) in *Phaseolus vulgaris* L. plants grown with magnetically treated water. * indicates statistically significant differences ($p < 0.05$). Control: plants grown without magnetically treated water, MTW: plants grown with magnetically treated water under a static magnetic field between 100 and 160 mT.

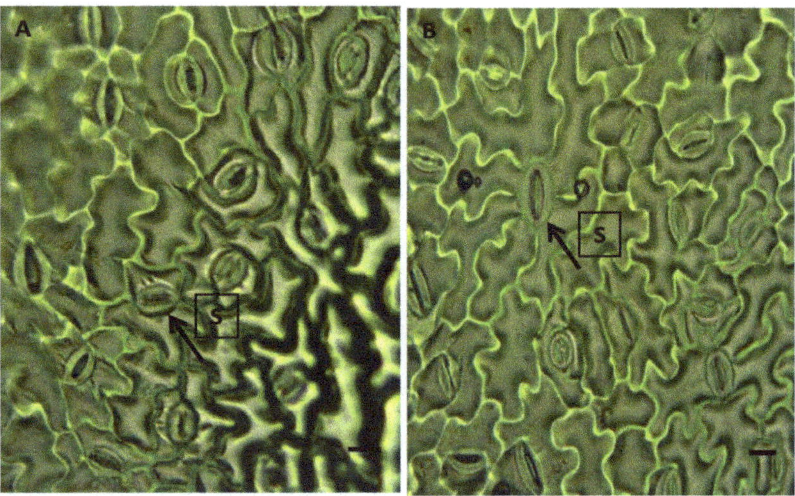

Figure 3. Microphotograph of stomata (S) in a leaf of *Phaseolus vulgaris* L. (**A**) Plants grown without magnetically treated water (control). (**B**) Plants grown with magnetically treated water (MTW). A 40× magnification. Bright field optical microscopy.

Magnetically treated water under an induction range of 100 to 150 mT increased water absorption and enhanced the transpiration process. It was confirmed that the stomal structural characteristics of *Phaseolus vulgaris* L. grown with MTW remained unchanged

(Figure 3). This study demonstrated that magnetically treated water increased absorption and transpiration processes in *Phaseolus vulgaris* L.

The mean values of the stem length in *Phaseolus vulgaris* L. did not show statistically significant differences (Figure 4A). From a biological standpoint, however, the higher values correspond to the plant group irrigated with MTW (5.50 ± 0.68 cm) compared to the control group (4.98 ± 0.83 cm). In contrast, the leaf area values showed statistically significant differences, with a significance level of 95%. The leaf area value was 88.12 ± 17.83 μm^2 for the treated group and 42.12 ± 15.51 μm^2 for the control group (Figure 4B).

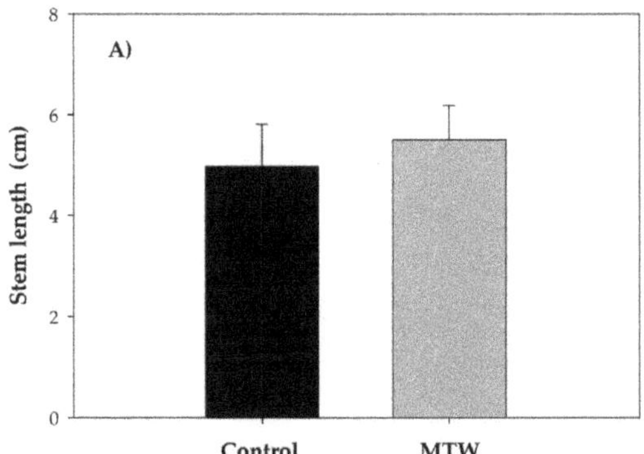

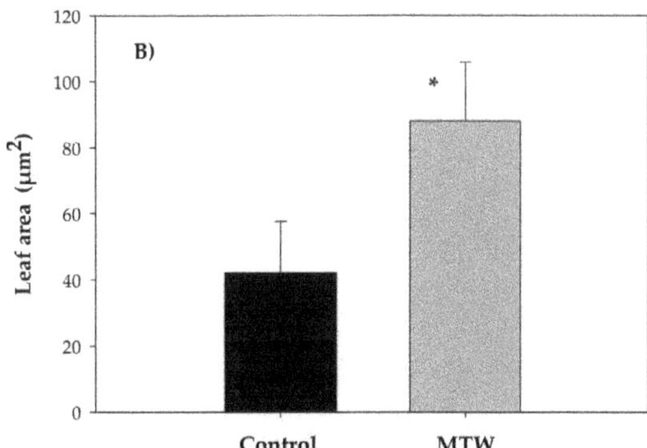

Figure 4. Mean values of stem length (**A**) and leaf area (**B**) in *Phaseolus vulgaris* L. grown with magnetically treated water. * indicates statistically significant differences ($p < 0.05$). Control: plants grown without magnetically treated water, MTW: plants grown with magnetically treated water under a static magnetic field between 100 and 160 mT.

Magnetically treated water impacts seed length, width and weight (Figure 5). Seed weight was the variable that showed statistically significant differences with respect to the control, with a significance level of 95% (Figure 5C). The treated group showed the highest seed weight values (31.73 ± 1.39 g) compared to the control plants (28.62 ± 1.51 g). Although there were no statistically significant differences ($p < 0.05$) in seed length (Figure 5A), the group of plants grown with magnetically treated water exhibited higher values (1.494 ± 0.070 cm) than the control plants (1.382 ± 0.047 cm). Similar results were obtained for the seed width, where magnetically treated plants were 0.71 ± 0.049 cm wide, and seeds in the control group were 0.68 ± 0.034 cm wide (Figure 5B).

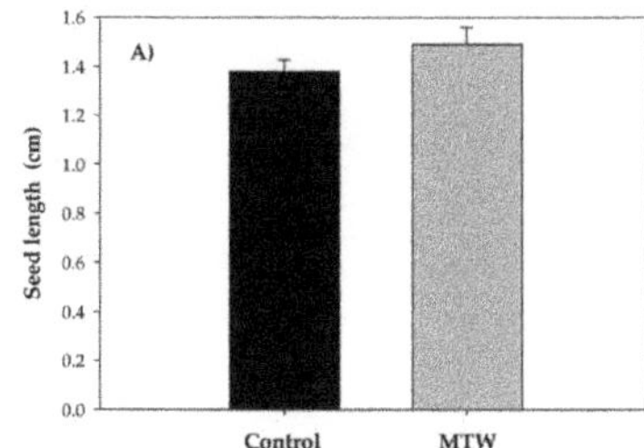

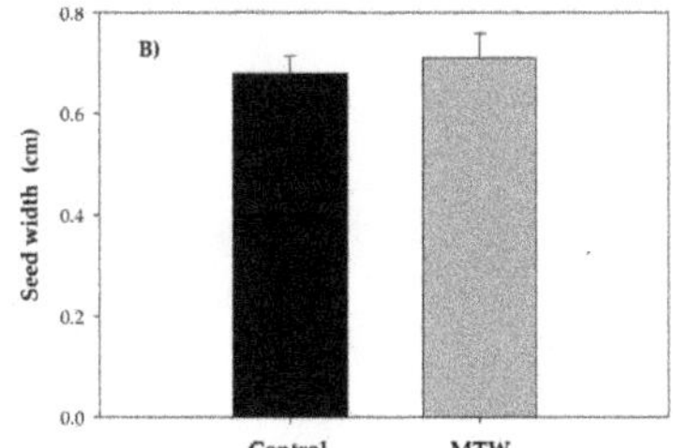

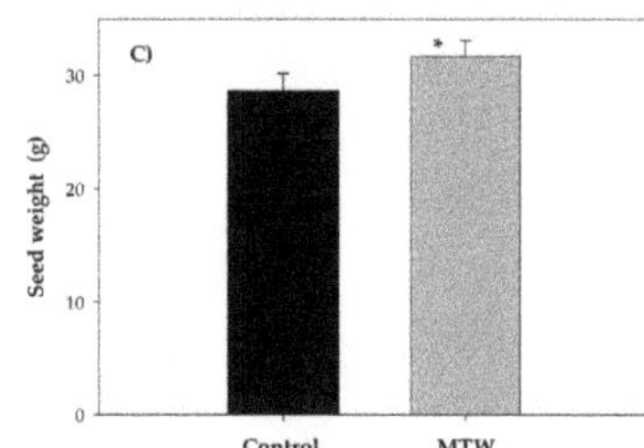

Figure 5. Mean values of seed length (**A**), seed width (**B**) and seed weight (**C**) in *Phaseolus vulgaris* L. grown with magnetically treated water. * indicates statistically significant differences ($p < 0.05$). Control: plants grown without magnetically treated water, MTW: plants grown with magnetically treated water under a static magnetic field between 100 and 160 mT.

In general, the results obtained in this experiment show that magnetically treated water had a positive impact on the anatomical and morphological features associated with the productivity of *Phaseolus vulgaris* L. seeds treated with an induction range of 100 to 150 mT.

2.3. Determination of the Effect of Irrigation with Magnetically Treated Water on the Physiological Characteristics of Phaseolus vulgaris L.

Chlorophyll pigments were determined for *Phaseolus vulgaris* L. species, in which chlorophyll a absorbs violet, blue, orange—reddish and red wavelengths of radiation and a few intermediate wavelengths of radiation (green–yellow–orange) (Figure 6). Accessory pigments include chlorophyll b and carotenoids, which act as antennas, conducting the energy they absorb toward the reaction center. Chlorophyll molecules are found in the reaction center and can transfer their excitation as useful energy in biosynthetic reactions.

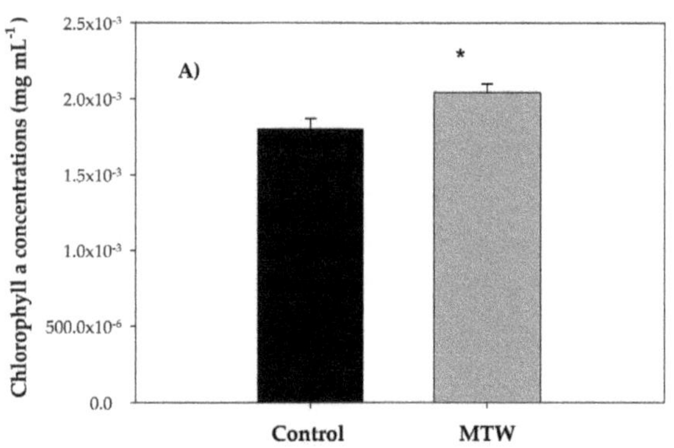

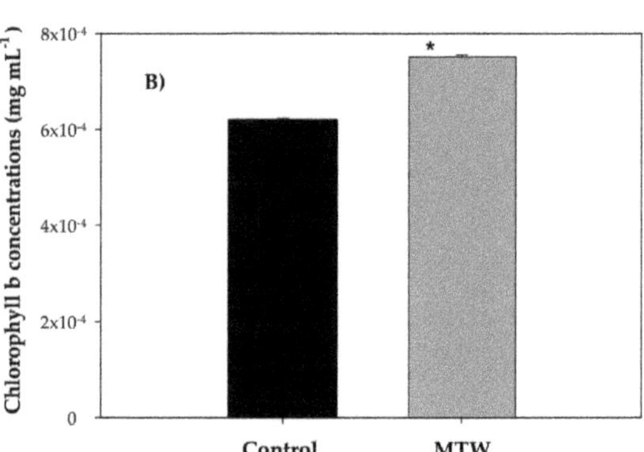

Figure 6. Mean values of chlorophyll concentrations in *Phaseolus vulgaris* L. grown with magnetically treated water. * indicates statistically significant differences ($p < 0.05$). Control: plants grown without magnetically treated water, MTW: plants grown with magnetically treated water under a static magnetic field between 100 and 160 mT.

The plants grown with MTW showed higher chlorophyll a content than the control group, with values of $2.04 \times 10^{-3} \pm 5.74 \times 10^{-5}$ mg mL^{-1} and $1.8 \times 10^{-3} \pm 6.9 \times 10^{-5}$ mg mL^{-1}, respectively. Statistically significant differences, with a significance level of 95%, were found for the mean values of chlorophyll a and b content in both experimental groups. The chlorophyll b content values in plants irrigated with magnetically treated water and the control group plants were $7.5 \times 10^{-4} \pm 4.6 \times 10^{-5}$ mg mL^{-1} and $6.2 \times 10^{-4} \pm 2.5 \times 10^{-5}$ mg mL^{-1}, respectively.

Figure 7 presents the mean values of the carbohydrate concentration in the experimental groups of *Phaseolus vulgaris* L., where the MTW group exhibited higher values (2350 ± 290 mgmL^{-1}) compared to the control group (1860 ± 260 mg mL^{-1}). Statistically significant differences with a probability of $p < 0.05$ were observed.

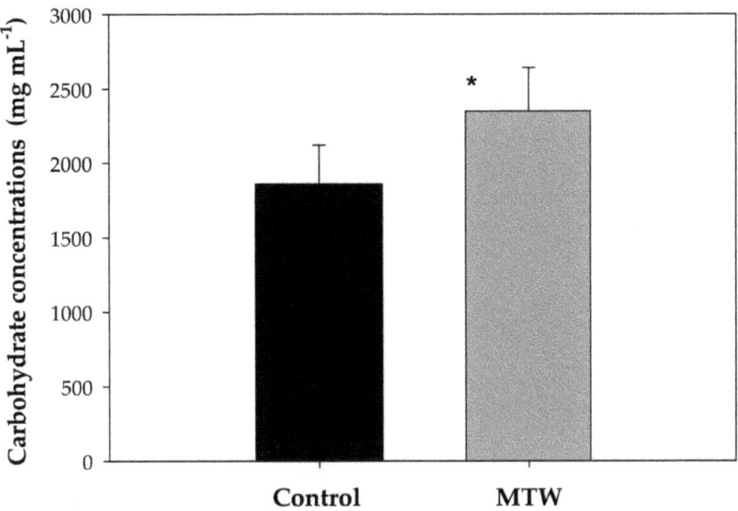

Figure 7. Carbohydrate concentrations in *Phaseolus vulgaris* L. grown with magnetically treated water * indicates statistically significant differences ($p < 0.05$). Control: plants grown without magnetically treated water, MTW: plants grown with magnetically treated water under a static magnetic field between 100 and 160 mT.

3. Discussion

The results obtained agree with the findings of Torres et al. [29], who reported a higher germination percentage for *Oryza sativa* L. (rice) and *Solanum lycopersicum* L. (tomato) after using magnetic fields of 10 and 15 mT. In the studies conducted by Grewal and Maheshwari [43] with *Pisum sativum* L. (pea) and *Cicer arietinum* L. (chickpea), magnetically treated water produced a significant increase ($p < 0.05$) in the germination percentage under an induction range between 3.5 and 136 mT. Selim et al. [27] also found that magnetically treated water increased the vigor index of *Solanum lycopersicum* L. (tomato), *Triticum aestivum* L. (chickpea) and *Pisum sativum* L. (pea) seedlings by 18.3, 81.9 and 65.1%, respectively, when compared to control plants.

The impact of magnetic field application on the germination process of plant species varies according to the magnetic inductions used. Plants grow under the effect of treated water that has different characteristics owing to changes in the water molecule structure, hydrogen bridges, pH variations and electrical conductivity [26,38]. These distinctive features of irrigation water favor the physiological and metabolic processes that occur during the germination of *Phaseolus vulgaris* L. from hypocotyl elongation to leaf appearance. This explains why the germination percentage and vigor index obtained in the group of plants grown with MTW were higher than those of control plants. Mroczek-Zdyrska et al. [42] found that the stimulation of plants by a weak permanent magnetic field (130 mT) increased the mitotic activity in the meristematic cells of the common bean. There was no influence of the 130 mT magnetic field stimulation on the development of aboveground plant parts.

In both experimental groups, stomata exhibited normal anatomical features, with two occlusive cells and three adjacent epidermal cells (Figure 3). Epidermal tissue has structures that are made up of joined cells, including stomata, formed by an ostiole, surrounded by two kidney-shaped occlusive cells which may or may not be accompanied by other epidermal cells that are structurally and physiologically associated with the stomata [44]. Cell wall thickening was observed in the inside rather than the outside. This structural adaptation of stomata confers on them the elasticity that enables them to open or close stomatal pores depending on turgid cell pressure [37].

Stomatal area values indicate that plants grown with magnetically treated water showed a larger stomatal aperture, which is proportional to an increased transpiration process. Certain environmental conditions, such as high relative humidity, have an impact on plant growth and can affect stomatal area and density. Although this may cause higher plant water consumption, a balanced use of water can be attained. As *Phaseolus vulgaris* L. shows a better water absorption rate, allowing its optimization, hydric equilibrium and plant growth are enhanced. Cultivation of this plant species is possible under extreme environmental conditions. Similar stem length values have been reported by Ahamed et al. [33] for *Capsicum annuum* L. (sweet pepper) seeds irrigated with magnetically treated water with an induction of 57–60 mT.

The lack of differences in stem length may be associated with meteorological conditions and the sowing time of *Phaseolus vulgaris* L. Khattab et al. [41] did not find significant differences in the stem length of *Phaseolus vulgaris* L. when compared with two varieties (Goya and Hama) of the species; however, this variable did show a significant increase due to the influence of weather conditions when varieties of the species were sown at different times of the year.

Leaves are the most active organs in processes such as photosynthesis, respiration and perspiration. The estimation of leaf area is of great importance for the analysis of plant growth, as it is closely related to the photosynthetic efficiency of plants [45]. Plants irrigated with magnetically treated water showed an increase in leaf growth and development, which could enhance photosynthesis (Figure 4B). Selim et al. [27] reported similar results for *Solanum lycopersicum* L. (tomato) and *Pisum sativum* L. (pea) irrigated with magnetically treated water. These results are in agreement with those obtained in leaf and stomatal areas and stomatal density estimations, as under normal conditions, stomata play a fundamental role in the water loss process in the form of steam from leaf surfaces during transpiration and stomatal opening and closure movements, which relates to the water supply of plants and conservation of plant homeostasis. This plant organ is directly involved in CO_2 uptake and, together with chlorophyll pigments, enables good plant development and growth [46].

Plant growth does not take place uniformly; instead, it concentrates in specific and distinctive areas. Initial plant growth is essential in certain parts of the plant showing active cell division known as meristems [45]. This behavior is expressed in the same way in the growth and development of the stem, where axillary and apical shoots have meristematic tissue capable of differentiating and starting their elongation. Leaves also play a key role in plant growth, as different metabolic processes take place in leaves with the presence of minerals, growth regulators and enzymes codified by a specific gene in each type of plant.

Water plays a pivotal role in seed growth and development. Water absorption is the first process that occurs during seed germination. All hydrolysis products feed the embryo so it can start growing [47]. The biological function of water is related to the hydration of the new plant, its nutrition and the production of essential compounds necessary for the germination process, thus enhancing embryo cell elongation and radicle emergence.

These results match those obtained for other species of leguminous plants reported by Moussa [40], who found a significant increase in the crop yield and weight of *Phaseolus vulgaris* L. cv. Master. However, for the production of the seeds of *Phaseolus vulgaris* L., the type of water used for magnetic treatment, exposure times and magnetic induction must be taken into consideration [26]. Other elements to be considered include endogenous factors such as genotype and species varieties, which have a different response depending on the sowing cycle and time [42].

Chlorophylls are pigments that enable light absorption in a very effective way as they possess conjugated double bond systems, according to Azcón-Bieto and Talón [48], and enhance photosynthesis [39]. Latef et al. [49] showed that in lettuce seeds (*Lactuca sativa* var. *cabitat* L.) seedlings, they used various intensities of the static magnetic field, and the best result was obtained in the treatment group with a magnetic induction of 770 m, specifically in terms of the growth parameters and metabolism compared to the control group. Photosynthetic pigments were induced markedly in the static magnetic field group, especially the chlorophyll a pigment.

Solar energy is absorbed by the photosynthetic pigments found in plant leaf chloroplasts; chlorophyll a and b are the most abundant in plants. All chlorophylls have a complex ring chemically related to the porphyrin-like groups found in the hemoglobin in animals and cytochromes. In addition, a long hydrocarbon tail is attached to the structure of the ring, which also contains weakly bound electrons and is part of the molecule involved in electron transfer and redox reactions [50]. In this case, as the content of chlorophyll pigments in the seedlings treated with magnetized water increased, the capacity to carry out photosynthesis was enhanced, thus providing greater growth vigor and development.

Selim et al. [27] reported that magnetized water increased photosynthetic pigments in pea plants (*Pisum sativum* L.) and chickpea (*Cicer arietinum*) and *Solanum lycopersicum* L. (tomato). The results obtained for these species are similar to those reported in the literature [37], where it is stated that magnetic field intensity and time of exposure influence the concentration of photosynthetic pigments, which may increase, decrease or maintain their composition with respect to plants irrigated with unmagnetized water. In this case, a decrease in the concentration of these chemical substances (statistically significant difference with a 95% confidence level) is observed, probably as a response to magnetic field overexposure.

The biological effect of the magnetic field depends on the magnetic field intensity. Depending on the magnetic field intensity, the accumulation of chemical substances can increase, decrease or remain unchanged [51]. Moussa [40] determined the concentration of photosynthetic pigments in *Phaseolus vulgaris* L. cv. Master and reported higher values in plants irrigated with magnetically treated water. They also found that carotenoid content and photosynthetic activity increased compared to the control group of the same species.

The results reported in this study are similar to those obtained by these authors, even though they used a 30 mT magnetic induction. Their research also suggests that magnetic fields applied to irrigation water can stimulate plant defensive systems, provided that photosynthesis and the translocation of the products obtained from this process are more efficient due to the mobilization of some growth regulators and enzymes determined in their studies. The results obtained in this paper with the use of a different magnetic induction range of 100–150 mT could lead to processes similar to those already described by these authors.

Magnetic treatment applied to irrigation water has led to increased carbohydrate concentrations, thus improving the photosynthesis and nutritional value of *Phaseolus vulgaris* L. Several authors have reported increments in the content of the total available carbohydrates (monosaccharides, disaccharides, polysaccharides) of *Vicia faba* L. plants irrigated with magnetically treated water compared to plants irrigated with water without magnetic treatment [36]. The increase in the concentration of carbohydrates was possible due to the close relationship between stomatal conductance and photosynthesis; this could be the result of bioenergetic structural excitation with increased cell pumping and enzymatic stimulation. During this metabolic process, a huge variety of chemical compounds are obtained, including indole acetic acid growth regulators (IAA) and proteins, as the expression of genes that encode them has been affected [36].

These results may be related to the effect of the static magnetic field on irrigation water, which generates chemical–physical changes. Pang and Deng [25] claim that the magnetic treatment of water and its effects on the chains of hydrogen bonds forming water molecules can generate a transfer of protons from the hydrogen bonds of closed chains, which may interact with the magnetic field applied externally. At the same time, due to these magnetic interactions, closed chains can be ordered. Therefore, for the purpose of this research, this could explain why the treatment of water used for the irrigation of *Phaseolus vulgaris* L. with an induction range of 100–150 mT (0.1–0.15 T) has also undergone some changes in its solubility, surface tension, electrical conductivity and pH, leading to changes in its physical–chemical parameters. In addition, it is argued that magnetic treatment can affect the cluster composition that structures water molecules. These changes lead to a reduction in the molecular diameter and complexity; plants assimilate water more easily, and the transport of nutrients necessary for metabolism occurs more efficiently.

When the seeds of *Phaseolus vulgaris* L. are irrigated with magnetized water, they capture a signal induced by the behavior of excited water that presents in the form of physical stress at the cellular and molecular levels, causing changes in cell permeability, all of which have been reported. These interactions may have allowed soil water absorption and the transport of ions to be carried out more efficiently through the seed coat, thus starting the softening process.

The irrigation of *Phaseolus vulgaris* L. seeds with magnetically treated water under a magnetic induction of 100 and 150 mT had positive effects compared to the experimental group without magnetic treatment. The anatomical, morphological and physiological characteristics of the species, as well as the germination of its seeds, showed changes that enhance its primary and secondary metabolism. The technology employed in this study did not have negative effects on the plant nor did it affect the presence of structures or the net content of the assessed compounds. Its use in the cultivation of *Phaseolus vulgaris* L. might represent a viable alternative for the improvement of the plant and may lead to the development of state and nonstate organic farming productions.

4. Materials and Methods

4.1. Plant Material

Experiments were performed in the Plant Biotechnology Laboratory of the National Center for Applied Electromagnetism (CNEA), Santiago de Cuba, Cuba. *Phaseolus vulgaris* L. seeds were obtained from the seed laboratory attached to the Provincial Division of the Ministry of Agriculture (MINAG). During the experiments, the temperature values was in the range of 32 ± 2 °C; the average mean relative humidity was between 70 and 80%, and the rainfall was between 60 and 65 mm^3. The Eastern Center for Ecosystems and Biodiversity (BIOECO), holder of registration identification No. 21842, was in charge of plant species classification.

4.2. Crop Conditions

A substrate composed of a mixture of soil and organic matter (3:1) was used. In addition, biological control of pathogens, bacteria and fungi was carried out in compliance with standards set out by the Center for Industrial Biotechnology Studies (CEBI) of the University of Oriente (Table 2).

Table 2. Physico-chemical analysis of soil in CNEA's experimental plot farm for the cultivation of *Phaseolus vulgaris* L.

Characteristics	Mean Values	Characteristics	Mean Values
Electrical conductivity (μScm^{-1})	397	Aluminum (%)	12.67
Potassium oxide (%)	0.30	Magnesium oxide (%)	8.06
Phosphorous (%)	0.940	Cobalt (%)	<0.005
Calcium oxide (%)	7.29	pH	7.84
Iron (%)	3.9	Manganese (%)	0.20

4.3. Irrigation Water Characteristics

Irrigation was performed twice a day through an aerial microjet system for 30 min. The accessories used were an ITUR pump and valve-controlled distributor system that ensures irrigation is run in sections. Water tests were performed at CNEA's LESA Lab. In the water sample analyzed, pH, total hardness and total dissolved solids values were below the maximum admissible limits (MAL) for chemical components, which may affect the organoleptic quality of drinking water (Cuban Standard 827:2012) (Table 3).

Table 3. Physico-chemical analysis of irrigation water at CNEA's experimental plot farm for the cultivation of *Phaseolus vulgaris* L. *: expressed as $CaCO_3$.

Parameters	Control	MTW
Total hardness * (mg L−1)	129.28 ± 0.00	162.95 ± 2.33
Calcium hardness * (mg L−1)	100.80 ± 0.00	92.74 ± 0.00
Magnesium hardness * (mg L−1)	28.48 ± 0.00	70.21 ± 2.33
Phenolphthalein alkalinity * (mg L−1)	0.00 ± 0.00	0.00 ± 0.00
Total alkalinity * (mg L−1)	149.18 ± 0.00	145.15 ± 0.00
Bicarbonate * (mg L−1)	149.18 ± 0.00	145.15 ± 0.00
Chlorides (mgL−1)	23.24 ± 0.54	24.38 ± 1.08
pH	7.71 ± 0.01	7.70 ± 0.00
Electrical conductivity µScm−1	301.00 ± 2.00	302.00 ± 0.00
Total dissolved solids (mg L−1)	150.53 ± 0.97	151.00 ± 0.00
Salinity (ppt)	0.15 ± 0.00	0.15 ± 0.00
Water speed (ms^{-1})	1.4–1.6	
Pump flow $m^3\ h^{-1}$	2.54–2.91	

4.4. Description of Static Magnetic Treatment of Irrigation Water

Seeds were planted in two 12 m long and 0.5 m wide beds at CNEA's experimental plot farm. The total sample size was 80 plants; each experimental group included 40 plants.

Treatments included two experimental groups: treatment 1 (control group): plants grown with water without magnetic treatment; treatment 2: plants grown with water magnetized with a static magnetic field (MTW) between 100 and 150 mT.

For the magnetic treatment, an external permanent magnet device designed, built and characterized at the National Center for Applied Electromagnetism was used. The device produced a magnetic induction range between 100 and 150 mT measured with a 192041 Soviet Microwebermeter (relative error less than 5%). These results were verified with nuclear magnetic resonance equipment and a 410 Gauss meter/Teslameter (relative error of 0.01 G) [52].

4.5. Determination of the Effect of Irrigation with Magnetically Treated Water on the Germination of Phaseolus vulgaris L. Seeds

Germination percentage: germinated seeds were counted for 24 days. Using these data, the seed germination percentage was calculated using the formula P = (seeds germinated/total seeds) × 100, according to Abdul-Baki and Andenson [53].

Hypocotyl length: measured from the base of seed hypocotyl to the starting point of epicotyl using a ruler (cm) following the methodology proposed by Abdul-Baki and Andenson [53].

Vigor index I: calculated using the formula suggested by Abdul-Baki and Andenson [53].

$$\text{Vigor index I} = \text{germination percentage} \times \text{hypocotyl length}$$

4.6. Determination of Irrigation with Magnetically Treated Water on the Anatomical and Morphological Characteristics of Phaseolus vulgaris L. Seeds

Anatomical characters comprise stomatal density and area, whereas morphological characters include stem length, leaf area and seed area and weight. These parameters were estimated as follows:

Stomatal density and area: calculated using the stomatal impression method. The sample was taken from the underside (abaxial side) of the leaf; the epidermal impression obtained was observed through an optical microscope (Olympus, Cx41, Japan) under 10× and 40× magnification. The number of stomata was calculated in three fields, and stomatal length and width were determined according to Ortega and Rodés [54]. The results were expressed as the number of stomata (μm^2), with three replications.

Stomatal area: Stomal length and width were measured and multiplied by $\pi = 3.14$, as suggested by Ortega and Rodés [54]. Values were expressed in μm^2 with three replications.

Stem length: The length of the stem was measured using a measuring tape (cm) in 40 plants of each experimental group from the base of the stem to the main root apex.

Length of leaf area: Leaf area was measured in the same plants of each experimental group. The following equation was used: LA= leaf length × leaf width × π (π = 3.14).

Seed length and width: The length and width of 100 seeds from each experimental group were measured using a calibrated ruler in centimeters (cm).

Seed weight: Seed weight was estimated with a Sartorious®® digital analytical balance (BS 124S, China; accuracy: 0.1 mg). Values were expressed in grams (g) with three replications for each experimental group.

4.7. Determination of the Effect of Irrigation with Magnetically Treated Water on the Physiological Characteristics of Phaseolus vulgaris L. Seeds

4.7.1. Concentration of Photosynthetic Pigments

To determine the concentration of photosynthetic pigments, 1 g of fresh leaf was weighed. Extracts were prepared with 50 mL of solvent (96% ethanol) for both the plant group irrigated with MTW and the control plant group. Leaves were macerated in a porcelain mortar with said solvent. They were filtered using filter paper, and ethanol was added to a volume of 50 mL; 5 mL of this solution was transferred to a 50 mL volumetric flask and brought up to the mark with 96% ethanol. Absorbance peaks of extracts were determined using a Genesys spectrophotometer (10 UV, United States) according to Ortega and Rodés [54]. Absorbance was obtained at wavelengths of 440, 472, 643, 645, 649, 654, 663 and 665 nm. The formulas shown below were employed:

$$\text{Ca} = 0.0127\ \text{A668} - 0.00269\ \text{A648}$$

$$\text{Cb} = 0.0229\ \text{A665} - 0.00468\ \text{A663}$$

$$\text{Ccarot} = 4.695\ \text{A440.5} - 0.268\ \text{Ca} + \text{b}$$

Values were expressed in g mL^{-1} with three repetitions.

4.7.2. Concentration of Carbohydrates

Estimation of total carbohydrates was carried out using the phenol-sulfuric colorimetric method described by Dubois et al. [55]. One milliliter of *Phaseolus vulgaris* L. was taken from the experimental groups. For the preparation of the glucose standard solution, 50 mg of glucose was diluted in 100 mL of distilled H_2O as a final volume. Absorbance was determined at 492 nm using a Genesys spectrophotometer (10 UV, United States). The results were expressed in D-glucose from a calibration curve between 0.6 and 1.6 $mgmL^{-1}$ obtained for this compound. The total carbohydrate content was expressed in mg of saccharides g^{-1} of dry weight. Distilled water was used as a blank in all three replicates. The mathematical equation used to determine the sample concentration was the following: "Extract Conc. = D-glucose Equivalent Conc.de * dilution math conc".

4.8. Statistical Analysis

A completely randomized experimental design with three replications was used. The Kolmogorov–Smirnov test was performed to check if the data had a normal distribution and verify homogeneity of variances. Descriptive statistics and simple classification analysis of variance (ANOVA) were performed. Student's t-test was used to compare two samples. All statistical analyses were performed with a significance value of $p < 0.05$ using Statgraphics Centurium XV for Windows (Graphics Software Systems, STCC, 2000, Inc., Rockville, MA, USA), Basic Statistics, Prisma 5.01 and Origen 6.0.

5. Conclusions

The magnetically treated water under a magnetic induction of 100 to 150 mT stimulated the germination, growth and biomass production of the bean plants. This positive impact of magnetically treated water on beans was associated with an increase in stomatal (38%)

and leaf area (109%), together with stem length (35%), vigor index (100%), leaf area (109%) and seed weight (16%), as well as the anatomical (chlorophyll a and b, with 13 and 21%, respectively) and carbohydrates (26%). These results can be translated as a greater seed recovery and better crop performance for food purposes. Thus, the consumption of beans treated as a food with a high nutritional value is possible. The data also demonstrate that no malformations or abnormal changes were detected in plants grown with magnetically treated water, so this technology is a commercially successful agricultural practice, and magnetic treatment can be used as a suitable environmentally friendly physical method capable of contributing to sustainable agriculture.

Author Contributions: Conceptualization, Y.F.B. and E.I.A.; data curation, Y.F.B., A.F.D., M.N.B. and L.M.-A.; formal analysis, Y.F.B., A.F.D., Y.P.Q., E.I.A., C.P.V., J.G.A., M.N.B. and L.M.-A.; funding acquisition, Y.F.B., J.G.A. and L.M.-A.; investigation, Y.F.B., A.F.D., Y.P.Q., E.I.A., C.P.V., J.G.A., M.N.B. and L.M.-A.; methodology, Y.F.B., A.F.D., Y.P.Q., E.I.A., J.G.A. and L.M.-A.; project administration, Y.F.B. and E.I.A.; resources, Y.F.B., J.G.A. and L.M.-A.; software, Y.F.B.; supervision, Y.F.B.; validation, Y.F.B., Y.P.Q., C.P.V., J.G.A. and L.M.-A.; visualization, Y.F.B., C.P.V. and L.M.-A.; writing—original draft, Y.F.B., A.F.D., Y.P.Q., E.I.A., C.P.V., J.G.A., M.N.B. and L.M.-A.; writing—review and editing, Y.F.B., A.F.D., Y.P.Q., E.I.A., C.P.V., J.G.A., M.N.B. and L.M.-A. All authors have read and agreed to the published version of the manuscript.

Funding: This research received no external funding.

Data Availability Statement: Not applicable.

Acknowledgments: This research is associated with grain production, one of CITMA´s prioritized programs for food production. The authors would like to thank the National Center for Applied Electromagnetism and the Cuban Ministry of Agriculture.

Conflicts of Interest: The authors declare no conflict of interest.

References

1. FAO STAT. Available online: http://www.fao.org/faostat/en/#compare (accessed on 22 April 2021).
2. Bitocchi, E.; Rau, D.; Bellucci, E.; Rodriguez, M.; Murgia, M.L.; Gioia, T.; Santo, D.; Nanni, L.; Attene, G.; Papa, R. Beans (*Phaseolus* ssp.) as a model for understanding crop evolution. *Front. Plant Sci.* **2017**, *8*, 722. [CrossRef] [PubMed]
3. Nadeem, M.A.; Yeken, M.Z.; Shahid, M.Q.; Habyarimana, E.; Yilmaz, H.; Alsaleh, A.; Hatipoglu, R.; Cilesiz, Y.; Khawar, K.M.; Ludidi, N.; et al. Common bean as a potential crop for future food security: An overview of past, current and future contributions in genomics, transcriptomics, transgenics and proteomics. *Biotechnol. Bitechnol. Equip.* **2021**, *35*, 758–786. [CrossRef]
4. Avican, O.; Bilgen, B.B. Investigation of the genetic structure of some common bean (*Phaseolus vulgaris* L.) commercial varieties and genotypes used as a genitor with SSR and SNP markers. *Genet. Resour. Crop Evol.* **2022**, *69*, 2755–2768. [CrossRef]
5. Assefa, T.; Mahama, A.A.; Brown, A.V.; Cannon, E.K.S.; Rubyogo, J.C.; Rao, I.M.; Blair, M.W.; Cannon, S.B. A review of breeding objectives, genomic resources and marker-assisted methods in common bean (*Phaseolus vulgaris* L.). *Mol. Breed.* **2019**, *39*, 20. [CrossRef]
6. Rodino, A.P.; Drevon, J.J. Migration of a grain legume *Phaseolus vulgaris* in Europe. In *Biological Resources and Migration*; Werner, D., Ed.; Springer: Berlin/Heidelberg, Germany, 2004.
7. Besaye, B.H.; Galgaye, G.G. Impact of common bean (*Phaseolus vulgaris* L.) genotypes on seed yield, and seed quality at different locations of Eastern Ethiopia. *Cogent Food Agric.* **2022**, *8*, 1. [CrossRef]
8. Calzada, K.; Fernández, J.; Sotolongo, M. Comportamiento productivo del frijol (*Phaseolus vulgaris* L.) ante la aplicación de un promotor del crecimiento activado molecularmente. *Avances* **2015**, *17*, 327–337. Available online: http://www.ciget.pinar.cu/ojs/index.php/publicaciones/article/view/130 (accessed on 16 September 2019).
9. Olivera, A.; Morales, A.; Batista, F.; Rodríguez, J.; Montero, M.; Alfonso, A. Comportamiento agroproductivo de un sistema de producción de frijol (*Phaseolus vulgaris* L.) con enfoque agroecológico en el municipio Primero de enero. *Univ. Cienc.* **2016**, *5*, 26–51. Available online: https://revistas.unica.cu/index.php/uciencia/article/view/211 (accessed on 20 September 2020).
10. Oficina Nacional de Estadística e Información. Anuario estadístico de Cuba. 2019. Available online: http://www.onei.gob.cu/sites/default/ (accessed on 14 September 2019).
11. Rodríguez, O.; Chaveco, O.; Ortiz, R.; Ponce, M.; Ríos, H.; Miranda, S.; Dias, O.; Portelles, Y.; Torres, R.; Cedeño, L. Líneas de frijol común (*Phaseolus vulgaris* L.) resistentes a la sequía. Evaluación de su comportamiento frente a condiciones de riego, sin riego y enfermedades. *Temas De Cienc. Y Tecnol.* **2009**, *38*, 17–26.
12. de la Fé, C.; Lamz, A.; Cárdenas, R.; Hernández, J. Respuesta agronómica de cultivares de frijol común (*Phaseolus vulgaris* L.) de reciente introducción en Cuba. *Cultiv. Trop.* **2016**, *37*, 102–107.

13. Aulakh, C.S.; Sharma, S.; Thakur, M.; Kaur, P. A review of the influences of organic farming on soil quality, crop productivity and produce quality. *J. Plant Nutr.* **2022**, *45*, 1884–1905. [CrossRef]
14. Morales-Soto, A.; Lamz-Piedra, A.; Chang-Sidorchuk, L.; Martínez-Zubiaur, Y. Common bean (*Phaseolus vulgaris*) genotypes with resistance to BGYMV in Cuba. *Agron. Mesoam.* **2022**, *33*, 3. [CrossRef]
15. Hampton, J.G. What is seed quality? *Seed Sci. Technol.* **2002**, *30*, 1–10. Available online: https://www.semanticscholar.org/paper/What-is-seed-quality-Hampton/e5cb4bc09e1bf6aa7653ad6f8b5601332b27cfe7 (accessed on 1 September 2018).
16. Egli, D.B.; TeKrony, D.M.; Heitholt, J.J.; Rupe, J. Air temperature during seed filling and soybean seed germination and vigor. *Crop Sci.* **2005**, *45*, 1329–1335. [CrossRef]
17. Baskin, C.C.; Baskin, J.M. *Seeds: Ecology, Biogeography, and Evolution of Dormancy and Germination*; Elsevier: Amsterdam, The Netherlands, 2014.
18. Acharya, P.; Jayaprakasha, G.K.; Crosby, K.M.; Jifon, J.L.; Patil, B.S. Nanoparticle-mediated seed priming improves germination, growth, yield, and quality of watermelons (*Citrullus lanatus*) at multilocations in Texas. *Sci. Rep.* **2020**, *10*, 5037. [CrossRef]
19. Accinelli, C.; Abbas, H.K.; Shier, W.T. A bioplastic-based seed coating improves seedling growth and reduces production of coated seed dust. *J. Crop Improv.* **2018**, *32*, 318–330. [CrossRef]
20. Amirkhani, M.; Netravali, A.N.; Huang, W.; Taylor, A.G. Investigation of Soy Protein–based Biostimulant Seed Coating for Broccoli Seedling and Plant Growth Enhancement. *HortScience* **2016**, *51*, 1121–1126. [CrossRef]
21. Ferrer, A.E.D.; Aguilera, J.G.; Fung, Y.B.; Issac, E.A.; Zuffo, A.M. Use of GREMAG® technology to improve seed germination and seedling survival. In *Ciência Em Foco*, 1st ed.; Zuffo, A.M., Aguilera, J.G., de Oliveira, B.R., Eds.; Pantanal Editora: Nova Xavantina, Brazil, 2019; Volume 1, pp. 138–149. [CrossRef]
22. Fung, Y.B.; Ferrer, A.E.D.; Issac, E.A.; Aguilera, J.G.; Victorio, C.P.; Cuypers, A.; Beernarts, N. Stimulation of physiological parameters of Rosmarinus officinalis L. with the use of magnetically treated water. In *Ciência Em Foco*, 1st ed.; Zuffo, A.M., Aguilera, J.G., de Oliveira, B.R., Eds.; Pantanal Editora: Nova Xavantina, Brazil, 2019; Volume 1, pp. 91–101. [CrossRef]
23. Aguilera, J.G.; Zuffo, A.M.; García, R.P.; Veitias, E.; Fung, Y.B. Magnetically treated irrigation water improved the adaptation of Spathoglottis plicata produced in vitro. *Amazon. J. Plant Res.* **2018**, *2*, 195–200. [CrossRef]
24. Ferrer-Dubois, A.E.; Zamora-Oduardo, D.; Rodríguez-Fernández, P.; Fung-Boix, Y.; Isaac-Aleman, E. Water treated with a static magnetic field on photosynthetic pigments and carbohydrates of *Solanum lycopersicum* L. *Rev. Cuba. De Química* **2022**, *34*, 34–48. Available online: http://scielo.sld.cu/scielo.php?script=sci_arttext&pid=S2224-54212022000100034&lng=es&tlng=en (accessed on 20 September 2022).
25. Pang, X.-F.; Deng, B. The changes of macroscopic features and microscopic structures of water under influence of magnetic field. *Phys. B* **2008**, *403*, 3571–3577. [CrossRef]
26. Pang, X.; Deng, B.; Tang, B. Influences of magnetic fiel on macroscopic properties of water. *Mod. Phys. Lett. B* **2012**, *26*, 1250069. [CrossRef]
27. Selim, A.; Zayed, M.; Zayed, M. Magnetic field treated water effects on germination, growth and physio-chemical aspects of some economic plants. *Acta Bot. Hung.* **2013**, *55*, 99–116. [CrossRef]
28. Sarraf, M.; Kataria, S.; Taimourya, H.; Santos, L.O.; Menegatti, R.D.; Jain, M.; Ihtisham, M.; Liu, S. Magnetic Field (MF) Applications in Plants: An Overview. *Plants* **2020**, *9*, 1139. [CrossRef] [PubMed]
29. Torres, C.; Díaz, J.; Cabal, P. Efecto de campos magnéticos en la germinación de semillas de arroz (*Oryza sativa* L.) y tomate (*Solanum lycopersicum* L.). *Agron. Colomb.* **2008**, *26*, 177–185. Available online: https://revistas.unal.edu.co/index.php/agrocol/article/view/13493 (accessed on 5 September 2018).
30. Aguilera, J.G.; Martin, R.M. Água tratada magneticamente estimula a germinação e desenvolvimento de mudas de *Solanum lycopersicum* L. *Brasilian J. Sustain. Agric.* **2016**, *6*, 47–53. [CrossRef]
31. Ferrer-Dubois, A.E.; Fung-Boix, Y.; Isaac-Alemán, E.; Beenaerts, N.; Cuypers, A. Determinación fitoquímica de frutos de *Solanum lycopersicum* L. Irrigados con agua tratada con campo magnético estático. *Rev. Cuba. De Química* **2018**, *30*, 232–242. Available online: http://scielo.sld.cu/scielo.php?script=sci_arttext&pid=S2224-54212018000200005&lng=es&nrm=iso (accessed on 13 September 2019).
32. Samarah, N.H.; Bany Hani, M.A.M.I.; Makhadmeh, I.M. Effect of Magnetic Treatment of Water or Seeds on Germination and Productivity of Tomato Plants under Salinity Stress. *Horticulturae* **2021**, *7*, 220. [CrossRef]
33. Ahamed, M.; Elzaawely, A.; Bayoumi, Y. Effect of magnetic field on seed germination, growth and yield of sweet pepper (*Capsicum annuum* L.). *Asian J. Crop Sci.* **2013**, *5*, 286–294. [CrossRef]
34. Aguilera, J.G.; Nogueira, J.C.; Dos Santos, R.G.; De Morais, K.A.D.; Lima, R.E.; Zuffo, A.M.; Ratke, R.F.; Boix, Y.F.; Argentel-Martínez, L. Efeito da água tratada magneticamente na emergência e desenvolvimento de mudas de pimentão amarelo. In *Pesquisas Agrárias e Ambientais: Volume IV*, 1st ed.; Zuffo, A.M., Aguilera, J.G., Eds.; Pantanal Editora: Nova Xavantina, Brazil, 2021; Volume 4, pp. 158–165. [CrossRef]
35. Elías-Vigaud, Y.; Rodríguez-Fernández, P.; Fung-Boix, Y.; Isaac-Aleman, E.; Ferrer-Dubois, A.; Asanza-Kindelán, G. Producción de pepino (*Cucumis sativus* L.) en casa de cultivo semiprotegido bajo riego con agua magnetizada. *Cienc. En Su PC* **2020**, *1*, 75–86. Available online: https://www.redalyc.org/articulo.oa?id=181363107006 (accessed on 13 September 2019).
36. El Sayed, H.; El Sayed, A. Impact of magnetic water Irrigation for Improve the growth, chemical composition and yield production of Broad Bean (*Vicia faba* L.) plant. *Am. J. Exp. Agric.* **2014**, *4*, 476–496. [CrossRef]

37. Ferrer, V. Efecto del tratamiento magnético sobre las características físico—Químicas del agua, el suelo y el cultivo de remolacha (*Beta vulgaris*). In *Tesis en Opción Al Título de Ingeniero Agrónomo*; Universidad de Oriente: Santiago de Cuba, Cuba, 2017; p. 74.
38. González, G. Efecto de la aplicación de agua y semilla magnetizada en el cultivo de rábano (*Raphanus sativus*). In *Tesis en Opción Al Título de Ingeniero en Ambiente Y Desarrollo*; Escuela Agrícola Panamericana: Zamorano, Honduras, 2016; p. 23.
39. Hozayn, M.; Qados, A.A. Irrigation with magnetized water enhances growth, chemical constituent and yield of chickpea (*Cicer arietinum* L.). *Agric. Biol. J. North Am.* **2010**, *1*, 671–676. Available online: http://scihub.org/ABJNA/PDF/2010/4/1-4-671-676.pdf (accessed on 13 September 2018).
40. Moussa, H.R. The impact of magnetic water application for improving common bean (*Phaseolus vulgaris* L.) production. *N. Y. Sci. J.* **2011**, *4*, 15–20.
41. Khattab, E.; Mona, M.; Abdel, W.; Hegazi, A.; Arafa, A. Magnetic treatment of irrigation water for improving vegetative growth, fresh and dry yield of bean (*Phaeolus vulgaris* L.) grown under plastic house. *Egypt. J. Agric. Res.* **2014**, *92*, 1395–1411.
42. Mroczek, M.; Tryniecki, L.; Kornarzynski, K.; Pietruszewski, S.; Gagos, M. Influence of magnetic field stimulation on the growth and biochemical parameters in *Phaseolus vulgaris* L. *J. Microbiol. Biotechnol. Food Sci.* **2016**, *5*, 548. [CrossRef]
43. Grewal, S.; Maheshwari, B. Magnetic treatment of irrigation water and snow pea and chickpea seeds enhances early growth and nutrient contents of seedlings. *Bioelectromagnetics* **2011**, *32*, 58–65. [CrossRef]
44. Taiz, L.; Zeiger, E.; Moller, I.A.; Murphy, A. *Fisiología e Desenvolvimento Vegetal*, 6th ed.; Artmed: Porto Alegre, Brazil, 2017; 888p.
45. Taiz, L.; Zeiger, E. *Plant Physiology*, 6th ed.; Oxford University: Oxford, UK, 2015.
46. Reyes, D.; Quiroz, J.; Kelso, H.; Huerta, M.; Avendaño, C.; Lobato, R. Caracterización estomática de cinco especies del género Vanilla. *Agron. Mesoam.* **2015**, *26*, 237–246. [CrossRef]
47. Sánchez, R. Efecto de un campo magnético de 60HZ en la germinación de semillas de especies de hortícolas. In *Tesis en Opción Al Título de Ingeniero Agrónomo*; Universidad de Oriente: Santiago de Cuba, Cuba, 2017; p. 62.
48. Azcón-Bieto, J.; Talón, M. *Fundamentos de Fisiología Vegetal*; Artmed Editora: Porto Alegre, Brazil, 2008.
49. Abdel Latef, A.A.H.; Dawood, M.F.; Hassanpour, H.; Rezayian, M.; Younes, N.A. Impact of the static magnetic field on growth, pigments, osmolytes, nitric oxide, hydrogen sulfide, phenylalanine ammonia-lyase activity, antioxidant defense system, and yield in lettuce. *Biology* **2020**, *9*, 172. [CrossRef]
50. Taiz, L.; Zeiger, E. *Plant Physiology*, 4th ed.; Sinauer Associates, Inc.: Sunderland, MA, USA, 2006.
51. Gallegos, J.S. Metodología de espectroscopía fotoacústica aplicada al estudio de los efectos producidos por tratamientos biofísicos sustentables a semillas y plántulas. In *Tesis en Opción Al Título de Maestro En Ciencias En Ingeniería De Sistemas*; Instituto Politécnico Nacional Escuela Superior De Ingeniería Mecánica Y Eléctrica, Unidad Zacatenco, D.F.: Mexico City, México, 2007; p. 166.
52. Gilart, F.; Deas, D.; Ferrer, D.; Lopez, P.; Ribeaux, G.; Castillo, J. High flow capacity devices for anti-scale magnetic treatment of water. *Chem. Eng. Process.* **2013**, *70*, 211–216. [CrossRef]
53. Abdul-Baki, A.A.; Anderson, J.D. Relationship Between Decarboxylation of Glutamic Acid and Vigor in Soybean Seed 1. *Crop Sci.* **1973**, *13*, 227–232. [CrossRef]
54. Ortega, E.; Rodés, R. *Manual de Prácticas de Laboratorio de Fisiología Vegetal*; Universidad de La Habana: La Habana, Cuba, 1986; pp. 147–152.
55. Dubois, M.; Gilles, K.A.; Hamilton, J.K.; Rebers, P.A.; Smith, F. Colorimetric method for determination of sugars and related substances. *Anal. Chem.* **1956**, *28*, 350–356. [CrossRef]

 plants

MDPI

Article

Growth and Biochemical Composition of Microgreens Grown in Different Formulated Soilless Media

Roksana Saleh [1], Lokanadha R. Gunupuru [1], Rajasekaran Lada [1], Vilis Nams [1], Raymond H. Thomas [2] and Lord Abbey [1,*]

[1] Department of Plant, Food, and Environmental Sciences, Faculty of Agriculture, Dalhousie University, 50 Pictou Road, Bible Hill, NS B2N 5E3, Canada

[2] Biotron Experimental Climate Change Research Centre, Department of Biology, University of Western Ontario, London, ON N6A 5B7, Canada

* Correspondence: loab07@gmail.com

Abstract: Microgreens are immature young plants grown for their health benefits. A study was performed to evaluate the different mixed growing media on growth, chemical composition, and antioxidant activities of four microgreen species: namely, kale (*Brassica oleracea* L. var. *acephala*), Swiss chard (*Beta vulgaris* var. *cicla*), arugula (*Eruca vesicaria* ssp. *sativa*), and pak choi (*Brassica rapa* var. *chinensis*). The growing media were T1.1 (30% vermicast + 30% sawdust + 10% perlite + 30% PittMoss (PM)); T2.1 (30% vermicast + 20% sawdust + 20% perlite + 30% PM); PM was replaced with mushroom compost in the respective media to form T1.2 and T2.2. Positive control (PC) was Pro-mix BX™ potting medium alone. Root length was the highest in T1.1 while the shoot length, root volume, and yield were highest in T2.2. Chlorophyll and carotenoid contents of Swiss chard grown in T1.1 was the highest, followed by T2.2 and T1.1. Pak choi and kale had the highest sugar and protein contents in T2.2, respectively. Consistently, total phenolics and flavonoids of the microgreens were increased by 1.5-fold in T1.1 and T2.2 compared to PC. Antioxidant enzyme activities were increased in all the four microgreens grown in T1.1 and T2.2. Overall, T2.2 was the most effective growing media to increase microgreens plant growth, yield, and biochemical composition.

Keywords: microgreens; natural amendment; soil health; phytochemicals; healthy food

Citation: Saleh, R.; Gunupuru, L.R.; Lada, R.; Nams, V.; Thomas, R.H.; Abbey, L. Growth and Biochemical Composition of Microgreens Grown in Different Formulated Soilless Media. *Plants* **2022**, *11*, 3546. https://doi.org/10.3390/plants11243546

Academic Editor: Maurizio Cocucci

Received: 20 October 2022
Accepted: 10 December 2022
Published: 15 December 2022

1. Introduction

Microgreens are immature greens harvested from tender young plants that are grown for their high health-promoting compounds and biological properties [1,2]. Previous researchers reported high amounts of phytochemicals such as ascorbic acid, α-tocopherol, β-carotene, phylloquinone, vitamins, and minerals in different species of microgreens [3–5]. Kale (*Brassica oleracea* L. var. *acephala*), Swiss chard (*Beta vulgaris* var. *cicla*), and arugula (*Eruca vesicaria* ssp. *sativa*), as microgreens, possess high levels of vitamins A, C, and K, essential lipids, carotenoids, and mineral nutrients [5,6]. Microgreens are delicate and are prone to various stress factors that can adversely affect the edible quality and bio-functional properties. Like all plants, the key preharvest factors that can affect a microgreen's edible quality are genotypic characteristics, growing media, climate, and management practices [7–10]. Hence, the presented study focuses on the impact of various growing media amendments on the quality of different microgreens. Natural amendments are organic substrates added to a growing medium to improve plant productivity and harvest quality, through enhancement of the physiochemical properties and functional activities of the media [11–13]. These amendments include compost, vermicast, humates, manures, and sawdust. They supply macro- and micro-nutrients, support beneficial microbes, improve water-holding capacity and gas exchange, and promote nutrient availability required for plant growth and development [13–15].

Vermicast (earthworm excreta or castings) is a humus-like material rich in beneficial microbiome and humic and non-humic substances such as mineral elements, amino acids, plant hormones, and other macromolecules that promote to plant growth and development [16,17]. According to Karthikeyan et al. [18], vermicast enhanced the seed germination rate and plant growth parameters, leaf pigmentation, root nodulation, and the yield of Lantana (*Lantana camara*) and cluster bean (*Cyamopsis tetragonoloba*), compared to inorganic fertilizer. In addition, adding vermicast to a growing media ameliorated soil physiochemical properties, leading to improved aeration, media porosity, field capacity, and microbial activity [19,20]. Similarly, Abbey et al. [14] showed that morphological indices of kale and postharvest essential fatty acids, mineral nutrients, phenolic compounds, and antioxidant capacity were increased by the application of dry vermicast, potassium humate, and volcanic minerals.

Sawdust is another potential growing medium substrate that is a waste from the forestry and wood industries. Currently, sawdust is burned or taken to landfills. There is a growing concern over the mining and use of Sphagnum peat moss. Therefore, sawdust can be used as an environmentally friendlier alternative or supplement to traditional substrates such as peat moss or can be used in combination with other substrates. Maharani et al. [21] showed that sawdust can improve the porosity and drainage of a growing medium. A study showed that sawdust delayed the initial growth of tomato seedlings (*Solanum lycopersicum*), but that the plant growth soared seven weeks after planting when the seedlings were established, and that the yield was higher than that of the control [22]. This delay can be attributed to toxic compounds from the wood such as lignin, cellulose, hemicellulose, and terpenes, which probably leached out, decomposed, or diluted by reaction with other amendments after seven weeks of planting [22–24].

Chang [25] showed that the combination of sawdust with 30% soil, plus nitrogen (N), phosphorus (P), and potassium (K) compound fertilizers gave rise to a higher productivity of the tomato plant compared to sawdust alone. Plant growth components, yield index, and nutritional values of *Syngonium podophyllum* were drastically increased following the application of vermicompost-sawdust extract [26]. A recent study showed that the combination of different proportions of vermicast and sawdust improved plant growth and biochemical compounds in Swiss chard, pak choi, and kale microgreens [23]. The authors found that 40% vermicast + 60% sawdust, or 60% vermicast + 40% sawdust improved the physiochemical properties of the growing media and enhanced the active microbial activity and nutrient mineralization necessary to meet potential plant growth requirements.

Therefore, amendments such as vermicast and compost can be added to sawdust to improve both the nutrient status and functionality of the growing medium. A study by Hernánde et al. [27] showed that the application of spent mushroom compost increased the seed germination percentage, fresh shoot weight, and yield of red baby leaf lettuce (*Lactuca sativa* L.) by up to 7-fold, compared to peat alone. Few studies on the effects of individual amendments on plants have been reported, but not on their combining effect on microgreen plant growth and chemical composition. Therefore, the objective of the present study was to evaluate the physiochemical properties of different proportions of mixed media and their effects on the growth and biochemical composition of four different plant species (kale, Swiss chard, arugula, and pak choi) that can be grown and harvested as microgreens.

2. Results

2.1. Growing Media Properties

The different additives in the growing media significantly affected the physiochemical properties (Table 1). It was found that T1.1 and T2.1 had a significantly ($p < 0.05$) low bulk density of an average of 0.07 g/cm^3 compared to an average of 0.10 g/cm^3 for T1.2, T2.2, PC, and NC. The highest porosity was observed in PC, followed by T1.2, and T2.2 compared to the other treatments. Porosity and field capacity of media T1.1, T2.1, and NC were significantly ($p < 0.05$) lower than the other media.

Table 1. Physiochemical properties of growing media affected by different proportions of mixed amended.

Treatment	Bulk Density (g/cm^3)	Porosity (%)	Field Capacity (%)	pH	Salinity (mg/L)	Electric Conductivity (μS/cm)	Total Dissolved Solids (mg/L)
T1.1	0.07 b	31.3 c	29.2 c	5.7 b	1299.7 c	2260.0 c	1719.6 c
T1.2	0.12 a	35.0 b	34.2 a	6.4 a	1689.7 a	2570.0 b	2139.7 b
T2.1	0.07 b	26.6 d	25.5 d	5.8 ab	1319.7 c	1233.0 e	1709.6 c
T2.2	0.10 ab	35.7 b	33.2 ab	6.3 ab	1494.9 b	2205.5 c	2028.4 b
PC	0.10 ab	37.8 a	30.2 bc	6.1 ab	802.9 d	1486.0 d	1233.5 d
NC	0.09 ab	27.6 d	24.5 d	5.9 ab	1861.2 a	3412.5 a	2479.7 a
p-value	0.015	0.000	0.000	0.029	0.000	0.001	0.001

T1.1: 30% vermicast + 30% sawdust + 10% perlite + 30% PittMoss (PM); T1.2: 30% vermicast + 30% sawdust + 10% perlite + 30% mushroom compost (MC); T2.1: 30% vermicast + 20% sawdust + 20% perlite + 30% PM; T2.2: 30% vermicast + 20% sawdust + 20% perlite + 30% MC; negative control (NC): 60% sawdust + 40% PittMoss; and positive control (PC): Pro-mix BX™ potting medium alone; significant at $p < 0.05$. Treatment means followed by a common letter are not significantly different.

The different growing media had pH values ranging from 5.7 to 6.4. The pH for T1.1 was significantly ($p < 0.05$) lower than that of T1.2. The overall trend for salinity, electrical conductivity, and total dissolve solids of the growing media was similar among the treatments (Table 1). NC had the highest salinity, electrical conductivity, and total dissolved solids followed by T1.2, then T1.1, and T2.1, and the least by NC.

2.2. Plant Growth and Yield

The growing media, plant species, and the interaction of growing media × plant species influenced plant growth components significantly ($p < 0.01$).

Total root lengths of arugula, pak choi, kale, and Swiss chard were increased by ca.79%, 83%, 61%, and 62% in T1.1, respectively, compared to the average for their counterparts grown in the PC and NC (Figure 1A). T1.1, T2.1, and T2.2 similarly had the highest effect on total root length compared to the others. Total shoot length of arugula, pak choi, kale, and Swiss chard were increased by ca. 99%, 105%, 62%, and 115%, respectively, in T2.2, compared to their counterparts in the PC (Figure 1B).

Consistently, the PC and the NC significantly ($p < 0.01$) reduced the total length of the roots and shoots of all the microgreen plants. Furthermore, the root volume was increased by ca. 67% to 143% in plants grown in T2.2, compared to those grown in the PC (Figure 1C). Consistently, the root volume of each plant was significantly ($p < 0.01$) reduced in T1.1, followed by NC and then PC (Figure 1C). The plant yield of the microgreens was significantly ($p < 0.01$) increased by ca. 230% in T2.2 and 160% in T1.2, respectively, compared to their PC counterparts (Figure 1D). Consistently, PC and T1.1 significantly ($p < 0.01$) reduced the yield of all the microgreens.

2.3. Microgreens Biochemical Composition

The ANOVA demonstrated that variations in the mixed media, plant species, and their interaction, significantly ($p < 0.01$) affected the biochemical compositions of the microgreens (Figure 2A–D). Total carotenoids, Chl a, Chl b, and Chl t of all the microgreens were increased significantly ($p < 0.05$) by T1.1 and T2.2, except Chl b in the pak choi, which was increased by T2.2 (Figure 2B). T1.2 had a similar effect to T1.1 and T2.2 in increasing Chl a, Chl b, Chl t and the total carotenoids in arugula and kale microgreens, but the effect varied for pak choi and Swiss chard (Figure 2A–D). Total chlorophyll and carotenoids were approximately 1.5-fold higher in T1.1 and T2.2 compared to their PC counterparts. Moreover, among the different plant species, kale and Swiss chard exhibited the highest Chl t by 67% in T1.2 and by 116% in T1.1 compared to PC (Figure 2C).

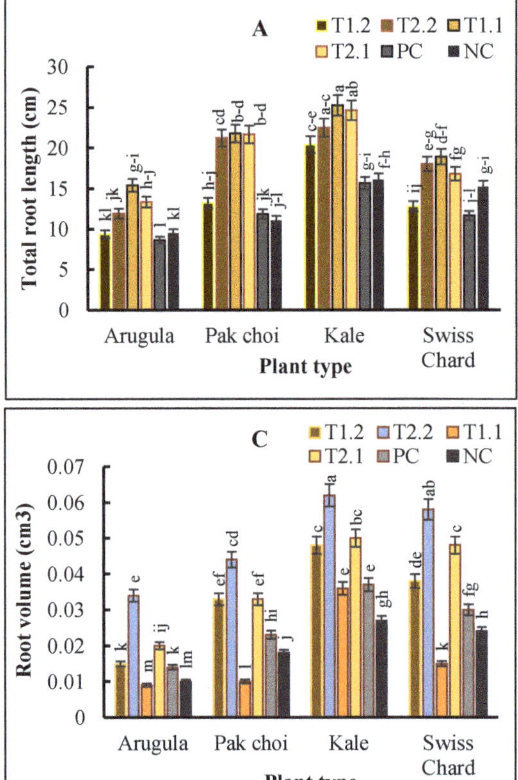

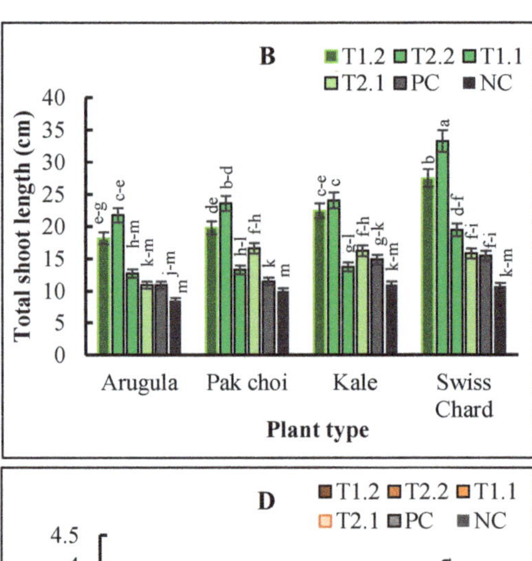

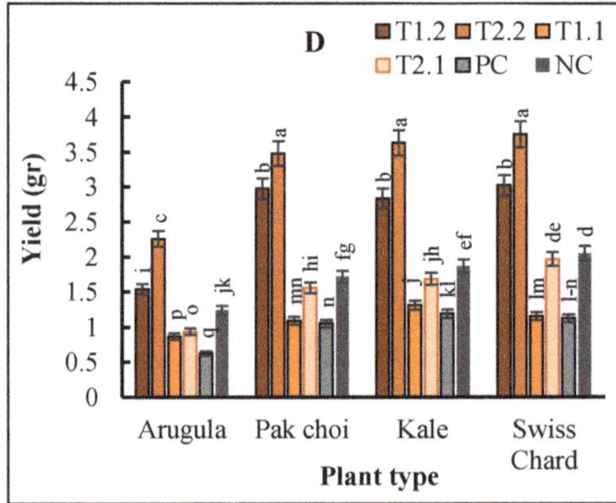

Figure 1. Total root length (**A**); total shoot length (**B**); root volume (**C**); yield (**D**) of arugula (*Eruca vesicaria* ssp. sativa), pak choi (*Brasica rapa* var. chinensis), kale (*Brassica oleracea* L. var. acephala) and Swiss chard (*Beta vulgaris* var. cicla) microgreens as affected by different growing media comprised of T1.1: 30% vermicast + 30% sawdust + 10% perlite + 30% PM; T1.2: 30% vermicast + 30% sawdust + 10% perlite + 30% MC; T2.1: 30% vermicast + 20% sawdust + 20% perlite + 30% PM; T2.2: 30% vermicast + 20% sawdust + 20% perlite + 30% MC; NC: 60% sawdust + 40% PittMoss; and PC: Pro-mix BX™ potting medium alone. Vertical bars represent standard errors of the means (N = 3). Bars with a common lower-case letter signifies treatment means that were not significantly different at $p < 0.05$.

Likewise, the highest total carotenoid content was about 72% higher for both kale and Swiss chard in T2.2 and T1.1, respectively, compared to their PC counterpart (Figure 2D). The total carotenoid content of arugula and pak choi was increased by ca. 15% and 24% in T2.2, respectively, compared to plants grown in the PC. Consistently, the lowest total carotenoid content was observed in all the microgreens grown in the T2.1, except for kale, which was lowest in the PC (Figure 2D). The overall trend for total carotenoid was arugula (562.35 µg/g FW) > Swiss chard (518.02 µg/g FW) > kale (472.69 µg/g FW) > pak choi (391.68 µg/g FW) (Figure 3D).

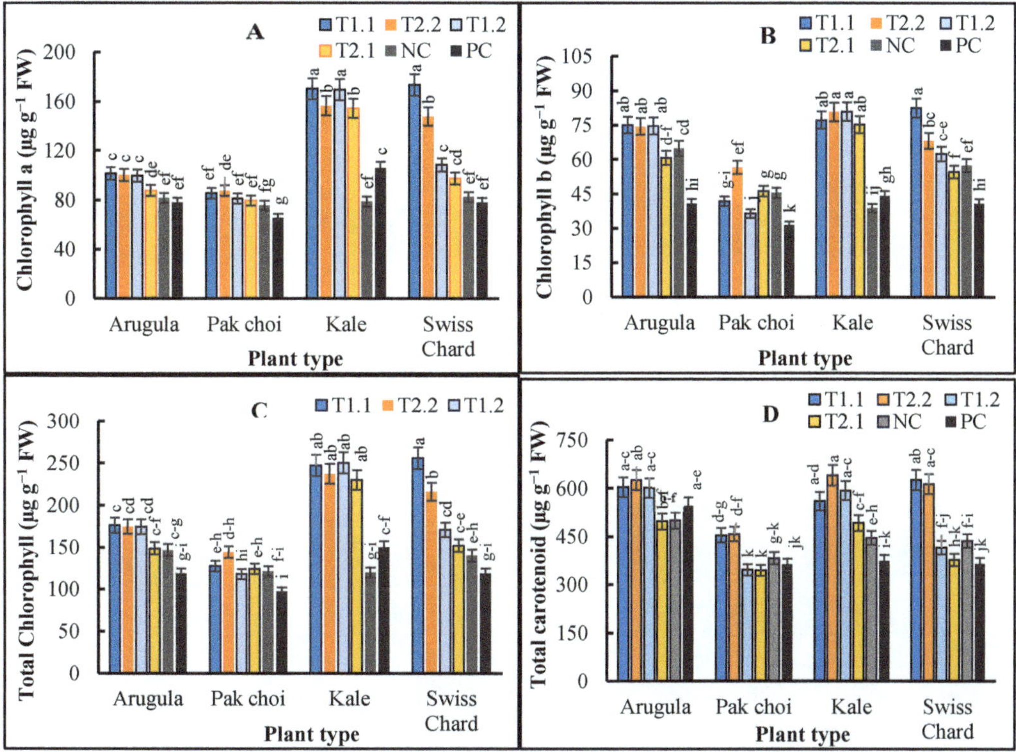

Figure 2. Chlorophyll a (**A**) and b (**B**), total chlorophyll (**C**) and carotenoid (**D**) contents of arugula (*Eruca vesicaria* ssp. sativa), pak choi (*Brasica rapa* var. chinensis), kale (*Brassica oleracea* L. var. acephala) and Swiss chard (*Beta vulgaris* var. cicla) microgreens as affected by different growing media comprised of T1.1: 30% vermicast + 30% sawdust + 10% perlite + 30% PM; T1.2: 30% vermicast + 30% sawdust + 10% perlite + 30% MC; T2.1: 30% vermicast + 20% sawdust + 20% perlite + 30% PM; T2.2: 30% vermicast + 20% sawdust + 20% perlite + 30% MC; NC: 60% sawdust + 40% PittMoss; and PC: Pro-mix BX™ potting medium alone. Vertical bars represent standard errors of the means (N = 3); significant at $p < 0.01$. Bars with a common lower-case letter signifies treatment means that were not significantly different at $p < 0.05$.

The highest sugar content was recorded by arugula microgreens grown in the PC, followed by T2.2 compared to other treatments (Figure 3A). On the contrary, the sugar content of pak choi was increased by 73% in T2.2 while T1.1 increased the sugar content of kale and Swiss chard by ca. 23% and 65%, respectively, compared to the PC (Figure 3A). Consistently, T2.1 significantly ($p < 0.01$) reduced the sugar content of all the four different microgreens. Among the microgreen plant species, the overall trend for the sugar content was arugula (3624.40 μg glucose/g) > kale (3204.99 μg glucose/g) > pak choi (3118.44 μg glucose/g) > Swiss chard (1944.46 μg glucose/g) (Figure 3A). As shown in Figure 3B, T1.1 significantly ($p < 0.01$) increased the protein content in arugula and Swiss chard by ca. 37% and 55%, respectively; while T2.2 significantly ($p < 0.01$) increased the protein content in pak choi and kale by ca. 23% and 105%, respectively, compared to their counterparts grown in the PC. The other media had similar effects on the total protein content of the microgreens. Overall, the trend for the protein content was Swiss chard (6372.85 μg Bovine/g) > kale (4941.84 μg Bovine/g) > arugula (4782.70 μg Bovine/g) > pak choi (3901.83 μg Bovine/g) (Figure 3B).

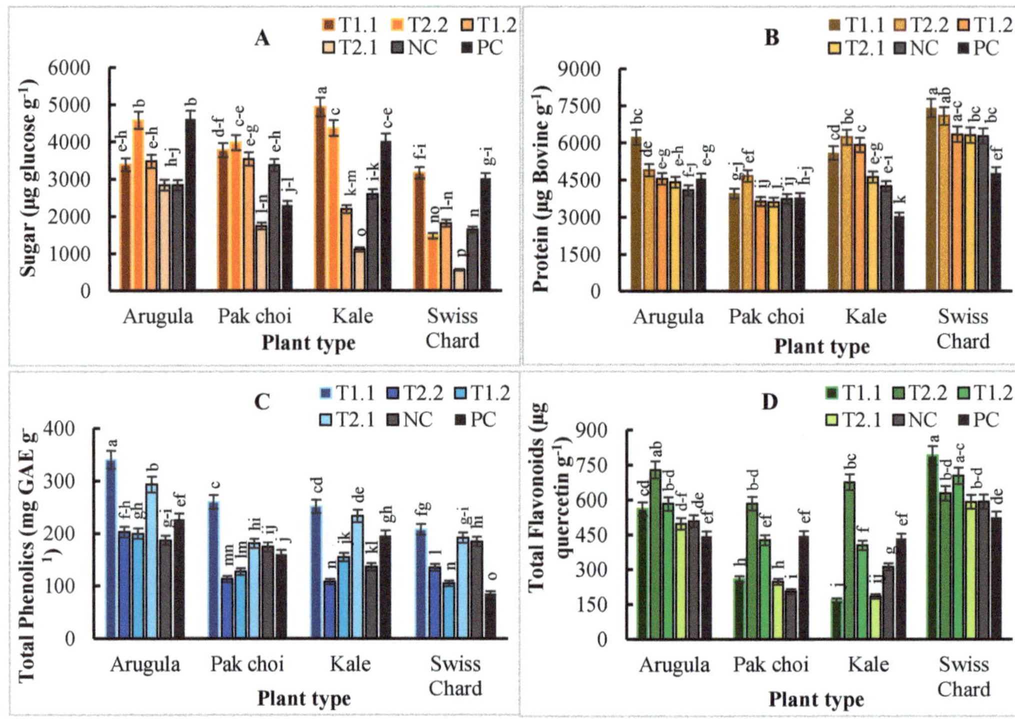

Figure 3. Sugar (**A**); protein (**B**); total phenolics (**C**); total flavonoids (**D**) contents of arugula (*Eruca vesicaria* ssp. sativa), pak choi (*Brasica rapa* var. chinensis), kale (*Brassica oleracea* L. var. acephala) and Swiss chard (*Beta vulgaris* var. cicla) microgreens as affected by different growing media comprised of T1.1: 30% vermicast + 30% sawdust + 10% perlite + 30% PM; T1.2: 30% vermicast + 30% sawdust + 10% perlite + 30% MC; T2.1: 30% vermicast + 20% sawdust + 20% perlite + 30% PM; T2.2: 30% vermicast + 20% sawdust + 20% perlite + 30% MC; NC: 60% sawdust + 40% PittMoss; and PC: Pro-mix BX™ potting medium alone. Vertical bars represent standard errors of the means (N = 3); significant at $p < 0.01$. Bars with a common lower-case letter signifies treatment means that were not significantly different at $p < 0.05$.

Total phenolics were significantly ($p < 0.01$) increased in all the plants grown in T1.1, followed closely by T2.1, which were not significantly ($p > 0.05$) different for Swiss chard (Figure 3C). The increase in total phenolics in arugula, pak choi, and kale by T1.1 and T2.1 were on the average, 1.5- and 1.2-fold higher than their counterparts that were grown in the PC. Interestingly, Swiss chard, followed by pak choi, and then arugula and kale had phenolics contents of ca. 144%, 63%, 50%, and 29% in T1.1, respectively, compared to their counterparts that were grown in the PC. Comparatively, the trend for the phenolics content in the microgreens was arugula (241.76 mg GAE/g) > kale (180.08 mg GAE/g) > pak choi (169.18 mg GAE/g) > Swiss chard (151.44 mg GAE/g) (Figure 3C). Total flavonoids in all the microgreens grown in T2.2, except for Swiss chard, increased by 1.5-fold compared to the microgreens grown in PC (Figure 3D). Total flavonoids in Swiss chard increased by 51% in T1.1 compared to PC. Total flavonoids in arugula, kale, and pak choi increased by 65%, 56%, and 31%, respectively, in T2.2 compared to PC. Among the microgreen plant species, the overall trend for the flavonoid was Swiss chard (638.34 µg quercetin/g) > arugula (553.84 µg quercetin/g) > kale (362.50 µg quercetin/g) > pak choi (360.96 µg quercetin/g) (Figure 3D).

The total ascorbate was increased by 57%, 64%, and 51% in arugula, pak choi, and kale grown in T1.2, respectively, compared to PC (Table 2). Furthermore, Swiss chard ascorbate content was significantly ($p < 0.01$) increased by 83% and 73% in T2.2 and T1.2, respectively, compared to PC. On the contrary, ascorbate was significantly ($p < 0.01$) reduced in microgreens grown in the T2.1 (Table 2). The overall trend for the microgreens' ascorbate content was kale (25.90 µmol/g FW) > Swiss chard (24.22 µmol/g FW) > arugula (23.41 µmol/g FW) > pak choi (22.40 µmol/g FW) (Table 2). Peroxidase was significantly ($p < 0.01$) increased in arugula and Swiss chard by T1.1 and T2.2 while T1.2 significantly ($p < 0.01$) increased POD in pak choi and kale.

Table 2. The effects of mixed growing media on total ascorbate, peroxidase activity and ascorbate peroxidase activity.

Treatment	Total Ascorbate (µmol g^{-1} FW)				Peroxidase Activity (Unit mg^{-1} FW)				Ascorbate Peroxidase Activity (Unit mg^{-1} FW)			
	Arugula	Pak Choi	Swiss Chard	Kale	Arugula	Pak Choi	Swiss Chard	Kale	Arugula	Pak Choi	Swiss Chard	Kale
T1.1	24.0 de	20.4 fgh	25.4 cd	23.6 de	0.94 bc	0.50 ij	0.96 bc	0.56 gh	0.23 a	0.08 fg	0.07 fg	0.06 g
T2.2	24.2 cde	28.0 bc	32.0 a	29.7 ab	0.67 ef	1.05 b	1.02 b	0.54 hij	0.15 cd	0.07 g	0.19 ab	0.19 ab
T1.2	32.2 a	29.5 ab	30.0 ab	32.0 a	0.60 fgh	1.25 a	0.50 ij	0.64 f	0.11 e	0.11 e	0.16 bcd	0.17 bcd
T2.1	21.5 efg	22.4 ef	21.9 ef	28.9 b	0.50 ij	0.94 bc	0.40 k	0.48 j	0.18 abc	0.04 j	0.11 e	0.05 hi
NC	17.9 hij	16.0 j	18.6 ghij	19.8 fgh	0.29 k	0.75 de	0.51 ij	0.17 m	0.06 g	0.06 gh	0.07 j	0.08 f
PC	20.6 fgh	18.1 hij	17.4 ij	21.4 efg	0.50 ij	0.87 cd	0.62 fg	0.50 ij	0.15 d	0.06 gh	0.05 de	0.07 fg
p value												
G	0.001	0.000	0.000	0.000	0.001	0.001	0.000	0.000	0.000	0.000	0.000	0.000
P	0.000	0.000	0.000	0.000	0.000	0.000	0.001	0.001	0.000	0.000	0.000	0.000
G × P	0.000	0.000	0.000	0.000	0.000	0.000	0.000	0.000	0.000	0.000	0.000	0.000

T1.1: 30% vermicast + 30% sawdust + 10% perlite + 30% PM; T1.2: 30% vermicast + 30% sawdust + 10% perlite + 30% MC; T2.1: 30% vermicast + 20% sawdust + 20% perlite + 30% PM; T2.2: 30% vermicast + 20% sawdust + 20% perlite + 30% MC; negative control (NC): 60% sawdust + 40% PittMoss; and positive control (PC): Pro-mix BX™ potting medium alone.; significant at $p < 0.01$. Treatment means followed by a common letter are not significantly different. G, growing media; P, plant species; G × P, interaction of growing media and plant species (N = 4).

Comparatively, Swiss chard followed by arugula had the highest POD activity and pak choi followed by kale had the lowest. The overall trend for the microgreens POD activity was pak choi (0.88 Unit/mg FW) > Swiss chard (0.66 Unit/mg FW) > arugula (0.58 Unit/mg FW) > kale (0.48 Unit/mg FW). Furthermore, APEX activity increased by 77% in Swiss chard and by 68% in kale when grown in T2.2, compared to PC. APEX activity of arugula and pak choi were increased by 55% and 54% in T1.1 and T1.2, respectively, compared to those grown PC (Table 2). Among the microgreen plant species, the overall trend for the microgreens APEX activity was arugula (0.146 Unit/mg FW) > Swiss chard (0.103 Unit/mg FW) = kale (0.103 Unit/mg FW) > pak choi (0.066 Unit/mg FW) (Table 2). Consistently, PC and NC significantly ($p < 0.01$) reduced the biochemical composition of the different microgreens (Table 2).

2.4. Association among Media, Plants, and Biochemical Composition

A multivariate two-dimensional PCA biplot was used to assess the association between the microgreens plant yield and biochemical parameters, as influenced by variations in growing media formulations (Figure 4). The PCA explained 80% of the total variations in the dataset. Treatments that are close to the origin of the PCA axes show a high association and stability than those on the periphery. The PCA demonstrated that treatment T2.2 can be associated with an improved plant yield and biochemical composition of the microgreens. The interaction of the growing media and plant species can be closely associated with the microgreens' yield and total ascorbates. Furthermore, kale carotenoid content was strongly influenced by the interaction between the growing media × plant species compared to the other plant species. APEX activities were associated with the interaction between the growing media and plant species in all the microgreens except arugula (Figure 4). Overall, the interaction between the growing media and plant species can be associated with the kale yield and its biochemical parameters compared to the other plant species.

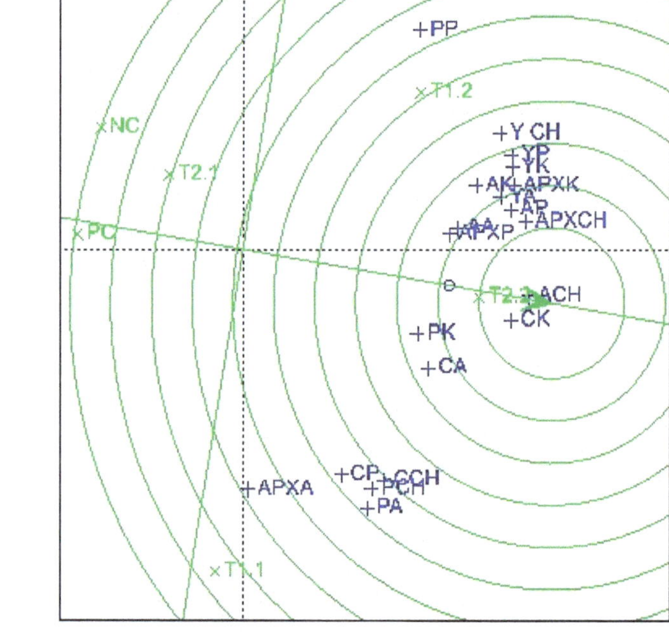

Figure 4. Ranking total × total biplot for comparison of treatment × plant species interaction effects on biochemical variations in all microgreens. Arugula (*Eruca vesicaria* ssp. sativa) yield (YA), arugula ascorbate (AA), arugula carotenoids (CA), arugula POD Activity (PA), arugula APEX Activity (APXA); Swiss chard (*Beta vulgaris* var. cicla) yield (YCH), Swiss chard ascorbate (ACH), Swiss chard carotenoids (CCH), Swiss chard POD Activity (PCH), Swiss chard APEX Activity (APXCH); kale (*Brassica oleracea* L. var. acephala) yield (YK), kale ascorbate (AK), kale carotenoids (CK), kale POD Activity (PK), kale APEX Activity (APXK); pak choi (*Brasica rapa* var. chinensis) yield (YP), pak choi ascorbate (AP), pak choi carotenoids (CP), pak choi POD Activity (PP), pak choi APEX Activity (APXP). T1.1: 30% vermicast + 30% sawdust + 10% perlite + 30% PM; T1.2: 30% vermicast + 30% sawdust + 10% perlite + 30% MC; T2.1: 30% vermicast + 20% sawdust + 20% perlite + 30% PM; T2.2: 30% vermicast + 20% sawdust + 20% perlite + 30% MC; NC: 60% sawdust + 40% PittMoss; and PC: Pro-mix BX™ potting medium alone.

3. Discussion

The effects of different substrates on the physiochemical characteristics of the formulated growing media and the differential response of the four different microgreens plant species were investigated under greenhouse conditions. The growing medium T1.2, followed by T2.2, had the highest effect on most of the plant growth components, except for root length. The growing media T1.1 and T2.1 contained PittMoss, which was made from mainly shredded cardboard, and T1.2 and T2.2 contained mushroom compost. The results show that mushroom compost was more beneficial than PittMoss. Similarly, Renaldo et al. [28] reported that mushroom compost increased the shoot and root dry mass in cucumber (*Cucumis sativus*) compared to biochar and corn stalks but had no effect on lettuce (*Lactuca sativa*), probably due to lettuce intolerance of the high salt content in the mushroom compost. Furthermore, Vahid Afagh et al. [15] reported that a 15% mushroom compost mixed in sandy loam soil increased both the plant growth and yield of German chamomile (*Matricaria recutita* L.) due to the improved medium structure, increased nutrient availability, and beneficial microbial activity [15,29]. Furthermore, the results also suggested that the variations in response of the microgreen plants to the different media were dependent on genotypic differences.

It was obvious that the improved structure and functionality of growing media T1.2 and T2.2 improved plant growth in all the plant species except for pak choi, as previously explained by Emami and Astaraei [30] and Vahid Afagh et al. [15]. The root lengths of all the microgreens were significantly increased in T1.1 and T2.1 compared to the other media. According to Vahid Afagh et al. [15], an addition of 15% mushroom compost to a medium increased aeration and water-holding capacity, leading to an improved crop productivity. The addition of PittMoss in T1.1 and T2.1 reduced the growing media bulk density, which in turn promoted root growth compared to the mushroom compost. A previous study using a high bulk density of (i.e., 1.35 g/cm^3) growing medium led to a reduction in lettuce root growth and yield [31]. In the present study, the bulk density ranged between 0.07 and 0.12 g/cm^3, which was below the root-restriction threshold bulk density of 1.6 g/cm^3, especially in T1.2 and T2.2. This may be the reason for the enhanced plant growth and yield of microgreens grown in T1.2 and T2.2. Moreover, Gillespie et al. [32] stated that the optimum range of pH for leafy greens growth is a 5.5 to 6.5 range, at which more nutrients become available to plants. However, it does not seem that the pH was a limitation in the present study, since all the media pH fell within the sufficiency range for the microgreen plants. Nevertheless, Ur Rahman et al. [33] reported that pH variation of the medium (from 5 to 9) significantly influenced the yield and biochemical constitutions in wheat (*Triticum aestivum* L.). The highest yield, total chlorophyll, and carotenoid contents were observed in seedlings grown in media with a neutral pH (6.5–7), while the lowest one was obtained in acidic (pH 5) and alkaline (pH 9) media that correspond with the results of this study.

Notably, there was a significant positive association between the yield, salinity, and TDS, suggesting sufficient growing medium fertility levels in particularly, T1.2 and T2.2, which were the only media with mushroom compost. Previous studies showed that high electric conductivity and salinity can reduce plant growth [34,35], which can be managed by adding perlite and wood-based substrates into the growing media to improve texture, structure, and porosity [35–37]. However, T1.2 and T2.2 had acceptable ranges of salinity thresholds between 640 and 1600 mg/L, as recommended for most vegetable crops [38]. Generally, NC recorded the highest salinity and the lowest yield, as previously reported by Shannon et al. [39], for kale and Swiss chard grown in media with excess salinity levels > 3.0 dS/m. Lin et al. [23] reported an increase in the plant growth and yield components of Swiss chard, pak choi, and kale in a medium consisted of 60% vermicast and 40% sawdust, with a considerably high electric conductivity of 1450 μS/cm and a pH of 7.3. Furthermore, Hernández et al. [27] attributed increased germination rate, fresh shoot weight, and yield in red baby leaf lettuce to mushroom compost, with a pH of 7 and an electric conductivity of > 4000 μS/cm. There was no significant correlation between EC and the measured growth components, but there was a strong relationship between pH and growth plant components in all the plants.

The microgreens' biochemical composition was significantly altered by the different mixed growing media. There are very few documented reports on the effect of different mixed growing media on biochemical quality of microgreens. Previous studies have demonstrated that vermicast and mushroom compost are well known to be rich in macro- and micro-elements including N, which is essential for chlorophyll and carotenoid synthesis as well as photosynthesis [40,41]. In this study, total flavonoids and ascorbates ranged from 404.1 to 653.7 μg quercetin/g, and 18.1 to 30.9 μmol/g FW, respectively. Media T1.2 and T2.2 impacted the highest amount of microgreen flavonoids and ascorbate contents, respectively, that most likely can be associated with media nutrient availability and a balance in C/N ratio, due to the added mushroom compost as explained by Hernández et al. [27]. Moreover, it was demonstrated that mushroom compost may be chitin-rich, which can be a significant source of plant growth stimulants and elicitors for the biosynthesis of secondary metabolites [42,43]. Therefore, a significant amount of chitin might be present in T1.2 and T2.2, leading to the high microgreen plants content of total carotenoid, flavonoids, and ascorbate, compared to media without mushroom compost. Treatments T1.1 and T2.1 improved phenolics content in all the microgreens irrespective of plant species. This can be ascribed to the high-carbon input from the thermally treated sawdust and PittMoss. This carbon might have improved the carbon-based phenolic compounds and their precursors involved in plant defense mechanisms and responses to environmental stress [44]. Contrary to this, the total phenolics was lower in T2.2, which suggested that the probably high N content in T1.2 and T2.2, due to the addition of N-rich vermicast and mushroom compost, might have reduced phenolic content in the microgreens as previously reported [14,44,45]. The difference in growing media had a significant effect on POD and APEX enzymes activities in the microgreens. Several studies have reported a strong correlation between bioactive phytochemicals and antioxidant properties [10,46]. Besides the increased ascorbate and flavonoids contents, POD and APEX were highly increased in the microgreens grown in T1.2 and T2.2. Our results are consistent with findings obtained by Shiri et al. [19], who reported a significant increase in antioxidant capacity with an elevated ascorbic acid content in plants.

4. Materials and Methods

4.1. Plant Material and Growing Condition

The experiment was carried out in July 2020 and repeated in December 2020 in the Department of Plant, Food, and Environmental Sciences greenhouse (45°23′ N, 63°14′ W), Dalhousie University, Truro, NS, Canada. The microgreens were kale (*Brassica oleracea* L. var. *acephala*), Swiss chard (*Beta vulgaris* var. *cicla*), arugula (*Eruca vesicaria* ssp. *sativa*), and pak choi (*Brasica rapa* var. *chinensis*), purchased from Halifax Seed Co., Halifax, NS, Canada. The growing media were PittMoss, vermicast, sawdust, mushroom compost, perlite and Pro-mix BX™. PittMoss® is a soilless potting mix made from recycled paper (Ambridge soil company, PA, USA). It is expected that the PittMoss will improve aeration and water retaining potential, resulting in the better delivery of nutrients to the root-zone environment. Vermicast, sawdust, and shiitake (Lentinula edodes) mushroom compost were obtained from Modgarden Company, Toronto, ON, Canada. Perlite and Pro-mix BX™ potting medium were purchased from Halifax Seed Company, NS, Canada. Kale, Swiss chard, arugula, and pak choi seeds were sown in flat plastic cell trays, measuring 19 cm length × 12 cm width × 2.5 cm deep, each containing a different mixed medium. The trays were kept in the greenhouse under a 16/8-hr day/night light regime (from high pressure sodium lamp) at a 24°/22 °C day/night temperature cycle with a 71% mean relative humidity. A 600 W HS2000 high-pressure sodium lamp with NAH600.579 ballast (P.L. Light Systems, Beamsville, ON, Canada) supplied the supplementary lighting. Air distribution in the greenhouse was distributed by a horizontal air-flow ventilation system. Watering was carried out every two days with 200 mL of tap water for each pot until the final harvest at 15 days after sowing. No additional fertilizer was applied.

4.2. Experimental Treatment and Design

The 2-factor experiment (i.e., plant species x growing media) was arranged in a completely randomized design with three replications. Seeds were sown in six different proportions of mixed media (Table 3). Pots were rearranged weekly on the growth shelf to offset microclimate variations in the greenhouse. The entire study was repeated twice. The data from the two studies were merged because the coefficient of variation was less than 5%. Seed germination, plant growth, yield, and various biochemical characteristics were measured.

Table 3. Proportions of mixed growing media.

Treatment	Formulation
T1.1	30% vermicast + 30% sawdust + 10% perlite + 30% PittMoss (PM)
T1.2	30% vermicast + 30% sawdust + 10% perlite + 30% mushroom compost (MC)
T2.1	30% vermicast + 20% sawdust + 20% perlite + 30% PittMoss (PM)
T2.2	30% vermicast + 20% sawdust + 20% perlite + 30% mushroom compost (MC)
NC	60% sawdust + 40% PittMoss
PC	Pro-mix BX™ potting medium alone

NC and PC are negative and positive control, respectively.

4.3. *Growing Media Physicochemical Properties*

To evaluate chemical properties of the growing media, 50 g of each media was added to 50 mL of deionized water and was thoroughly mixed before the determination of chemical properties. pH, salinity, electrical conductivity, and total dissolved solids were measured using an ExStik® II EC500 waterproof pH/conductivity meter (Extech ITM Instruments Inc., Newmarket, ON, Canada). The growing media physical properties and water retention characteristics were determined in triplicate as described by Armah [47], with slight modifications. Bulk density (D_b) was determined from the weight (M) and volume (V_1) of the soil core, using a graduated glass cylinder after continuous tapping, until there was no observable change in soil volume.

$$\textbf{Bulk density} = \mathrm{M}/\mathrm{v}_1 \tag{1}$$

$$\textbf{Porosity} = \mathrm{Ms}/\mathrm{v}_2 \tag{2}$$

Water saturation, field capacity, and wilting point were determined after the soil was air-dried under ambient conditions (*ca.* 22 °C). A known mass of the fresh soil sample (M_s) was placed in a 15.24 cm plastic pot with drainage holes and was weighed (M_{sp}). The potted soil was placed in a saucer and was saturated with distilled water, and the saturated soil weight (M_{sat}) was recorded after 48 h. Then, the saucer was removed so that the free water could drain out under atmospheric pressure for 72 h and was then weighed ($M_{drained}$). The drained soil was spread evenly in a flat aluminum tray and air-dried under ambient conditions for 72 h and then weighed (M_{dried}).

$$\textbf{Field capacity } (\mathbf{F_c}) = \frac{\mathbf{M_{drained}} - \mathbf{M_{sp}}}{\mathbf{M_s}} \times 100 \tag{3}$$

4.4. *Plant Growth and Yield Components*

Data on seedling growth indices were collected 14 days after sowing the seeds. Plant samples (n = 15) were randomly and gently uprooted from the middle section of the growing trays for each treatment per replicate using a spatula. The seedlings were placed on tissue paper before carefully removing chunks of loosely attached media from the roots. The roots were then thoroughly washed under a gentle running deionized with minimum root loss (i.e., ca. < 2%). After drying with a blotting paper, the total lengths of roots and shoots and root volume were determined using a Perfection V800 Photo Color Scanner Digital ICE® Technologies (Epson America Inc., Los Alamitos, CA, USA). The shoots of the remaining microgreens were cut with a pair of scissors at the growing media surface after 14 days of sowing, and the fresh weights were recorded as the estimated yield per treatment. At the final harvest, there was no seed residue on the shoots that we had to worry about.

4.5. *Microgreen Quality and Phytochemical Analysis*

4.5.1. Chlorophylls a and b, Total Chlorophyll, and Total Carotenoid

Samples of the microgreens per treatment from the final harvest in Section 4.4 above were immediately frozen in liquid N to avoid changes in the biochemical compounds present in the plants. Pooled samples of the microgreens frozen in liquid N were ground to fine powder and stored in −20 °C until analyzed. Briefly, 0.2 g of each ground microgreen was separately dissolved in 10 mL of 80% acetone. After centrifuging at 12,000 rpm for 15 min, the supernatant was collected and transferred into 96 micro-well plates to measure the absorbance at 646.8 nm and 663.2 nm wavelength, using a UV-Vis spectrophotometer (Evolution™ Pro, Thermo Fisher scientific, Waltham, MA, USA) against acetone as blank, using the method described by Lightenthaler [48]. Chlorophyll and carotenoid concentrations were obtained by the following formula.

$$\mathbf{Chla}\ (\mu g/mL) = 12.25 \times \mathbf{A663.2} - 2.79 \times \mathbf{A646.8} \quad (4)$$

$$\mathbf{Chlb}\ (\mu g/mL) = \mathbf{21.50 \times A646.8 - 5.1 \times A663.2} \quad (5)$$

$$\mathbf{Chlt}\ (\mu g/mL) = \mathbf{chla + chlb} \quad (6)$$

$$\mathbf{Car}\ (\mu g/mL) = \mathbf{(1000 \times A470 - 1.8 \times chla - 85.02 \times chlb)/198} \quad (7)$$

Finally, the calculated value was multiplied by the total volume (10 mL) and then divided by the total fresh weight (0.2 g), which was expressed as μg/g FW.

4.5.2. Total Sugar

The total sugar content of the microgreens was measured using the method described by Mohammadkhani and Heidari [49], with some modifications. Firstly, 0.2 g of powder was dissolved in 10 mL of 90% ethanol and was incubated in a water bath for 60 min. The mixture was topped with up to 25 mL with 90% ethanol and centrifuged at 4000 rpm for 3 min. An amount of 1 mL of the supernatant was transferred into a glass test tube and 1 mL of 5% phenol was added and vortexed. Subsequently, 5 mL of sulfuric acid was added and incubated in the dark for 15 min. The mixture was cooled, and the absorbance was measured at 490 nm using a UV-Vis spectrophotometer against a blank made up of deionized water, phenol, and sulfuric acid. The total sugar was obtained by a standard sugar curve prepared by dissolving sucrose in distilled water at different concentrations, from 0 to 300 μg. Then, 1 mL of 5% phenol and 5 mL of sulfuric acid was added to the mixture and the absorbance was recorded at 490 nm. The sugar content was expressed as μg glucose/g FW.

4.5.3. Total Protein

The total protein content was measured using the Bradford assay, as described by Hammond and Kruger [50]. In brief, 0.2 g of the ground microgreen tissue samples was transferred into a test tube, added with 5 mL ice-cold extraction buffer (i.e., 50 mM potassium phosphate buffer at pH 7.0) and 0.1 mM EDTA. The mixture was vortexed for 30 s before centrifugation at 15,000 rpm for 20 min. The supernatant was collected and kept on ice. Subsequently, the supernatant was mixed with 100 μL of enzyme extract and 1 mL of Bradford reagent, before recording the absorbance against a blank (Bradford reagent) at 595 nm after a 5 min incubation. The protein concentration was determined by the regression equation obtained from a Bovine serum albumin at different concentrations (200–900 μg mL^{-1}) and was expressed as μg Bovine/g.

4.5.4. Total Phenolics

The total phenolic (TPC) was measured using the Folin–Ciocalteu method, as described by Alothman et al. [51]. Briefly, 0.2 g of the ground microgreens was dissolved in ice-cold 80% methanol and incubated at an ambient temperature (approximately, 22 °C) for 48 h in the dark. The mixture was then centrifuged at 13,000 rpm for 5 min. A 100 μL sample of the supernatant, the standard at different concentrations (i.e., 0, 5, 10, 15, 20, 25 mg/L), and a methanol blank were added into distinct tubes before adding 200 μL Folin-Ciocalteu reagent and 800 μL of Na_2CO_3 and then incubating it for 2 h in the dark. Eventually, 200 μL of the mixture, the standard, and the blank were individually transferred into a microplate to measure the absorbance at 765 nm by UV-vis spectrophotometer. TPC concentration was determined by the standard curve obtained from Gallic acid equivalents and expressed as mM Gallic acid per g of fresh sample (mg GAE/g).

4.5.5. Total Flavonoids

The total flavonoid was measured using the method described by Chang et al. [25]. Ground samples of each microgreen (0.2 g) and 2.5 mL of 95% methanol was mixed and vortexed before centrifugation at 13,000 rpm for 10 min. The supernatant (500 μL) standard (1 mg quercetin dissolved in 95% methanol at 5, 10, 15, 25, 50, 100, 150, 200 μg/mL concentrations), and 95% methanol were transferred into separate tubes. Then, 1.5 mL 95% methanol, 0.1 mL 10% $AlCl_3$, 0.1 mL 1 M potassium acetate, and 2.8 mL distilled water were added to each tube. Afterward, the mixture was incubated at an ambient temperature for 30 min, and the absorbance was recorded at 415 nm against a blank using a UV-Vis spectrophotometer. The flavonoids content was measured by the standard curve obtained from the quercetin standard curve. The total flavonoids content was expressed as μg quercetin/g of plant fresh weight.

$$\text{Total flavonoid} = \frac{(\mathbf{[flavonoids]}\,(\boldsymbol{\mu}\mathbf{g/mL}) \times \mathbf{total\ volume\ of\ methanolic\ extract\ (mL)})}{\mathbf{mass\ of\ extract\ (g)}} \tag{8}$$

4.5.6. Total Ascorbate

The total ascorbate was measured using the method described by Ma et al. [52]. In brief, 0.2 g of the ground microgreens was mixed with 1.5 mL ice-cold 5% trichloroacetic acid (TCA) and centrifuged for 15 min at 4 °C. Then, 100 μL of the supernatant was collected and added to 400 μL phosphate buffer (150 mM KH_2PO_4), 5 mM EDTA, and 100 μL10 mM dithiothreitol and vortexed. Following the incubation of the mixture, 0.5% N-ethylmaleimide was added to the mixture and vortexed. To obtain color, 400 μL 10% TCA, 400 μL 44% orthophosphoric acid, 400 μL4% dipyridyl and 200 μL 30 g/L $FeCl_3$ was added to the mixture and incubated at 40 °C for 1 h before recording the absorbance at 525 nm using a UV-Vis spectrophotometer against a blank. The standard was prepared from L-ascorbic acid in 5% TCA (0–5 mM). Total ascorbate content was expressed as μmol/g FW.

4.5.7. Antioxidant Enzyme Activity

The peroxidase (POD) and ascorbate peroxidase enzyme activities (APEX) were measured using the method described by Patterson et al. [53]. Briefly, 0.2 g of the ground microgreens was mixed with 5 mL ice-cold extraction buffer and centrifuged at 15,000 rpm for 20 min. The extraction buffer contained mM potassium_phosphate buffer (pH 7.0), 1% polyvinylpyrrolidone, and 0.1 mM EDTA. The supernatant (i.e., enzyme extract) was collected for POD and APEX assays. For POD, the reaction mixture was prepared from the combination of 100 mM potassium-phosphate buffer (pH 7.0), 0.1 mM pyrogallol, and 5 mM H_2O_2. Then, 10 μL of the supernatant was added to the mixture and incubated for 5 min at room temperature. To stop any enzyme reaction in the mixture, 0.1 mL of NH_2SO_4 was added. Finally, the absorbance was recorded at 420 nm using a UV-Vis spectrophotometer against a blank (Milli-Q water). The enzyme activity was calculated by the following formula and expressed as unit/mg FW.

$$\text{POD} = \text{A}_{420} \times 3/(12 \times 0.1))/0.2 \tag{9}$$

To assay APEX, 100 μL of the supernatant was added to the reaction mixture, i.e., 1372 μL of 50 mM potassium_phosphate buffer (pH 7.0), 75 μL of 10 mM ascorbate, and 3 μL of 100 mM H_2O_2. The mixture was incubated for 1 min before reading the absorbance at 290 nm using a UV-Vis spectrophotometer against a blank. The enzyme activity in unit/mg FW was obtained by:

$$\text{APEX} = (\text{A } 290 \times 1/(2.8 \times 0.1))/0.2 \tag{10}$$

4.6. Statistical Analysis

All the data were subjected to a two-way analysis of variance (ANOVA) using Minitab version 18.3. Fisher method was used to separate treatment means when the ANOVA showed a significant difference at $p < 0.05$. Furthermore, a multivariate analysis using a two-dimensional principal component analysis (PCA) was carried out using GenStat software.

5. Conclusions

Global warming and climate change have had adverse impacts on plant production and food security. During the last decade, synthetic chemical fertilizers and pesticides have been extensively used in conventional agriculture to meet global food and nutrition demand. However, their application negatively affects the environment and human health. Therefore, the development of an innovative and climate-smart approach to food production is of high importance. In the present study, the effect of different mixed natural growing media on the growth and biochemical properties of different microgreen plant species was investigated. Overall, our results showed that variations in the growing media characteristics had a significant effect on the studied traits of the microgreens. Overall, growing media containing mushroom compost, i.e., T2.2, was found to be the most favorable. The efficacy of T2.2 on the assessed growth, yield, and quality traits was further confirmed through the PCA analysis. The ingredients used to make the mixed growing media in this study are reasonably inexpensive and locally available. Therefore, they can be used as an alternative to conventional media such as Pro-mix BX™ potting medium for growing microgreens to improve productivity and nutrient and non-nutrient bioactive compounds.

Author Contributions: Conceptualization: L.A., L.R.G. and R.S.; Formal analysis: R.S., L.A. and V.N.; Funding acquisition: L.A. and R.H.T.; Investigation: R.S. and L.R.G.; Methodology: R.S., L.R.G. and L.A.; Project administration: L.A. Resources: L.A.; Supervision: R.H.T. and L.A.; Validation: R.S., L.A. and R.H.T.; Writing—original draft: R.S.; Writing—review and editing: L.R.G., R.S., R.L., V.N., R.H.T. and L.A. All authors have read and agreed to the published version of the manuscript.

Funding: This work was financially supported by the Natural Sciences and Engineering Research Council of Canada (NSERC), Grant #CRDPJ 523129-17.

Data Availability Statement: Not applicable.

Acknowledgments: The lead author wishes to thank Samuel K. Asiedu and her laboratory mates for their generous assistance and support. We also thank Jason Giffin of Maritime Gourmet Mushrooms Inc., Great Village, NS, Canada for the supply of the mushroom compost.

Conflicts of Interest: The authors declare no conflict of interest.

References

1. Bulgari, R.; Negri, M.; Santoro, P.; Ferrante, A. Quality evaluation of indoor-grown microgreens cultivated on three different substrates. *Horticulturae* **2021**, *7*, 96. [CrossRef]
2. Kyriacou, M.C.; El-Nakhel, C.; Graziani, G.; Pannico, A.; Soteriou, G.A.; Giordano, M.; Rouphael, Y. Functional quality in novel food sources: Genotypic variation in the nutritive and phytochemical composition of thirteen microgreens species. *Food Chem.* **2019**, *277*, 107–118. [CrossRef] [PubMed]
3. Bajaj, S.; Khan, A. Antioxidants and diabetes. *Indian J. Endocrinol. Metab.* **2012**, *16* (Suppl. 2), S267. [CrossRef] [PubMed]
4. Kyriacou, M.C.; Rouphael, Y.; Di Gioia, F.; Kyratzis, A.; Serio, F.; Renna, M.; Santamaria, P. Micro-scale vegetable production and the rise of microgreens. *Trends Food Sci. Technol.* **2016**, *57*, 103–115. [CrossRef]
5. Pinto, E.; Almeida, A.A.; Aguiar, A.A.; Ferreira, I.M. Comparison between the mineral profile and nitrate content of microgreens and mature lettuces. *J. Food Compos. Anal.* **2015**, *37*, 38–43. [CrossRef]
6. Di Noia, J. Peer Reviewed: Defining powerhouse fruits and vegetables: A nutrient density approach. *Prev. Chronic Dis.* **2014**, *11*, 57–76. [CrossRef]
7. Aftab, T. A review of medicinal and aromatic plants and their secondary metabolites status under abiotic stress. *J. Med. Plants* **2019**, *7*, 99–106.
8. Nguyen, D.T.; Lu, N.; Kagawa, N.; Takagaki, M. Optimization of Photosynthetic Photon Flux Density and Root-zone Temperature for Enhancing Secondary Metabolite Accumulation and Production of Coriander in Plant Factory. *Agronomy* **2019**, *9*, 224. [CrossRef]

9. Zhang, F.; Tiyip, T.; Ding, J.L.; He, Q.S. The effects of the chemical components of soil salinity on electrical conductivity in the region of the delta oasis of Weigan and Kuqa Rivers, China. *Agric. Sci. China* **2009**, *8*, 985–993. [CrossRef]
10. Zhang, B.; Wijesundara, N.M.; Abbey, L.; Rupasinghe, H.V. Growing medium amendments effect on growth, secondary metabolites, and anti-streptococcal activity of two species of Plectranthus. *J. Appl. Res. Med. Aromat. Plants* **2017**, *5*, 53–59. [CrossRef]
11. Abbott, L.K.; Macdonald, L.M.; Wong, M.T.F.; Webb, M.J.; Jenkins, S.N.; Farrell, M. Potential roles of biological amendments for profitable grain production—A review. *Agric. Ecosyst. Environ.* **2018**, *256*, 34–50. [CrossRef]
12. da Costa Jaeggi, M.E.P.; Rodrigues, R.R.; Pereira, I.M.; do Carmo Parajara, M.; Rocha, R.S.; da Cruz, D.P.; de Araújo Capetini, S. Vegetative Development of Radish Seedlings in Different Organic Substrates. *J. Exp. Agric. Int.* **2019**, *10*, 1–8. [CrossRef]
13. Treutter, D. Managing phenol contents in crop plants by phytochemical farming and breeding—Visions and constraints. *Int. J. Mol. Sci.* **2010**, *11*, 807–857. [CrossRef] [PubMed]
14. Abbey, L.; Pham, T.H.; Annan, N.; Leke-Aladekoba, A.; Thomas, R.H. Chemical composition of kale as influenced by dry vermicast, potassium humate and volcanic minerals. *Food Res. Int.* **2018**, *107*, 726–737. [CrossRef]
15. Vahid Afagh, H.; Saadatmand, S.; Riahi, H.; Khavari-Nejad, R.A. Influence of spent mushroom compost (SMC) as an organic fertilizer on nutrient, growth, yield, and essential oil composition of German chamomile (*Matricaria recutita* L.). *Commun. Soil Sci. Plant Anal.* **2019**, *50*, 538–548. [CrossRef]
16. Abbey, L.; Udenigwe, C.; Mohan, A.; Anom, E. Microwave irradiation effects on vermicasts potency, and plant growth and antioxidant activity in seedlings of Chinese cabbage (*Brassica rapa* subsp. *pekinensis*). *J. Radiat. Res. Appl. Sci.* **2017**, *10*, 110–116. [CrossRef]
17. Iheshiulo, E.M.A.; Abbey, L.; Asiedu, S.K. Response of kale to single-dose application of K humate, dry vermicasts, and volcanic minerals. *Int. J. Veg. Sci.* **2007**, *23*, 135–144. [CrossRef]
18. Karthikeyan, M.; Hussain, N.; Gajalakshmi, S.; Abbasi, S.A. Effect of vermicast generated from an allelopathic weed lantana (*Lantana camara*) on seed germination, plant growth, and yield of cluster bean (*Cyamopsis tetragonoloba*). *Environ. Sci. Pollut. Res.* **2014**, *21*, 12539–12548. [CrossRef]
19. Shiri, M.A.; Ghasemnezhad, M.; Bakhshi, D.; Saadatian, M. Effects of ascorbic acid on phenolic compounds and antioxidant activity of packaged fresh cut table grapes. *Electron. J.Environ. Agric. Food Chem.* **2011**, *10*, 27–39.
20. Trivedi, P.; Singh, K.; Pankaj, U.; Verma, S.K.; Verma, R.K.; Patra, D.D. Effect of organic amendments and microbial application on sodic soil properties and growth of an aromatic crop. *Ecol. Eng.* **2017**, *102*, 127–136. [CrossRef]
21. Maharani, R.; Yutaka, T.; Yajima, T.; Minoru, T. Scrutiny on physical properties of sawdust from tropical commercial wood species: Effects of different mills and sawdust's particle size. *Indones. J. For. Res.* **2010**, *7*, 20–32. [CrossRef]
22. Agboola, O.O.; Oseni, O.M.; Adewale, O.M.; Shonubi, O. Effect of the use of sawdust as a growing medium on the growth and yield of tomato. *Ann. West Univ. Timisoara. Ser. Biol.* **2018**, *21*, 67–74.
23. Lin, S. Combined Effect of Vermicast-Trichoderma-Sawdust on Kale, Swiss Chard, and Pak Choi Growth. Master's Thesis, Faculty of Agriculture, Department of Plant, Food, and Environmental Sciences, Dalhousie University, Halifax, NS, Canada, 2020. Available online: http://hdl.handle.net/10222/79587 (accessed on 2 October 2022).
24. Mohan, G.; Johnson, R.L.; Yu, J. Conversion of Pine Sawdust into Polyhydroxyalkanoate Bioplastics. *ACS Sustain. Chem. Eng.* **2021**, *9*, 8383–8392.
25. Chang, C.C.; Yang, M.H.; Wen, H.M.; Chern, J.C. Estimation of total flavonoid content in propolis by two complementary colorimetric methods. *J. Food Drug Anal.* **2002**, *10*, 3–7.
26. Mahboub-Khomami, A.; Padasht, M.N.; Ajili Lahiji, A.; Shirinfekr, A. The effect of sawdust vermicompost extract on Syngonium podophyllum growth and nutrition. *J. Plant Nutr.* **2019**, *42*, 410–416. [CrossRef]
27. Hernández, D.; Ros, M.; Carmona, F.; Saez-Tovar, J.A.; Pascual, J.A. Composting Spent Mushroom Substrate from *Agaricus bisporus* and *Pleurotus ostreatus* Production as a Growing Media Component for Baby Leaf Lettuce Cultivation under Pythium irregulare Biotic Stress. *Horticulturae* **2021**, *7*, 13. [CrossRef]
28. Renaldo, B.; Gaius, E.; Gregory, G.; Paul, V. Effects of organic amendments on early plant growth. In Proceedings of the Conference: 20th World Congress of Soil Science, Jeju, Korea, 8–13 June 2014; Available online: https://www.researchgate.net/publication/263656746_Effects_of_organic_amendments_on_early_plant_growth (accessed on 2 October 2022).
29. Demir, H. The effects of spent mushroom compost on growth and nutrient contents of pepper seedlings. *Mediterr. Agric. Sci.* **2017**, *30*, 91–96.
30. Emami, H.; Astaraei, A.R. Effect of organic and inorganic amendments on parameters of water retention curve, bulk density, and aggregate diameter of a saline-sodic soil. *J. Agric. Sci. Technol.* **2012**, *14*, 1625–1636.
31. Chen, Z.; Han, Y.; Ning, K.; Luo, C.; Sheng, W.; Wang, S.; Wang, Q. Assessing the performance of different irrigation systems on lettuce (*Lactuca sativa* L.) in the greenhouse. *PLoS ONE* **2019**, *14*, e0209329. [CrossRef]
32. Gillespie, D.P.; Kubota, C.; Miller, S.A. Effects of low pH of hydroponic nutrient solution on plant growth, nutrient uptake, and root rot disease incidence of basil (*Ocimum basilicum* L.). *HortScience* **2020**, *55*, 1251–1258. [CrossRef]
33. Ur Rahman, S.; Xuebin, Q.; Riaz, L.; Yasin, G.; Noor Shah, A.; Shahzad, U.; Du, Z. The interactive effect of pH variation and cadmium stress on wheat (Triticum aestivum L.) growth, physiological and biochemical parameters. *PLoS ONE* **2021**, *16*, e0253798. [CrossRef] [PubMed]

34. Vidal, N.P.; Pham, H.T.; Manful, C.; Pumphrey, R.; Nadeem, M.; Cheema, M.; Thomas, R. The use of natural media amendments to produce kale enhanced with functional lipids in controlled environment production system. *Sci. Rep.* **2018**, *8*, 14771. [CrossRef]
35. Warrence, N.J.; Bauder, J.W.; Pearson, K.E. Basics of salinity and solidity effects on soil physical properties. *Land Resour. Environ. Sci.* **2002**, *7*, 129.
36. Lee, Y.H.; Ahn, B.K.; Sonn, Y.K. Effects of electrical conductivity on the soil microbial community in a controlled horticultural land for strawberry cultivation. *Korean J. Soil Sci. Fertil.* **2011**, *44*, 830–835. [CrossRef]
37. Meyer, M.H.; Cunliffe, B.A. Effects of media porosity and container size on overwintering and growth of ornamental grasses. *HortScience* **2004**, *39*, 248–250. [CrossRef]
38. Machado, R.M.A.; Serralheiro, R.P. Soil salinity: Effect on vegetable crop growth. Management practices to prevent and mitigate soil salinization. *Horticulturae* **2017**, *3*, 30. [CrossRef]
39. Shannon, M.C.; Grieve, C.M.; Lesch, S.M.; Draper, J.H. Analysis of salt tolerance in nine leafy vegetables irrigated with saline drainage water. *J. Am. Soc. Hortic. Sci.* **2000**, *125*, 658–664. [CrossRef]
40. Gonani, Z.; Riahi, H.; Sharifi, K. Impact of using leached spent mushroom compost as a partial growing media for horticultural plants. *J. Plant Nutr.* **2011**, *34*, 337–344. [CrossRef]
41. Zietz, M.; Weckmuller, A.; Schmidt, S.; Rohn, S.; Schreiner, M.; Krumbein, A.; Kroh, L.W. Genotypic and climatic influence on the antioxidant activity of flavonoids in kale (*Brassica oleracea* var. *sabellica*). *J. Agric. Food Chem.* **2010**, *58*, 2123–2130. [CrossRef]
42. Sharp, R.G. A review of the applications of chitin and its derivatives in agriculture to modify plant-microbial interactions and improve crop yields. *Agronomy* **2013**, *3*, 757–793. [CrossRef]
43. Li, H.; Yoshida, S.; Mitani, N.; Egusa, M.; Takagi, M.; Izawa, H.; Ifuku, S. Disease resistance and growth promotion activities of chitin/cellulose nanofiber from spent mushroom substrate to plant. *Carbohydr. Polym.* **2022**, *284*, 119–233. [CrossRef] [PubMed]
44. Singh, S.; Singh, J.; Vig, A.P. Earthworm as ecological engineers to change the physico-chemical properties of soil: Soil vs vermicast. *Ecol. Eng.* **2016**, *90*, 1–5. [CrossRef]
45. Ibrahim, M.H.; Jaafar, H.Z.; Rahmat, A.; Rahman, Z.A. The relationship between phenolics and flavonoids production with total non-structural carbohydrate and photosynthetic rate in *Labisia pumila* Benth. under high CO_2 and nitrogen fertilization. *Molecules* **2011**, *16*, 162–174. [CrossRef] [PubMed]
46. Arumugam, G.; Swamy, M.; Sinniah, U. *Plectranthus amboinicus* (Lour.) Spreng: Botanical, phytochemical, pharmacological and nutritional significance. *Molecules* **2016**, *21*, 369. [CrossRef]
47. Armah, A. The Use of Rock Dust as a Natural Media Amendment for the Production of Horticultural Crops in Controlled Environments. Master's Thesis, Memorial University, Grenfell Campus, Corner Brook, NL, Canada, 2021.
48. Lightenthaler, H.K. Chlorophylls and carotenoids: Pigments of photosynthetic biomembranes. *Methods Enzymol.* **1987**, *148*, 350–382.
49. Mohammadkhani, N.; Heidari, R. Drought-induced accumulation of soluble sugars and proline in two maize varieties. *World Appl. Sci. J.* **2008**, *3*, 448–453.
50. Hammond, J.B.; Kruger, N.J. The Bradford method for protein quantitation. In *New Protein Techniques*; Humana Press: Clifton, NJ, USA, 1988; pp. 25–32.
51. Alothman, M.; Bhat, R.; Karim, A.A. Antioxidant capacity and phenolic content of selected tropical fruits from Malaysia, extracted with different solvents. *Food Chem.* **2009**, *115*, 785–788.
52. Ma, Y.H.; Ma, F.W.; Zhang, J.K.; Li, M.J.; Wang, Y.H.; Liang, D. Effects of high temperature on activities and gene expression of enzymes involved in ascorbate–glutathione cycle in apple leaves. *Plant Sci.* **2008**, *175*, 761–766. [CrossRef]
53. Peterson, J.M. Soils-Part 2: Physical Properties of Soil and Soil Water. 1999. Available online: https://passel2.unl.edu/view/lesson/0cff7943f577 (accessed on 2 October 2022).

 plants MDPI

Article

Irrigation Effect on Yield, Skin Blemishes, Phellem Formation, and Total Phenolics of Red Potatoes

Manlin Jiang [1], Tracy Shinners-Carnelley [2], Darin Gibson [3], Debbie Jones [3], Jyoti Joshi [1] and Gefu Wang-Pruski [1,*]

1 Faculty of Agriculture, Dalhousie University, Truro, NS B2N 5E3, Canada
2 Research, Quality & Sustainability, Peak of the Market, Winnipeg, MB R3H 0R5, Canada
3 Gaia Consulting Ltd., Newton, MB R0H 0X0, Canada
* Correspondence: gefu.wang-pruski@dal.ca

Abstract: Dark Red Norland is an important potato cultivar in the fresh market due to its attractive bright, red colour, and good yield. However, skin blemishes such as silver patch, surface cracking, and russeting can negatively influence the tuber skin quality and marketability. It is well known that potato is a drought-sensitive plant. This study was conducted to determine whether irrigation would affect Dark Red Norland's yield and skin quality. A three-year field trial was conducted by Peak of the Market in Manitoba, Canada. Plants were treated under both irrigation and rainfed conditions. The results show that irrigation increased the total yield by 20.6% and reduced the severity of surface cracking by 48.5%. Microscopy imaging analysis demonstrated that tubers from the rainfed trials formed higher numbers of suberized cell layers than those of the irrigated potatoes, with a difference of 0.360 to 0.652 layers in normal skins. Surface cracking and silver patch skins had more suberized cell layers than the normal skins, with ranges of 7.805 to 8.333 and 7.740 to 8.496, respectively. A significantly higher amount of total polyphenols was found in the irrigated samples with a mean of 77.30 mg gallic acid equivalents (GAE)/100 g fresh weight (fw) than that of the rainfed samples (69.80 mg GAE/100 g fw). The outcome of this study provides a better understanding of the water regime effect causing these skin blemishes, which could potentially be used to establish strategies to improve tuber skin quality and minimize market losses.

Keywords: anthocyanin; russeting; silver patch; *Solanum tuberosum*; suberized cell layer; surface cracking

Citation: Jiang, M.; Shinners-Carnelley, T.; Gibson, D.; Jones, D.; Joshi, J.; Wang-Pruski, G. Irrigation Effect on Yield, Skin Blemishes, Phellem Formation, and Total Phenolics of Red Potatoes. *Plants* **2022**, *11*, 3523. https://doi.org/10.3390/plants11243523

Academic Editor: José M. Escalona Lorenzo

Received: 10 November 2022
Accepted: 12 December 2022
Published: 14 December 2022

Publisher's Note: MDPI stays neutral with regard to jurisdictional claims in published maps and institutional affiliations.

1. Introduction

Potato (*Solanum tuberosum* L.) is the fourth most important food crop in terms of volume and consumption after maize, wheat, and rice in the world [1]. It is also one of the most studied crops in much of the latest literature [1–3]. About one-fifth of the potatoes grown in Canada are for the fresh table market [4]. Canada ranks as the fifth biggest fresh potato exporter in the world [4]. During 2020 and 2021, Canada exported CAD 319 million's worth of fresh potatoes [4]. Norland cultivar was one of the top three registered seed potato varieties grown in Canada in 2020 [4]. It is a common fresh market cultivar and popular because of its bright red colour. Dark Red Norland is a developed strain of the Norland cultivar, which has darker red skin colour and high yield but the same weaknesses: skin discoloration and skin blemish defects. Skin blemishes can badly affect tuber appearance and marketability.

Potato tubers are covered with a protective corky skin tissue called periderm. Periderm has a complex structure that is made up of three types of cells. The visible outermost layer of the skin is called the phellem layer, which is composed of many layers of suberized cells. Under the phellem is the phellogen, which is made of layers of meristematic cells. Under the phellogen is the phelloderm, which is made of layers of parenchyma-like cells [5]. Phellogen cells divide outwards to make suberized phellem; phellogen cells divide inwards to make phelloderm [6].

Potato periderm contains enzymes and metabolites that can respond to biotic and abiotic stresses. Phenolic compounds are the most abundant secondary metabolites in plants. Potato skin has higher amounts of phenolic compounds than potato flesh [7]. These compounds are important in plant defence mechanisms as antioxidants. The synthesis of phenolic compounds is induced in response to biotic and abiotic stimulation, such as drought, chilling, pathogens, or nutrient deficiency [8]. As a group of phenolic compounds, anthocyanin synthesis and accumulation in potato tissues are also considered indicators of stress resistance [9,10]. Anthocyanin discoloration in *Solanaceae* is more likely due to a change in the balance between anthocyanin biosynthesis and degradation [11]. The ability to enhance skin-set development and suberization can greatly reduce surface blemishes, shrinkage and flaccidity, blemishes, and infections [7].

'Skin blemishes' are those defects on tuber skin that can badly influence the tuber's appearance and grading. Surface cracking, silver patch, and russeting are the three major skin blemishes found on Dark Red Norland tubers in the field. Surface cracking is seen as shallow, corky cracks on the tuber skin, normally presenting as rough, latticed areas of tuber skin (Figure 1a). Most cracks are generated when the internal pressure exceeds the tensile strength of the surface tissues during tuber enlargement, and the outer periderm bursts [12]. Silver patch is a defect that appears as silvery, smooth patches on the tuber skin (Figure 1b). This defect has not yet been described in the literature but has been given the name silver patch by Dr. Tracy Shinners-Carnelley (personal communication). Russeting presents as protruding dark-brown patches on the surface of tubers and is considered a defect when it occurs on tubers of smooth-skinned cultivars (Figure 1c). It negatively affects the protective functions of the skin, including the prevention of water loss and resistance to pathogen invasion [13]. It is believed that these defects are not associated with any disease-causing pathogens, since no pathogens have been isolated and identified in these defective tissues (Dr. Tracy Shinners-Carnelley, personal communication).

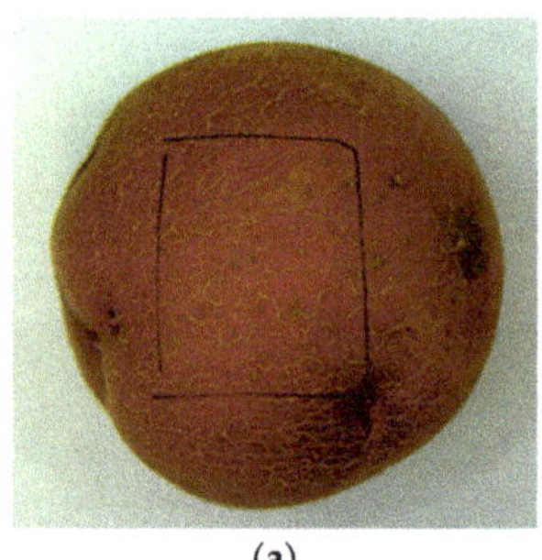
(**a**)

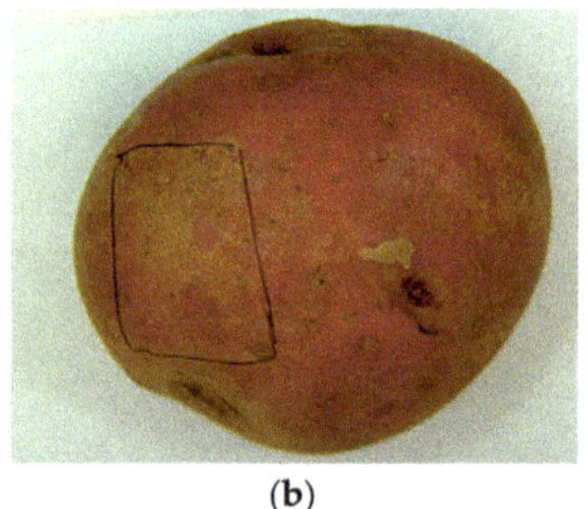
(**b**)

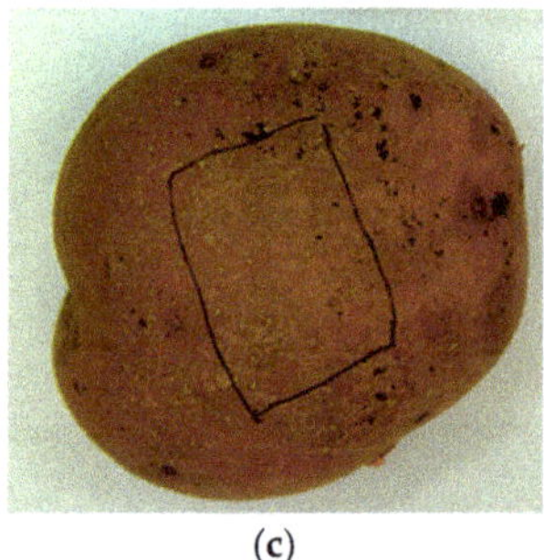
(**c**)

Figure 1. Tuber skin blemishes found on Dark Red Norland, as shown in the marked areas. (**a**) Surface cracking (SC); (**b**) silver patch (SP); (**c**) russeting (R).

As a drought-sensitive plant, adequate soil moisture is suggested to be maintained at all stages of potato development [14]. Water stress may inhibit or even completely stop one or more physiological processes, such as transpiration, photosynthesis, cell enlargement, and enzymatic activities [15]. Limited irrigation at different stages of potato growth results in earlier crop maturity and decreases plant growth, tuber yield, the number of tubers per plant, and tuber size and quality [16,17]. Drought during the periods of tuber initiation and bulking has the most drastic effect on the yield [16]. Smaller tuber sizes and increased external defects were found in a previous study when the irrigation gradually declined and no irrigation occurred during the tuber initiation [18]. The effects of drought stress on tuber physiological development could include decreases in tuber number, increases in misshapen tubers, reduced tuber dry matter, and reduced water content [19].

Skin blemishes, such as surface cracking, silver patch, and russeting, significantly showed up on Dark Red Norland tubers for unknown reasons in 2019–2021 field trials in Manitoba. It is proposed that these blemishes were caused by environmental factors. This study determined the relationships between the water regime and the yield of Dark Red Norland and their tuber skin blemishes. In addition, phellem structure, total phenolics, and anthocyanin were studied to obtain a better understanding of these skin blemishes.

2. Results

2.1. Total Yield

The three-year data show that the highest yield of Dark Red Norland was in the medium size with a range of 2.25–3.0″ in diameter for approximately 71.2% of the total yield (Figure 2). The proportions of less than 2″, 2–2.25″, and 3–3.5″ of the total yield were 4.8%, 8.2%, and 14.7%, respectively. When the total yields (tubers of all sizes) were compared, higher yields were found in the irrigated plots, with a mean of 429.032 cwt/ac for the three years. Irrigation improved the proportion of the yield from medium- and large-size tubers (by 8.1% and 125.6%) and increased the proportion of the total yield by 20.6% but did not significantly change the yields of less than 2″ and between 2 and 2.25″.

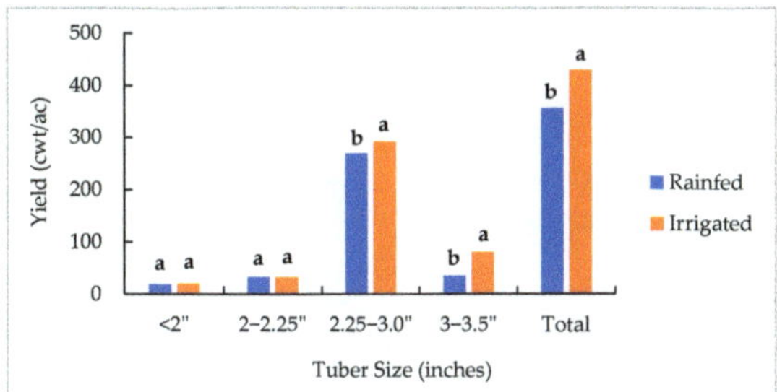

Figure 2. Tuber yields (cwt/ac) under rainfed and irrigated plots in different tuber sizes (in inches) of the three-year trial. Means not sharing a common letter in two adjacent columns are significantly different at $p < 0.05$ according to the Fisher LSD method.

2.2. Skin Blemishes

Silver patch appeared with the highest percentage of occurrence (nearly 50%) in both the rainfed and irrigated plots (Figure 3). A significantly higher percentage of surface cracking was found in the rainfed plots (27.39%) compared to the irrigated (18.45%), which increased by 48.5%. A lower percentage of russeting was found in the rainfed plots (10.06%) compared to the irrigated plots. (17.98%) (Figure 3). Normal tubers accounted for only 22.72% and 24.79% of the rainfed and irrigated treatments, respectively.

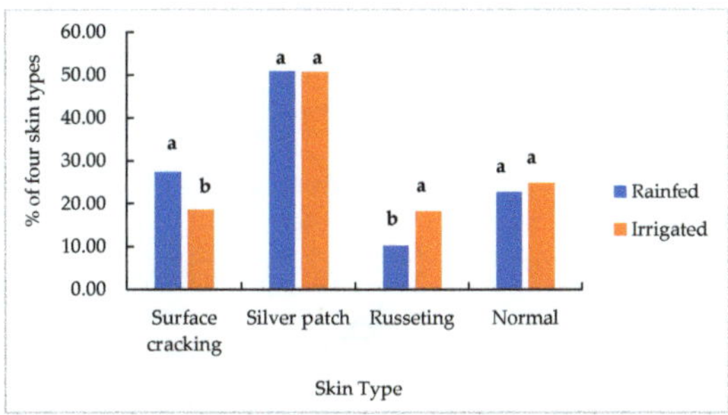

Figure 3. Percentages of the four skin types including 'surface cracking', 'silver patch', 'russeting', and 'normal' in all three years from rainfed and irrigated trials. Means not sharing a common letter in two adjacent columns are significantly different at $p < 0.05$ according to the Fisher LSD method.

2.3. Suberized Cell Layer

The representative samples of suberized skins in the normal skin type and the three defected skin types of surface cracking, silver patch, and russeting are shown in Figure 4. Normal skin had organized cells, which were well-packed (Figure 4a); russeting skin had rough skin surface and irregular cells (Figure 4d); silver patch and surface cracking had obviously more suberized cell layers (Figure 4b,c). Silver patch skin had well-arranged cells (Figure 4c), while surface cracking had cracks between the cells (Figure 4b).

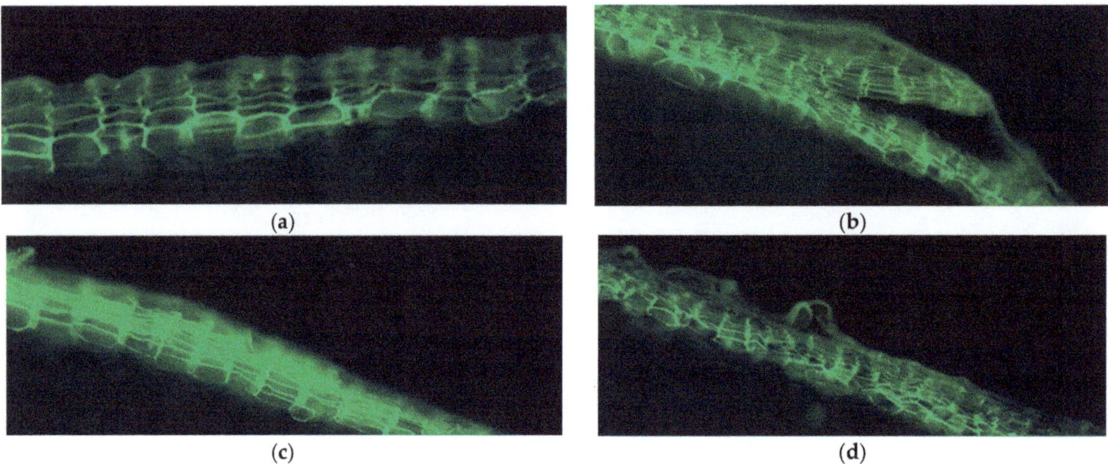

(a) (b) (c) (d)

Figure 4. Representative samples of suberized cell layers in the (**a**) normal (N) skin, as well as the three defected skin types, including (**b**) surface cracking (SC), (**c**) silver patch (SP), and (**d**) russeting (R) of Dark Red Norland tubers under fluorescence. Photos were taken by Manlin Jiang using Leica LAS X Imaging and Analysis Software.

In regard to normal skins, the tubers of the rainfed plots had more layers of suberized cells compared to those of the irrigated plots in all three years (Table 1). The differences in the suberized cell layers in the normal skins between the two water regimes were in a range of 0.360–0.652 during the three years. Among the four skin types, surface cracking and silver patch skins had significantly more suberized cell layers than those of normal and russeting skins. This situation occurred in most of the plots in all three years (Table 2). Table 2 lists eight treatments with heat stress applied at different growth stages, including 'Tuber Initiation', 'Tuber Bulking', 'Tuber Skinset', and 'No Stress' in both the rainfed and irrigated plots. We do not discuss the heat treatment in this paper because the heat treatment did not have significant effects on the soil temperature in the field trial. However, we could clearly distinguish the differences in the number of suberized cell layers among the four skin types. Surface cracking and silver patch skins had the most suberized cell layers with ranges of 7.805 to 8.333 and 7.740 to 8.496, respectively.

Table 1. Two-sample t-test comparisons of suberized cell layer in normal skin samples between two water regimes in each year.

Water Regime	2019	2020	2021
Rainfed	7.327 ^{a}	6.730 ^{a}	7.110 ^{a}
Irrigated	6.675 ^{b}	6.370 ^{b}	6.735 ^{b}

Means that do not share a letter within a column are significantly different, $p = 0.05$.

Table 2. Pairwise comparisons of suberized cell layer using Tukey's method among four skin types, including normal skin (N), surface cracking (SC), silver patch (SP), and russeting (R) in samples of 3 years.

Skin Type	Treatment							
	Rainfed				Irrigated			
	No Stress	Tuber Initiation	Tuber Bulking	Tuber Skinset	No Stress	Tuber Initiation	Tuber Bulking	Tuber Skinset
N	7.000 [c]	6.900 [b]	6.959 [b]	7.358 [b]	6.752 [c]	6.385 [c]	6.600 [c]	6.611 [b]
SC	7.856 [b]	7.970 [a]	7.805 [a]	8.330 [a]	8.329 [a]	8.170 [a]	8.333 [a]	8.031 [a]
SP	8.495 [a]	8.000 [a]	7.740 [a]	8.086 [a]	7.835 [b]	7.950 [a]	7.710 [b]	8.119 [a]
R	6.464 [c]	6.282 [c]	6.860 [b]	6.700 [c]	7.000 [c]	7.044 [b]	6.810 [c]	6.214 [b]

Means that do not share a letter within a column are significantly different, $p = 0.05$.

2.4. Total Phenolic Content

Tubers from 32 plots in the 2020 and 2021 field seasons were analyzed for the total phenolic content. In both years, silver patch skin and russeting skin showed more total phenolics than normal skin (Figure 5). The amounts of total phenolic content in the silver patch and russeting skin samples were 76.85 mg GAE/100 g fw and 77.63 mg GAE/100 g fw, respectively. The normal skin had the lowest amount of total phenolics, with an average of 69.12 mg GAE/100 g fw. In addition, the irrigated samples had more total phenolics (77.30 mg GAE/100 g fw) than the rainfed samples (69.80 mg GAE/100 g fw).

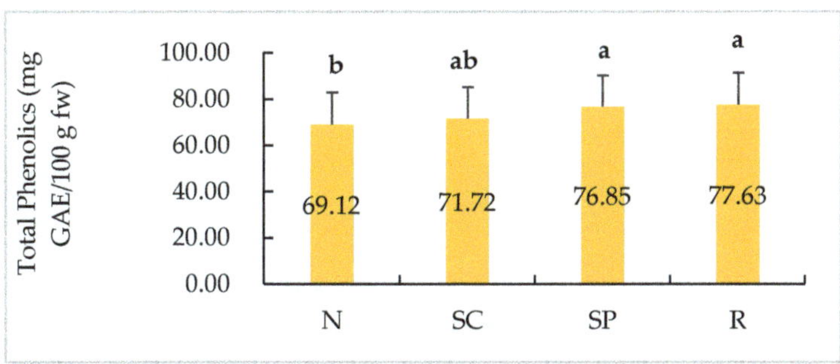

Figure 5. Comparisons of total phenolics (mg GAE/100 g fw) among four skin types including, normal (N), surface cracking (SC), silver patch (SP), and russeting (R) in two-year samples. Means that do not share a common letter are significantly different.

2.5. Anthocyanin Content

Thinly sliced 'normal' and 'silver patch' skin samples were observed under the bright field of a microscope (Figure 6). In Figure 6A–J, normal skin samples are shown in the left column, which clearly show more pinkish pigments in the periderm. Silver patch skin samples are shown in Figure 6K–T in the right column, showing a less red colour; instead, there is a layer of a brown-coloured compound in the skin cells. The comparisons between the normal and silver patch skin samples demonstrated that the loss of the reddish pigment in the silver patch skin was the reason for the blemish. Therefore, the anthocyanin contents were measured in the normal and silver patch skin tissues. The anthocyanin content was significantly lower in the silver patch skin tissues compared to the tissues of the normal skin type. Figure 7 shows a summarized analysis, including all the data from both the 2020 and 2021 seasons. Within two years, the normal skins had a higher total anthocyanin content, with an average of 0.0624 mg C3GE/100 mg fresh weight than that of the silver patch skins, which had an average of 0.0444 mg C3GE/100 mg fresh weight. These results demonstrate that the silver patch skins lost a significant amount of anthocyanin.

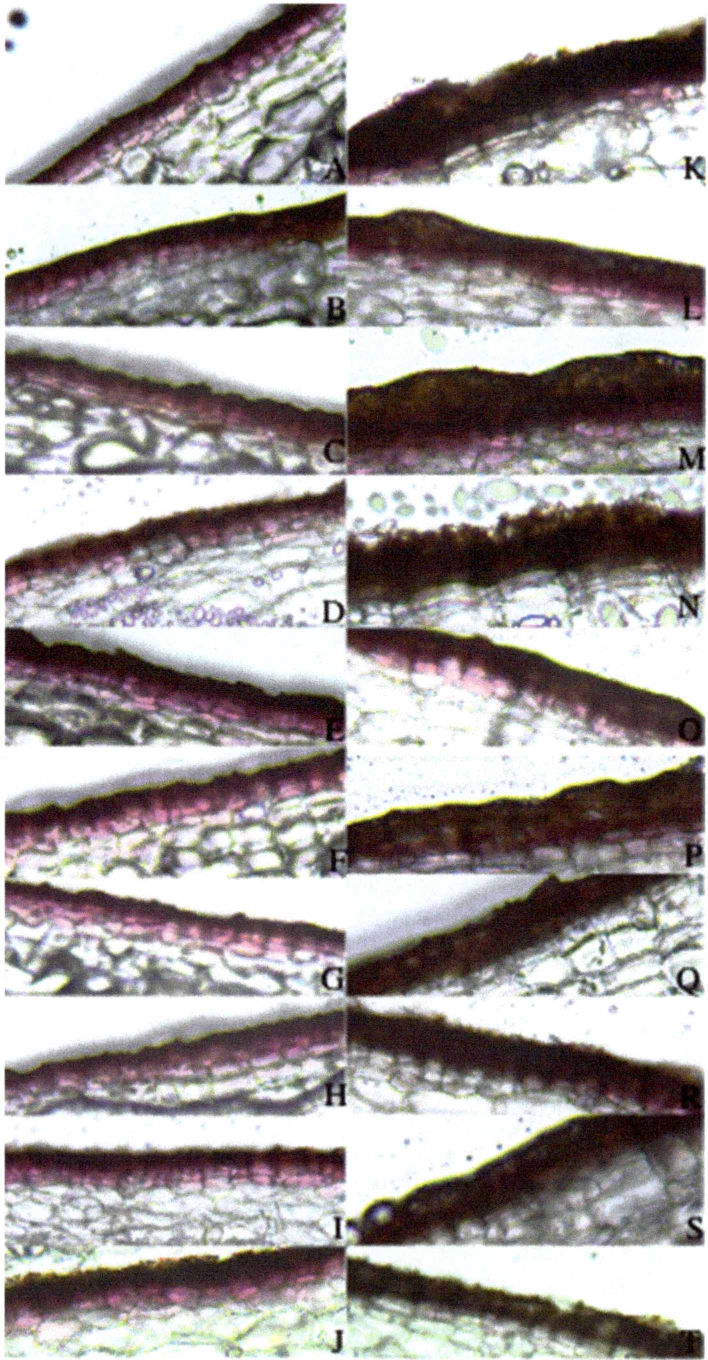

Figure 6. Photos comparison between normal skin (**A–J** on the left column) and silver patch skin (**K–T** on the right column) of Dark Red Norland tubers under bright field microscopy observation.

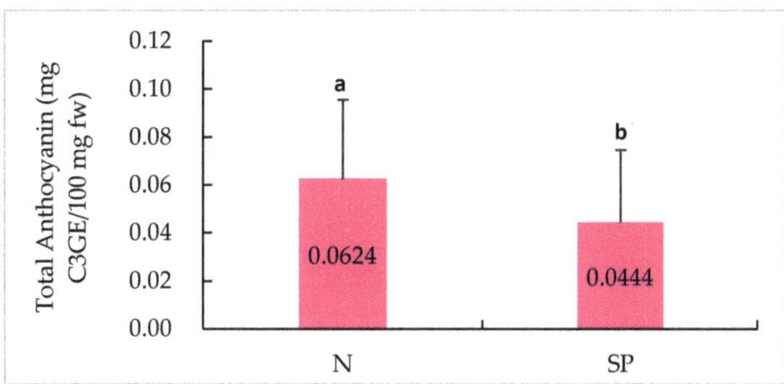

Figure 7. Comparisons of total anthocyanin contents between normal (N) and silver patch (SP) skins in all samples from 2020 and 2021. Means that do not share a common letter are significantly different at $p = 0.05$.

3. Materials and Methods

Three-year (2019–2021) field trials were conducted by Peak of the Market (POM) in Manitoba, Canada. All the laboratory experiments of this study were conducted at the Faculty of Agricultural, Dalhousie University located in Truro, Nova Scotia, Canada.

3.1. Three-Year Field Trial

The field trial to produce the potato cultivar, Dark Red Norland (*Solanum tuberosum* L.), for this study was conducted at the POM Research Site in Winkler, Manitoba, by Gaia Consulting (https://gaiaconsulting.mb.ca/) (accessed on 10 December 2022). There were 32 plots each year. Half of the plots were treated without irrigation, called 'rainfed' plots, while the others were treated with irrigation. The irrigation was applied using a lateral irrigation system (Figure 8). The hand-feel method was used to determine if the water holding capacity was close to or below 70%, which meant it was time to irrigate the plots. The irrigation schedule was different in each of the three years due to the local daily precipitation, soil moisture evaporation, and the amount of water storage for irrigation. The dates and the amount of applied irrigation each year are shown in Table 3. In 2019, irrigation was applied 11 times in the field trial, while there it was only applied 6 and 4 times in 2020 and 2021, respectively. It should be mentioned that the water reservoir ran dry on 9 July 2021 and no additional irrigation water could be applied after that in the 2021 field trial.

Figure 8. Irrigation system used in the field trials in Manitoba, Canada (photo was taken by Dr. Tracy Shinners-Carnelley).

Table 3. Irrigation dates and applied amount (inches) in each year from 2019 to 2021.

Year	Date of Irrigation	Amount Applied (Inches)
2019	3 June	0.50
	5 June	0.50
	12 June	0.50
	27 June	0.50
	3 July	0.50
	18 July	0.50
	24 July	0.50
	29 July	0.50
	5 August	0.75
	7 August	0.50
	13 August	0.75
2020	4 June	0.75
	17 June	0.75
	24 June	0.50
	29 June	0.50
	21 July	0.75
	29 July	1.20
2021	24 June	1.00
	30 June	0.75
	6 July	0.50
	9 July	0.50

The tuber yield and tuber blemish defects, including surface cracking, silver patch, and russeting, were recorded. After harvest, tubers from the differently treated plots were rated for skin colour, external blemishes, yield, and size. The harvested tubers were separated into 5 groups based on size, which were <2″, 2–2.25″, 2.25–3.0″, 3–3.5″, and >3.5″. Tubers with different skin blemishes were counted and transformed into percentage numbers in each size group by using the formula:

$$\text{Number of defected tuber/Total number of graded tubers} \times 100 \tag{1}$$

The grading rules were based on the Peak of the Market Pre-Pack Inspection Manual (POM, 2010). Selected tubers were shipped to Dr. Wang-Pruski's lab at Dal AC for all the lab analyses.

3.2. *Sampling*

3.2.1. Sampling for Suberin Analysis

The tubers were washed and graded after harvesting each year by Gaia Consulting. Four medium-sized tuber samples were randomly selected from each plot and sent to Dal AC for suberin analysis each year. Thirty-two bags of tubers were received each year and stored in a cooler at 4 °C and 90% relative humidity (RH). Each bag was checked and typical skin types, including normal, surface cracking, silver patch, and russeting, were marked on the tubers (Figure 1). Tubers were photographed under bright light before cutting. Skin samples were taken based on the four types of skin blemish occurrence in each bag and processed using the methods described later.

3.2.2. Sampling for Total Phenolics and Anthocyanin Analyses

Eight tubers were selected from each plot for total phenolics and anthocyanin analyses in both the 2020 and 2021 seasons. Half of the samples (four tubers) had bright red colour and relatively normal skins, while the other half (four tubers) were selected with one or more skin blemishes of surface cracking, silver patch, and/or russeting. These tubers normally had lighter skin colour (less red) (Figure 9a).

Figure 9. Photographs of sampling of selected tubers. (**a**) Eight tubers, including four tubers with relatively healthy skin type (left four showing red skin) and four tubers with blemishes of surface cracking, silver patch, and/or russeting (right four showing less reddish colour). (**b**,**c**) Skin samples were cut off from the eight tubers. (**d**) Skin samples were cut into small pieces.

Skin samples were collected from the eight tubers based on normal, surface cracking, silver patch, and russeting skin types for each bag of samples. Each type of skin sample was cut off from at least two tubers in a bag by a knife with a thickness of 2 mm to 5 mm, which included the whole periderm and partial cortex structure (Figure 9b,c). After that, the skin tissues were cut into small pieces (Figure 9d), wrapped in aluminum foil paper, cooled down in liquid nitrogen, put into a 50 mL polypropylene conical tube, and stored in a −80 °C freezer for further usage. The processes are shown in Figure 9a–d. The total phenolic content was measured for all the skin samples, while the anthocyanin content was tested for the normal and silver patch skin samples.

3.3. Evaluation of Suberized Cell Layer

After the visual assessments of the tubers were completed, 4 medium-sized tubers were randomly picked out from each treatment plot and used for suberized cell layer analysis based on the method published by Dr. Gefu Wang-Pruski's lab [20]. These tubers were washed, dried, and photographed on both sides using a digital camera (Sony DSC-F717) or a mobile phone (iPhone 13 Pro). Tuber skins were hand sliced into about 3 × 4 mm skin samples. The skin slices were stained with TBO solution (0.05% (*w*/*v*) Toluidine Blue O dissolved in 0.1 M sodium acetate (pH 4.5)), and then the samples were placed in complete darkness for 5–10 min. After that, the slices were washed with ddH_2O and post-stained by neutral red (NR) solution (0.1% (*w*/*v*) Neutral Red dissolved in 0.1 M potassium phosphate (pH 6.5)) for 1–5 min. The stained slices were washed with ddH_2O, de-stained by lactic acid (85% lactic acid and ddH_2O water at 1:1 (*v*/*v*) ratio) and washed with ddH_2O again. The prepared samples were observed under a microscope (Leica DMi8) and fluorescence light source (Leica EL6000) under a 10× objective lens. A Leica microscope and Leica DMC6200 camera were used to observe the samples and take images (Figure 10a). Based on the images, the number of suberized cell layers was counted (Figure 10b).

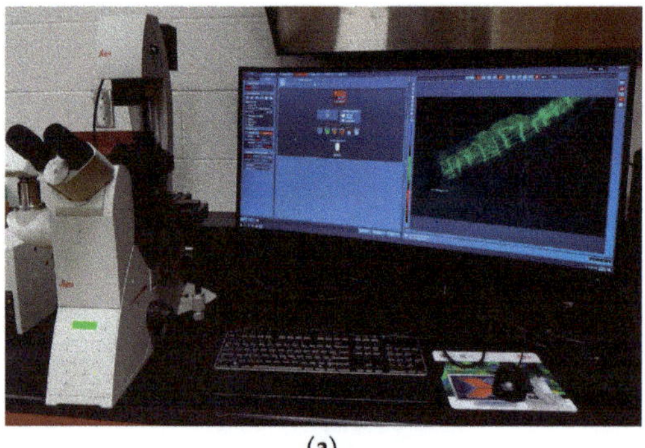

(a)

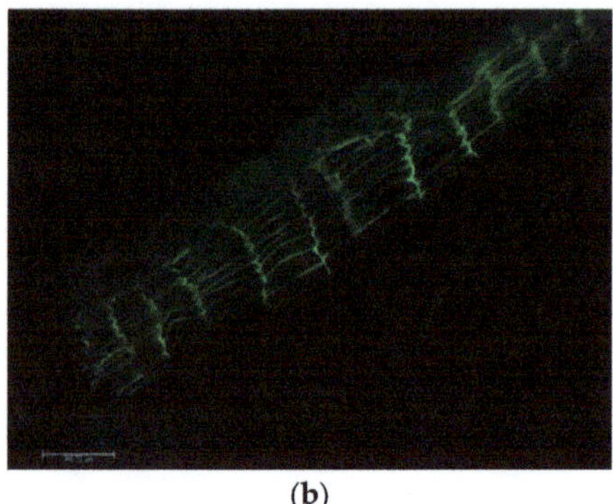

(b)

Figure 10. Analysis of the suberized cell layer. (**a**) The Lecia microscope (Leica DMi8) and Leica external light source for fluorescence excitation (Leica EL6000) and camera system (Leica DMC6200); (**b**) Image of the suberized cell layers under the fluorescence (captured by Leica LAS X Imaging and Analysis Software), showing the number of layers of suberized cells. (Leica system can be found in Leica Microsystems Inc., 71 Four Valley Drive, Concord, ON, Canada.)

3.4. Determination of Total Phenolics

The total phenolics were measured using the Folin–Ciocalteu (FC) method [21] with garlic acid as a standard. The absorbance against the prepared sample reagents was measured using a UV-VIS spectrophotometer (Ultrospec 3000, Biochrom, Unit 7, Enterprise Zone, 3970 Cambridge Research Park, Beach Drive, Waterbeach, Cambridge, UK). Approximately 0.5 g of each the tuber skin samples was weighed and recorded. The absorbance was measured against a prepared reagent blank (0 mg/L gallic acid) at 760 nm. All samples were analyzed in duplicate. The total phenolic content was expressed as 'mg gallic acid equivalents/100 g fresh weight' (mg GAE/100 g fw). Based on the skin blemish occurrence, at most, 32 samples were measured for each skin type from 256 tubers each year. Among all four skin types, at most, 128 samples were measured from both the rainfed and irrigated treatments each year.

3.5. Determination of Total Anthocyanin

The anthocyanin content was analyzed using both visual observation and biochemical analysis to show if the reduced redness in the tuber skin colour was related to a loss of anthocyanin. Brightfield microscopy observation was performed for the colour comparisons. For the biochemical analysis, the anthocyanin contents in the normal and silver patch skins were evaluated in all treatments. The extraction and quantification of anthocyanin were carried out by the pH differential method [22], with a few modifications as indicated below.

The amount of sample per extraction was 100 mg in this experiment. The density of the skin tissue was set as 1 g/mL. The total dilution factor (DF) was determined to be 100 as shown in the equation:

$$\text{Total/Final dilution factor (DF)} = DF_1 \times DF_2 = 10 \times 10 = 100 \tag{2}$$

The previously identified dilution factor (DF_1) was set to be 10:

$$DF_1 = V_{(\text{tissue + solvent})}/V_{(\text{tissue})} = 10 \tag{3}$$

The absorbance of the sample was read at 520 nm and 700 nm 3 times after zeroization.

The absorbance (A) and the total monomeric anthocyanin of each sample were calculated by using the equations:

$$A = (Abs_{520} - Abs_{700})_{pH\ 1.0} - (Abs_{520} - Abs_{700})_{pH\ 4.5} \tag{4}$$

$$\text{Total anthocyanin (mg C3GE/L)} = (\text{A} \times \text{MW} \times \text{DF} \times 1000)/(\varepsilon \times \text{L}) \quad (5)$$

where MW is the molecular weight of the predominant anthocyanin. In this experiment, cyanidin-3-O-glucoside (C3G) was used to express the total anthocyanin, since it is the most abundant anthocyanin in nature [23]. The MW of C3G is 449.2 g/mol. The molar extinction coefficient (ε) is 26,900. L is the path length (in cm), which is 1 cm. The conversion factor from g to mg is 1000. C3GE is the cyanidin-3-O-glucoside equivalent. The total anthocyanin (mg C3GE/L) was divided by 1000 to obtain a final unit of mg C3GE/100 mg fresh weight.

At most, 32 samples were measured for both the normal and silver patch skin types from 256 tubers each year.

3.6. Statistical Analysis

For total yield analysis, a two-sample *t*-test and Fisher's least significant difference (LSD) pairwise comparisons were used. Mood's Median test and Fisher's LSD pairwise comparisons were used for skin blemish analysis. The statistically significant level was set as $p = 0.05$.

For suberin analysis, the number of suberized cell layers of all normal skin samples was compared between the rainfed and irrigated plots using a two-sample *t*-test. The number of suberized cell layers was compared among the 4 skin types using one-way ANOVA and Tukey's pairwise comparison. A normality test was performed before the ANOVA and two-sample *t*-test. The statistically significant level was set as $p = 0.05$.

One-way ANOVA and Tukey's pairwise comparison were used for the comparison of the total phenolic contents among 4 skin types, including normal, silver patch, surface cracking, and russeting. A two-sample *t*-test was performed to compare the total phenolic content from all the skin samples between rainfed and irrigated plots. A normality test was performed before the ANOVA and two-sample *t*-test. The statistically significant level was set as $p = 0.05$.

A two-sample *t*-test was performed to compare the total anthocyanin content in the normal and silver patch skin samples for both the 2020 and 2021 samples. A two-sample *t*-test was performed to compare all the normal and silver patch skin samples extracted in these 2 years. A normality test was performed before the two-sample *t*-test. When the data did not fit the normality, the Mann–Whitney test was used. The statistically significant level was set as $p = 0.05$.

4. Discussion

Water regime is an important factor that can affect tuber yield and quality. Irrigation did increase the total tuber yield and decreased the occurrence of surface cracking skin defects. The data from the 3-year field trials show that irrigated plants had significantly higher total yields (Figure 2). The yields of medium (2.25–3.0″) and large (3–3.5″) tubers were increased with irrigation. This result agrees with those of many previous studies about the importance of water availability during the growing season, especially its significant effect on tuber yield [24–26].

Surface cracking defects were found to be induced by water deficit in the field trial (Figure 3). Irrigation significantly reduced the occurrence (%) of surface cracking defects, however, russeting defects were increased to some degree (Figure 3). The higher russeting defects in the irrigated plots may have been caused by the expansion of the tuber skin in the skin developmental process [27], which is similar to the out-of-step cell division speed due to a fluctuating moisture supply [15].

The rainfed normal tubers tended to form more suberized cell layers compared to the irrigated normal tubers. Based on the three years of suberin analysis data, we found that the rainfed normal samples had more suberized cell layers than those of the irrigated normal samples. This result demonstrates that the tubers grown without irrigation tended to form more suberized cell layers, which can result in a thicker phellem. Following suberization, phellem cells die and create an outer defensive layer, which possesses a waxy component that protects against cell desiccation, and a protective suberin biopolymer, which provides a barrier to pathogens and other intrusions [7,28,29]. Suberin serves as a protective barrier in the periderm tissue layers, controls water and ion transport, restricts infection, and maintains integrity [7]. It has been reported that suberization in potato tuber periderm is associated with protection against biotic and abiotic stresses [7,27–29]. It has been observed that in response to heat stress, there was increased production and accumulation of periderm cell layers to protect the tubers, and many transcriptional factors of periderm responded to heat stress [7,30]. An increased number of suberized phellem cell layers also provided resistance against tuber greening [7,31]. Our results demonstrate that suberization can also be responsive to water deficit stress, during which the rainfed tubers tended to form more suberized cell layers to protect the tubers from water loss and a drought environment.

Many studies have been performed to understand the molecular mechanism of potato periderm [5,7,32,33]. A gene called '*CYP86A33*' was proven to have a strong function in the formation of ω-functionalized monomers in aliphatic suberin, which are necessary for the suberin typical lamellar organization and the periderm resistance to water loss [34–36]. Another potato gene encoding a fatty ω-hydroxyacid/fatty alcohol hydroxycinnamoyl transferase (*FHT*) was reported to have significant effects on the anatomy, sealing properties, and maturation of the periderm [37]. When *FHT* was down-regulated, the tuber skin became thicker and russeted, water loss was greatly increased, and maturation was prevented [37]. It is suggested that future studies analyze the suberization-related gene expression under different water regimes, which can improve our understanding of the influence of water stress.

In addition to this, our results also show that the surface cracking and silver patch skins had more suberized cell layers than those of the normal and russeting skins. This result demonstrates that the normal skin had fewer suberized cell layers than skins with defects. However, this is opposite to a previous study that showed russeting had increased suberization and a thicker layer of phellem [27]. This difference could have been caused by the different observation methods and different potato varieties. Dark Red Norland is a smooth-skinned cultivar with a relatively thin phellem layer. The cells tend to be cracked when the suberization activity is increased. As the tuber skin expands during development, the thick part of the skin cracks away from the original thin skin and sloughs off, resulting in netted, rough skins. Surface cracking has a similar process of formation, which could also explain why the phellem of surface cracking seemed to be cracked and had more suberized cell layers (Figure 4 and Table 2). It is suggested that irrigation is applied throughout the tuber growing stage to reduce the soil temperature and create a good condition for Dark Red Norland tuber skin formation. Soil temperature is another important factor that can influence tuber growth, which is related to heat stress. A study showed that high temperature had negative effects on tuber yield and skin formation [38].

Significantly higher amounts of total phenolics were found in the irrigated treatments. Many studies have proven that environmental factors can profoundly influence the phenolic content in plants [8,21,39–43]. However, the results are often conflicting. Drought is likely to make plants accumulate phenolic compounds. The biosynthesis and accumulation of phenolic compounds during drought stress are regulated by enzymes of the phenylpropanoid pathway [42]. Studies on leafy lettuce, grapes, leaves of maize, and leaves of *Amaranthus tricolor* have observed a high accumulation of phenolic compounds in samples under drought stress [39,41,44,45]. In addition, cherry tomato, which also belongs to *Solanum* genus, was indicated to have decreased polyphenol content under irrigation. In contrast, contradictory results were obtained by Sánchez-Rodríguez et al. [46]. Studies on broccoli, sweet potato, and cauliflower have demonstrated that irrigation could have a positive effect on the phenolic content [43,47–50]. According to a review on the influence of water stress on the production of phenolic compounds in plants of medicinal interest [51], the widely accepted idea that there is a widespread increase in phenolic compounds in response to water stress is most often incorrect [51]. The total phenolic system is complicated and can be different for each plant species [51]. Our results show that irrigation had a positive effect on the phenolic content accumulation in Dark Red Norland tuber skins. This suggests that irrigation has an important role in regulating total phenolic biosynthesis in potato skins.

Significantly lower anthocyanin contents were found in silver patch skins (Figure 7). This finding is shown in Figure 6, in which less reddish pigmentation in the silver patch skins can be seen; instead, more brown-coloured compounds were found in the skin cells. It has been reported that phenolic extracts can strongly stimulate the oxidation of anthocyanin due to an anthocyanin-PPO (polyphenol oxidases)–phenol reaction that produces a brown by-product [52,53]. This explains why there were dark brown-coloured skin surfaces in the silver patch skin and not the pinkish-red skin colour; there was a higher concentration of total phenolics and less anthocyanin in the silver patch skins.

5. Conclusions

Irrigation plays an important role in Dark Red Norland potato production. It significantly improves the total yield of Dark Red Norland tubers, while reducing the occurrence of surface cracking skin blemishes and producing more good tubers. Our results also show that tubers grown without irrigation tended to form more suberized cell layers to protect the tubers from drought stress. Irrigation can increase the level of total phenolics in Dark Red Norland tuber skins. Different skin blemishes have different levels of suberization, total phenolics, and anthocyanins, in which surface cracking and silver patch form more suberized cell layers. Silver patch skins have higher contents of total phenolics but fewer anthocyanins. This study provides a better understanding of potato production and skin blemishes in Dark Red Norland tubers. Future studies can be conducted regarding drought stress effects on suberization-related genes in potato skins.

Author Contributions: Conceptualization, M.J., T.S.-C. and G.W.-P.; investigation, M.J.; resources, T.S.-C. and G.W.-P.; data curation, M.J., T.S.-C., D.G., D.J., J.J. and G.W.-P.; writing—original draft preparation, M.J.; writing—review and editing, T.S.-C. and G.W.-P. All authors have read and agreed to the published version of the manuscript.

Funding: This research was funded by the Canadian Agricultural Partnership AG Action Manitoba Project #1000219633.

Data Availability Statement: All the relevant data of the study are provided in the manuscript.

Acknowledgments: The authors are highly thankful to the Ag Action Manitoba Program, Peak of the Market, Gaia Consulting, and the Dalhousie Agricultural Campus for their great support and help in this project.

Conflicts of Interest: The authors declare no conflict of interest.

References

1. Stewart, D.; McDougall, G. Potato; A Nutritious, Tasty but Often Maligned Staple Food. Available online: https://www.hutton.ac.uk/webfm_send/743 (accessed on 10 December 2022).
2. Asim, A.; Öztürk Gökçe, Z.N.; Bakhsh, A.; Tindaş Çayli, İ.; Aksoy, E.; Çalişkan, S.; Çalişkan, M.E.; DemiRel, U. Individual and Combined Effect of Drought and Heat Stresses in Contrasting Potato Cultivars Overexpressing MiR172b-3p. *Turk. J. Agric. For.* **2021**, *45*, 651–668. [CrossRef]
3. Hossain, M.J.; Aksoy, E.; Öztürk Gökçe, N.Z.; Joyia, F.A.; Khan, M.S.; Bakhsh, A. Rapid and Efficient in Vitro Regeneration of Transplastomic Potato (*Solanum tuberosum* L.) Plants after Particle Bombardment. *Turk. J. Agric. For.* **2021**, *45*, 313–323. [CrossRef]
4. Agriculture and Agri-Food Canada. Potato Market Information Review, 2020–2021. Available online: https://agriculture.canada.ca/en/agriculture-and-agri-food-canada/canadas-agriculture-sectors/horticulture/horticulture-sector-reports/potato-market-information-review-2020-2021 (accessed on 5 December 2022).
5. Vulavala, V.K.R.; Fogelman, E.; Faigenboim, A.; Shoseyov, O.; Ginzberg, I. The Transcriptome of Potato Tuber Phellogen Reveals Cellular Functions of Cork Cambium and Genes Involved in Periderm Formation and Maturation. *Sci. Rep.* **2019**, *9*, 10216. [CrossRef]
6. Keren-Keiserman, A.; Baghel, R.S.; Fogelman, E.; Faingold, I.; Zig, U.; Yermiyahu, U.; Ginzberg, I. Effects of Polyhalite Fertilization on Skin Quality of Potato Tuber. *Front. Plant Sci.* **2019**, *10*, 1379. [CrossRef] [PubMed]
7. Singh, B.; Bhardwaj, V.; Kaur, K.; Kukreja, S.; Goutam, U. Potato Periderm Is the First Layer of Defence against Biotic and Abiotic Stresses: A Review. *Potato Res.* **2021**, *64*, 131–146. [CrossRef]

8. André, C.M.; Schafleitner, R.; Legay, S.; Lefèvre, I.; Aliaga, C.A.A.; Nomberto, G.; Hoffmann, L.; Hausman, J.-F.; Larondelle, Y.; Evers, D. Gene Expression Changes Related to the Production of Phenolic Compounds in Potato Tubers Grown under Drought Stress. *Phytochemistry* **2009**, *70*, 1107–1116. [CrossRef] [PubMed]
9. Seitz, H.U.; Hinderer, W. CHAPTER 3—Anthocyanins. In *Phytochemicals in Plant Cell Cultures*; Constabel, F., Vasil, I.K., Eds.; Academic Press: Cambridge, MA, USA, 1988; pp. 49–76, ISBN 978-0-12-715005-5.
10. Strygina, K.V.; Kochetov, A.V.; Khlestkina, E.K. Genetic Control of Anthocyanin Pigmentation of Potato Tissues. *BMC Genet.* **2019**, *20*, 27. [CrossRef]
11. Liu, Y.; Tikunov, Y.; Schouten, R.E.; Marcelis, L.F.M.; Visser, R.G.F.; Bovy, A. Anthocyanin Biosynthesis and Degradation Mechanisms in Solanaceous Vegetables: A Review. *Front. Chem.* **2018**, *6*, 52. [CrossRef] [PubMed]
12. Mikitzel, L. Tuber Physiological Disorders. In *The Potato: Botany, Production and Uses*; Botany, Production and Uses; CABI: Oxford, UK, 2014; pp. 237–254, ISBN 978-1-78064-280-2.
13. Ginzberg, I.; Minz, D.; Faingold, I.; Soriano, S.; Mints, M.; Fogelman, E.; Warshavsky, S.; Zig, U.; Yermiyahu, U. Calcium Mitigated Potato Skin Physiological Disorder. *Am. J. Pot. Res.* **2012**, *89*, 351–362. [CrossRef]
14. Lynch, D.R.; Foroud, N.; Kozub, G.C.; Fames, B.C. The Effect of Moisture Stress at Three Growth Stages on the Yield, Components of Yield and Processing Quality of Eight Potato Varieties. *Am. Potato J.* **1995**, *72*, 375–385. [CrossRef]
15. van Loon, C.D. The Effect of Water Stress on Potato Growth, Development, and Yield. *Am. Potato J.* **1981**, *58*, 51–69. [CrossRef]
16. Onder, S.; Caliskan, M.E.; Onder, D.; Caliskan, S. Different Irrigation Methods and Water Stress Effects on Potato Yield and Yield Components. *Agric. Water Manag.* **2005**, *73*, 73–86. [CrossRef]
17. Karafyllidis, D.I.; Stavropoulos, N.; Georgakis, D. The Effect of Water Stress on the Yielding Capacity of Potato Crops and Subsequent Performance of Seed Tubers. *Potato Res.* **1996**, *39*, 153–163. [CrossRef]
18. Miller, D.E.; Martin, M.W. Effect of Declining or Interrupted Irrigation on Yield and Quality of Three Potato Cultivars Grown on Sandy Soil. *Am. Potato J.* **1987**, *64*, 109–117. [CrossRef]
19. Hill, D.; Nelson, D.; Hammond, J.; Bell, L. Morphophysiology of Potato (*Solanum tuberosum*) in Response to Drought Stress: Paving the Way Forward. *Front. Plant Sci.* **2021**, *11*, 597554. [CrossRef]
20. Zhang, Z.; Huang, Z.; Lu, H.; Wang-Pruski, G. A Rapid and Effective Method for Observation of Suberized Cell Layers in Potato Tuber Skin. *Sci. Hortic.* **2017**, *224*, 215–218. [CrossRef]
21. Škerget, M.; Kotnik, P.; Hadolin, M.; Hraš, A.R.; Simonič, M.; Knez, Ž. Phenols, Proanthocyanidins, Flavones and Flavonols in Some Plant Materials and Their Antioxidant Activities. *Food Chem.* **2005**, *89*, 191–198. [CrossRef]
22. Giusti, M.M.; Wrolstad, R.E. Characterization and Measurement of Anthocyanins by UV-Visible Spectroscopy. *Curr. Protoc. Food Anal. Chem.* **2001**, F1.2.1–F1.2.13. [CrossRef]
23. Francis, F.J. Food Colorants: Anthocyanins. *Crit. Rev. Food Sci. Nutr.* **1989**, *28*, 273–314. [CrossRef]
24. Ávila-Valdés, A.; Quinet, M.; Lutts, S.; Martínez, J.P.; Lizana, X.C. Tuber Yield and Quality Responses of Potato to Moderate Temperature Increase during Tuber Bulking under Two Water Availability Scenarios. *Field Crops Res.* **2020**, *251*, 107786. [CrossRef]
25. Ojala, J.C.; Stark, J.C.; Kleinkopf, G.E. Influence of Irrigation and Nitrogen Management on Potato Yield and Quality. *Am. Potato J.* **1990**, *67*, 29–43. [CrossRef]
26. Wagg, C.; Hann, S.; Kupriyanovich, Y.; Li, S. Timing of Short Period Water Stress Determines Potato Plant Growth, Yield and Tuber Quality. *Agric. Water Manag.* **2021**, *247*, 106731. [CrossRef]
27. Kumar, P.; Ginzberg, I. Potato Periderm Development and Tuber Skin Quality. *Plants* **2022**, *11*, 2099. [CrossRef]
28. Barel, G.; Ginzberg, I. Potato Skin Proteome Is Enriched with Plant Defence Components. *J. Exp. Bot.* **2008**, *59*, 3347–3357. [CrossRef] [PubMed]
29. Woolfson, K.N.; Esfandiari, M.; Bernards, M.A. Suberin Biosynthesis, Assembly, and Regulation. *Plants* **2022**, *11*, 555. [CrossRef] [PubMed]
30. Ginzberg, I.; Barel, G.; Ophir, R.; Tzin, E.; Tanami, Z.; Muddarangappa, T.; de Jong, W.; Fogelman, E. Transcriptomic Profiling of Heat-Stress Response in Potato Periderm. *J. Exp. Bot.* **2009**, *60*, 4411–4421. [CrossRef]
31. Tanios, S.; Thangavel, T.; Eyles, A.; Tegg, R.S.; Nichols, D.S.; Corkrey, R.; Wilson, C.R. Suberin Deposition in Potato Periderm: A Novel Resistance Mechanism against Tuber Greening. *New Phytol.* **2020**, *225*, 1273–1284. [CrossRef] [PubMed]
32. Vulavala, V.K.R.; Fogelman, E.; Rozental, L.; Faigenboim, A.; Tanami, Z.; Shoseyov, O.; Ginzberg, I. Identification of Genes Related to Skin Development in Potato. *Plant Mol. Biol.* **2017**, *94*, 481–494. [CrossRef]
33. Neubauer, J.D.; Lulai, E.C.; Thompson, A.L.; Suttle, J.C.; Bolton, M.D. Wounding Coordinately Induces Cell Wall Protein, Cell Cycle and Pectin Methyl Esterase Genes Involved in Tuber Closing Layer and Wound Periderm Development. *J. Plant Physiol.* **2012**, *169*, 586–595. [CrossRef]
34. Serra, O.; Soler, M.; Hohn, C.; Sauveplane, V.; Pinot, F.; Franke, R.; Schreiber, L.; Prat, S.; Molinas, M.; Figueras, M. *CYP86A33* -Targeted Gene Silencing in Potato Tuber Alters Suberin Composition, Distorts Suberin Lamellae, and Impairs the Periderm's Water Barrier Function. *Plant Physiol.* **2009**, *149*, 1050–1060. [CrossRef]
35. Bjelica, A.; Haggitt, M.L.; Woolfson, K.N.; Lee, D.P.N.; Makhzoum, A.B.; Bernards, M.A. Fatty Acid ω-Hydroxylases from *Solanum tuberosum*. *Plant Cell. Rep.* **2016**, *35*, 2435–2448. [CrossRef] [PubMed]
36. Haggitt, M. Role of Fatty Acid Omega-Hydroxylase 1 and Abscisic Acid in Potato Tuber Suberin Formation. Available online: https://ir.lib.uwo.ca/cgi/viewcontent.cgi?article=5355&context=etd (accessed on 10 December 2022).

37. Serra, O.; Hohn, C.; Franke, R.; Prat, S.; Molinas, M.; Figueras, M. A Feruloyl Transferase Involved in the Biosynthesis of Suberin and Suberin-Associated Wax Is Required for Maturation and Sealing Properties of Potato Periderm: FHT Function in Potato Periderm. *Plant J.* **2010**, *62*, 277–290. [CrossRef] [PubMed]
38. Rykaczewska, K. The Effect of High Temperature Occurring in Subsequent Stages of Plant Development on Potato Yield and Tuber Physiological Defects. *Am. J. Potato Res.* **2015**, *92*, 339–349. [CrossRef]
39. Malejane, D.N.; Tinyani, P.; Soundy, P.; Sultanbawa, Y.; Sivakumar, D. Deficit Irrigation Improves Phenolic Content and Antioxidant Activity in Leafy Lettuce Varieties. *Food Sci. Nutr.* **2017**, *6*, 334–341. [CrossRef] [PubMed]
40. Payyavula, R.S.; Navarre, D.A.; Kuhl, J.C.; Pantoja, A.; Pillai, S.S. Differential Effects of Environment on Potato Phenylpropanoid and Carotenoid Expression. *BMC Plant Biol.* **2012**, *12*, 39. [CrossRef]
41. Pérez-Álvarez, E.P.; Intrigliolo, D.S.; Almajano, M.P.; Rubio-Bretón, P.; Garde-Cerdán, T. Effects of Water Deficit Irrigation on Phenolic Composition and Antioxidant Activity of Monastrell Grapes under Semiarid Conditions. *Antioxidants* **2021**, *10*, 1301. [CrossRef] [PubMed]
42. Chowdhary, V.; Alooparampil, S.; Pandya, R.V.; Tank, J.G.; Chowdhary, V.; Alooparampil, S.; Pandya, R.V.; Tank, J.G. *Physiological Function of Phenolic Compounds in Plant Defense System*; IntechOpen: London, UK, 2021; ISBN 978-1-83969-347-2.
43. Kałużewicz, A.; Lisiecka, J.; Gąsecka, M.; Krzesiński, W.; Spiżewski, T.; Zaworska, A.; Frąszczak, B. The Effects of Plant Density and Irrigation on Phenolic Content in Cauliflower. *Hortic. Sci.* **2017**, *44*, 178–185. [CrossRef]
44. Hura, T.; Hura, K.; Grzesiak, S. Contents of Total Phenolics and Ferulic Acid, and PAL Activity during Water Potential Changes in Leaves of Maize Single-Cross Hybrids of Different Drought Tolerance. *J. Agron. Crop Sci.* **2008**, *194*, 104–112. [CrossRef]
45. Sarker, U.; Oba, S. Drought Stress Enhances Nutritional and Bioactive Compounds, Phenolic Acids and Antioxidant Capacity of Amaranthus Leafy Vegetable. *BMC Plant Biol.* **2018**, *18*, 258. [CrossRef]
46. Sánchez-Rodríguez, E.; Ruiz, J.M.; Ferreres, F.; Moreno, D.A. Phenolic Profiles of Cherry Tomatoes as Influenced by Hydric Stress and Rootstock Technique. *Food Chem.* **2012**, *134*, 775–782. [CrossRef]
47. Robbins, R.J.; Keck, A.-S.; Banuelos, G.; Finley, J.W. Cultivation Conditions and Selenium Fertilization Alter the Phenolic Profile, Glucosinolate, and Sulforaphane Content of Broccoli. *J. Med. Food* **2005**, *8*, 204–214. [CrossRef] [PubMed]
48. Cogo, S.L.P.; Chaves, F.C.; Schirmer, M.A.; Zambiazi, R.C.; Nora, L.; Silva, J.A.; Rombaldi, C.V. Low Soil Water Content during Growth Contributes to Preservation of Green Colour and Bioactive Compounds of Cold-Stored Broccoli (*Brassica oleraceae* L.) Florets. *Postharvest Biol. Technol.* **2011**, *60*, 158–163. [CrossRef]
49. Fortier, E.; Desjardins, Y.; Tremblay, N.; Bélec, C.; Côté, M. Influence of Irrigation and Nitrogen Fertilization on Broccoli Polyphenolics Concentration. *Acta Hortic.* **2010**, *856*, 55–62. [CrossRef]
50. Mao, L.; Jett, L.E.; Story, R.N.; Hammond, A.M.; Peterson, J.K.; Labonte, D.R. Influence of Drought Stress on Sweetpotato Resistance to Sweetpotato Weevil, Cylas Formicarius (*Coleoptera: Apoinidae*), and Storage Root Chemistry. *Fla. Entomol.* **2004**, *87*, 261–267. [CrossRef]
51. The Effect of Water Deficit Stress on the Composition of Phenolic Compounds in Medicinal Plants. *South Afr. J. Bot.* **2020**, *131*, 12–17. [CrossRef]
52. Andersen, A.W.; Tong, C.B.S.; Krueger, D.E. Comparison of Periderm Color and Anthocyanins of Four Red Potato Varieties. *Am. J. Potato Res* **2002**, *79*, 249–253. [CrossRef]
53. Jiang, Y. Role of Anthocyanins, Polyphenol Oxidase and Phenols in Lychee Pericarp Browning. *J. Sci. Food Agric.* **2000**, *80*, 305–310. [CrossRef]

 *plants*

Article

Changes in Soil Characteristics, Microbial Metabolic Pathways, TCA Cycle Metabolites and Crop Productivity following Frequent Application of Municipal Solid Waste Compost

Lord Abbey [1,*], Svetlana N. Yurgel [1,2], Ojo Alex Asunni [3], Raphael Ofoe [1], Josephine Ampofo [4], Lokanadha Rao Gunupuru [1] and Nivethika Ajeethan [1]

1 Department of Plant, Food, and Environmental Sciences, Faculty of Agriculture, Dalhousie University, Halifax, NS B2N 5E3, Canada
2 United States Department of Agriculture, ARS, Grain Legume Genetics and Physiology Research Unit, 24106 N Bunn Road, Prosser, DC 99350-9687, USA
3 Department of Applied Disasters and Emergency Studies, Brandon University, Brandon, MB R7A 6A9, Canada
4 Department of Food Science and Technology, University of California, Davis, CA 95616, USA
* Correspondence: loab07@gmail.com

Abstract: The benefit sof municipal solid waste (MSW) compost on soil health and plant productivity are well known, but not its long-term effect on soil microbial and plant metabolic pathways. A 5-year study with annual (AN), biennial (BI) and no (C, control) MSW compost application were carried out to determine the effect on soil properties, microbiome function, and plantgrowth and TCA cycle metabolites profile of green beans (*Phaseolus vulgaris*), lettuce (*Latuca sativa*) and beets (*Beta vulgaris*). MSW compost increased soil nutrients and organic matter leading to a significant ($p < 0.05$) increase in AN-soil water-holding capacity followed by BI-soil compared to C-soil. Estimated nitrogen release in the AN-soil was ca. 23% and 146% more than in BI-soil and C-soil, respectively. Approximately 44% of bacterial community due to compost. Deltaproteobacteria, Bacteroidetes Bacteroidia, and Chloroflexi Anaerolineae were overrepresented in compost amended soils compared to C-soil. A strong positive association existed between AN-soil and 18 microbial metabolic pathways out of 205. Crop yield in AN-soil were increased by 6–20% compared to the BI-soil, and by 35–717% compared to the C-soil. Plant tricarboxylic acid cycle metabolites were highly ($p < 0.001$) influenced by compost. Overall, microbiome function and TCA cycle metabolites and crop yield were increased in the AN-soil followed by the BI-soil and markedly less in C-soil. Therefore, MSW compost is a possible solution to increase soil health and plants production in the medium to long term. Future study must investigate rhizosphere metabolic activities.

Keywords: organic amendment; plant metabolites; soil health; environmental health

Citation: Abbey, L.; Yurgel, S.N.; Asunni, O.A.; Ofoe, R.; Ampofo, J.; Gunupuru, L.R.; Ajeethan, N. Changes in Soil Characteristics, Microbial Metabolic Pathways, TCA Cycle Metabolites and Crop Productivity following Frequent Application of Municipal Solid Waste Compost. *Plants* **2022**, *11*, 3153. https://doi.org/10.3390/plants11223153

Academic Editor: Christian Dimkpa

Received: 15 October 2022
Accepted: 16 November 2022
Published: 18 November 2022

1. Introduction

The aim of the United Nations sustainable development goal #2 is to provide adequate and consistent nutritious food for the estimated 9.73 billion world population by 2050 [1]. Researchers are therefore, exploring different ways to meet this food gap without endangering agroecological systems and environment. One such approach is amendment of agricultural soils with municipal solid waste (MSW) compost to improve soil organic matter content and soil nutrient status, and to sustain soil ecosystem services and biodiversity with the added benefit of increasing food and nutrition security [2,3].

Global generation of municipal solid waste (MSW) is estimated at ca. 2 billion Mt per annum, and it is expected to rise to ca. 2.59 billion Mt by 2030 mainly due to increases in urbanization and changes in people's lifestyles [4,5]. In Canada like many other countries, at least 40% of the MSW from kitchen, yard, restaurants, hotels, and groceries are organic and compostable [6]. MSW compost is a significant source of nutrients, macromolecules, and other compounds essential for plant growth and development [7,8]. Nonetheless, anecdotal evidence suggests that farmers and the public are concerned about the use of MSW compost for food production due to possible contaminations with microbial pathogens like *Salmonella*, and *Escherichia coli* in addition to pharmaceutical residues, plastics, and other harmful physical objects. However, recent trend shows a gradual change in public perception towards acceptability and application of MSW compost. This trend can be attributed to sustained global climate change and deterioration of soil health, scarcity and increase cost of fertilizers, establishment of compost quality standards in many jurisdictions such as in the European Union, Canada and USA, and increase in global research and promotion of MSW compost [9–13].

By nature, compost is complex and comprises humic substances (i.e., fulvic acids, humic acids and humins), non-humic substances (i.e., nutrient elements, macromolecules, and plant growth promoters) and beneficial microbiome that can positively alter many plant structural, physiological, and biochemical functions [12–14]. Soil microbiome and function are the basics for the promotion of the stability, productivity, and sustainability of soil health that ensure essential nutrients cycling for optimum plant growth and development [11]. Previous studies demonstrated varied microbial community composition in mature compost including high diversity of nitrogen (N)-fixing, sulphur (S)-oxidizing and nitrifying bacteria, and the biomass of actinobacteria, but a remarkable reduction in Gram-negative bacteria compared to immature compost [15–17]. Furthermore, Kelly et al. [18] reported that long-term application of compost increased microbial communities and their functions that culminated in increased nutrient uptake and plant productivity. Besides, MSW compost was also found to increase the accumulation of plant primary metabolites such as organic acids, essential amino acids, and phospholipids in different crop species [19]. Despite these positive observations, the main limitation of MSW compost is the slow and irregular release of N for plant use [10]. Therefore, the appropriate frequency and rate of MSW compost application will be critical to the determination of N availability to plants, and aversion of a decline in crop productivity.

Many metabolic pathways have been studied in isolation [20], and there is not much literature to link MSW compost effect to plant metabolic pathways. A study by Neugart et al. [21] indicated that food waste compost increased carotenoids concentrations but reduced glucosinolates and phenolics concentrations in pak choi (*Brassica rapa* ssp. *Chinensis*). According to Zhou et al. [22], plants use their root exudates (e.g., oxalate, malate and citrate) as signals to mobilize specific microbial communities to combat disease pathogens, facilitate nutrient acquisition and crosstalk amongst various plant growth regulators. These mechanisms may influence various plant metabolic pathways but understudied. One such major metabolic pathway is the tricarboxylic acid (TCA) cycle, which involves the interconversion of cytosolic glucose, fatty acids and amino acids to acetyl-CoA or other intermediates for mitochondria energy generation [23,24]. Organic acids are the key intermediate metabolites of the TCA cycle, which is a major metabolic pathway for the adenosine triphosphate (ATP) and nicotinamide adenine dinucleotide phosphate (NADPH) syntheses in all organisms [24–26]. The precursor of the TCA cycle is pyruvic acid from cytosolic glucose via glycolysis from which acetyl-CoA is formed [27]. So far, literature on if and how compost regulate TCA cycle intermediate metabolites in plants is scarce.

A recent study by Rosa et al. [28] showed that as the concentration of water extract of compost applied to maize (*Zea mays*) was increased from 0 to 80 mg/L, root exudate of organic acids associated with the TCA cycle metabolites, i.e., oxalic acid, citric acid, malic acid and succinic acid were increased by more than 100%. This might be excess metabolites that were not used in the TCA cycle activities and may suggests that soils amended with compost can influence the former but understudied. Consequently, we postulate that variations in application frequency of MSW compost will differentially alter soil health, microbiome function, and TCA cycle pathway and crop productivity. The objective of the study was to determine how variations in application frequency of MSW compost affect soil health, soil microbiome, and crop productivity and TCA cycle intermediate metabolites. We previously investigated one plant species and did not determine microbiome function and plant metabolic response. In the present study, we were interested to understand the responses of different crop species, so three test crops were used; namely green beans (*Phaseolus vulgaris* cv. Golden Wax) (leguminous crop), lettuce (*Latuca sativa* cv. Grand Rapids) (leafy vegetable) and beets (*Beta vulgaris* cv. Detroit Supreme) (root vegetable).

2. Results and Discussion

2.1. Location Climate

Overall, the climatic conditions in Brandon, MB during the 5-year research were similar except the low minimum mean temperatures in May 2015 (i.e., 1.8 °C) and 2019 (i.e., 0.4 °C) while there was flood and drought conditions in June (i.e., 106 mm) and August 2016 (i.e., 0 mm), respectively (Supplementary Table S1). The annual mean temperature ranged from 15–17 °C and the annual mean precipitation ranged from 41 mm to 70 mm. Overall, the differences in climatic conditions did not adversely impact the research over the 5-year period.

2.2. Soil Physical Properties

Continuous application of MSW compost for 5 years remarkably altered soil structure and soil function such as increased soil water content and nutrient status and accessibility to plants compared to the control (Table 1) as previously reported by [29]. After Year 5, soil particle and bulk densities were significantly ($p < 0.05$) reduced in the annual plot (AN-soil) seconded by the biennial plot (BI-soil) and then the control plot (C-soil). The high soil organic matter (SOM) content of the AN-soil and the BI-soil enhanced their respective soil structural properties leading to a significant ($p < 0.001$) improvement in soil water retention, i.e., water-holding, water saturation, and field capacities (Table 1) as previously explained by Rawls et al. [30]. All the measured soil water indices were similar for AN-soil and the BI-soil except for water-holding capacity that was increased significantly ($p < 0.05$) in AN-soil by ca. 12% and ca. 27% compared to the BI-soil and the C-soil, respectively.

Table 1. Soil physical and chemical properties of experimental plots in Year 5.

Soil Properties	Compost Application		
	Annual	Biennial	Control
Saturation capacity (%)	49.33 ± 1.23 a	46.39 ± 1.91 a	39.65 ± 0.98 b
Field capacity (%)	41.24 ± 0.98 a	35.37 ± 1.00 ab	30.49 ± 1.12 b
Wilting capacity (%)	6.40 ± 1.09 a	4.69 ± 0.09 ab	4.16 ± 1.11 b
Water-holding capacity (%)	34.80 ± 2.42 a	30.68 ± 1.49 b	25.33 ± 2.01 c
Bulk density (g/cm^3)	0.99 ± 0.00 c	1.06 ± 0.01 b	1.22 ± 0.04 a
Particle density (g/cm^3)	1.22 ± 0.10 c	1.33 ± 0.09 b	1.41 ± 0.12 a
Turbidity (NTU)	584.33 ± 8.14 a	516.22 ± 7.80 ab	453.56 ± 6.77 c
Organic matter content (%)	6.11 ± 1.00 a	4.61 ± 1.23 b	1.93 ± 0.21 c
Total Dissolved Solids (mg/L)	1535.2 ± 10.1 a	498.1 ± 9.91 b	272.4 ± 8.71 b
Electric conductivity (µS/cm)	2175.0 ± 10.8 a	715.2 ± 12.1 b	377.5 ± 10.00 b
Salinity (mg/L)	1040.6 ± 9.90 a	334.8 ± 7.61 b	180.8 ± 6.97 b

NTU, Nephelometric turbidity unit; means sharing the same alphabetical letters within the same row are not significantly different at the 5% level.

2.3. Soil Chemical Properties

Total dissolved solids, which is usually used to estimate the proportion of dissolved organic materials including organic matter and salts, was significantly ($p < 0.001$) higher for the AN-soil by more than 208% and 463% compared to that of the BI-soil and the C-soil, respectively (Table 1). Electric conductivity and salinity, which are indicators of soil fertility status, were significantly ($p < 0.001$) high in the AN-soils., i.e., ca. 204% and ca. 211% compared to the BI-soil and ca. 476% and ca. 476% compared to the C-soil, respectively (Table 1). These were expected due to the high SOM in the AN-soil followed by the BI-soil.

This is because SOM is a reservoir of soil nutrients [14]. The chemistry of the soil was influenced by continuous and long-term MSW compost application (Figure 1A–H). The SOM in the C-soil increased slightly from Year 1 up to Year 3 before it declined from 2.4% to 1.9% (Figure 1A). There was a sharp increase in AN-soil and BI-soil SOM from Year 1 with a dip in Year 4 before rising again. The dip in Year 4 was due to the late application of compost in Year 3 because of a delay in MSW compost delivery for the study. SOM of the AN-soil at Year 5 was significantly ($p < 0.01$) increased by ca. 33% and ca. 217% compared to those of the BI-soil and the C-soil, respectively. High SOM is associated with high soil organic carbon (SOC) and ultimately, desirable environmental and soil health [31,32]. In 5 years, the AN-soil pH was in general higher (i.e., between lower-upper difference) then BI-soil and C-soil. Soil pH increased from 7.7 in Year 1 to a range between 8.4 (BI-soil) and 8.7 (AN-soil) in Year 3, before declining slightly to an average of 8.15 in Year 5 (Figure 1B). The increase in AN-soil and BI-soil pH could be due to the intrinsically high Na content of the compost, which is supported by the corresponding increases in AN-soil and BI-soil electric conductivity and salinity (Table 1). The increase in pH can also be attributed to the release of hydroxyl ions from the high organic matter AN-soil and BI-soil, which declined after Year 3. This is because MSW compost has high organic matter content with negatively charged sites that can bind or release hydroxyl ions in acidic and basic soils, respectively to buffer soil acidity [33]. There was a slight increase in C-soil pH from Year 4, which can be ascribed to possible base cations naturally associated with Orthic Black Chernozem solum on moderate to strong calcareous, loamy morainal till of limestone, granitic and shale origin (Newdale series) of the experimental site (MAFRD, 2010). This is evident in the highest Ca and Mg levels in C-soil (Figure 1F,G).

The inherent capacity of the soil particles to adsorb cations (i.e., CEC) was not altered by compost application within the first 3 years of the study (Figure 1C). The AN-soil had a higher CEC compared to the BI-soil in Year 5. There was a dip in AN-soil CEC in Year 4, which was not significantly ($p > 0.05$) different from those of BI-soil and C-soil and cannot be readily explained. Cations like K^+, Na^+, Ca^{2+} and Mg^{2+} are retained on negatively charged soil components such as organic matter. According to Solly et al. [34], exchangeable Ca contributes the most (i.e., 59–83%) to CEC at pH > 5.5 with a strong positive relationship existing between CEC and SOM. In the present study, the exchangeable Ca increased from 19.41 meq/100 g from Year 1 to 23.67 and 23.99 and 21.85 meq/100 g in the AN-, BI-, and C-soils, respectively, in Year 5 (data not presented). Therefore, the trend of the AN-soil CEC compared to the BI-soil can be ascribed to the pH range (Figure 1B) and its high SOM (Figure 1A) and exchangeable Ca (Table 2).

Total nitrogen (N) was highly increased in the AN-soil by ca. 149% and ca. 390% more than the BI-soil and the C-soil, respectively (Table 2). Of particular interest was the high estimated nitrogen release (ENR) in the AN-soil followed by the BI-soil (Figure 1D). ENR is a critical index for the estimation of N availability to plants in the next growing season. Typically, nutrients are slowly released from compost due to slow microbial decomposition and mineralization processes. ENR of the C-soil progressively declined while compost application increased ENR, especially in the AN-soil (Figure 1D). The AN-soil ENR was ca. 23% more than that of the BI-soil from Years 3–5; and ca. 69% and 146% more than the C-soil at Years 3–4 and Year 5, respectively.

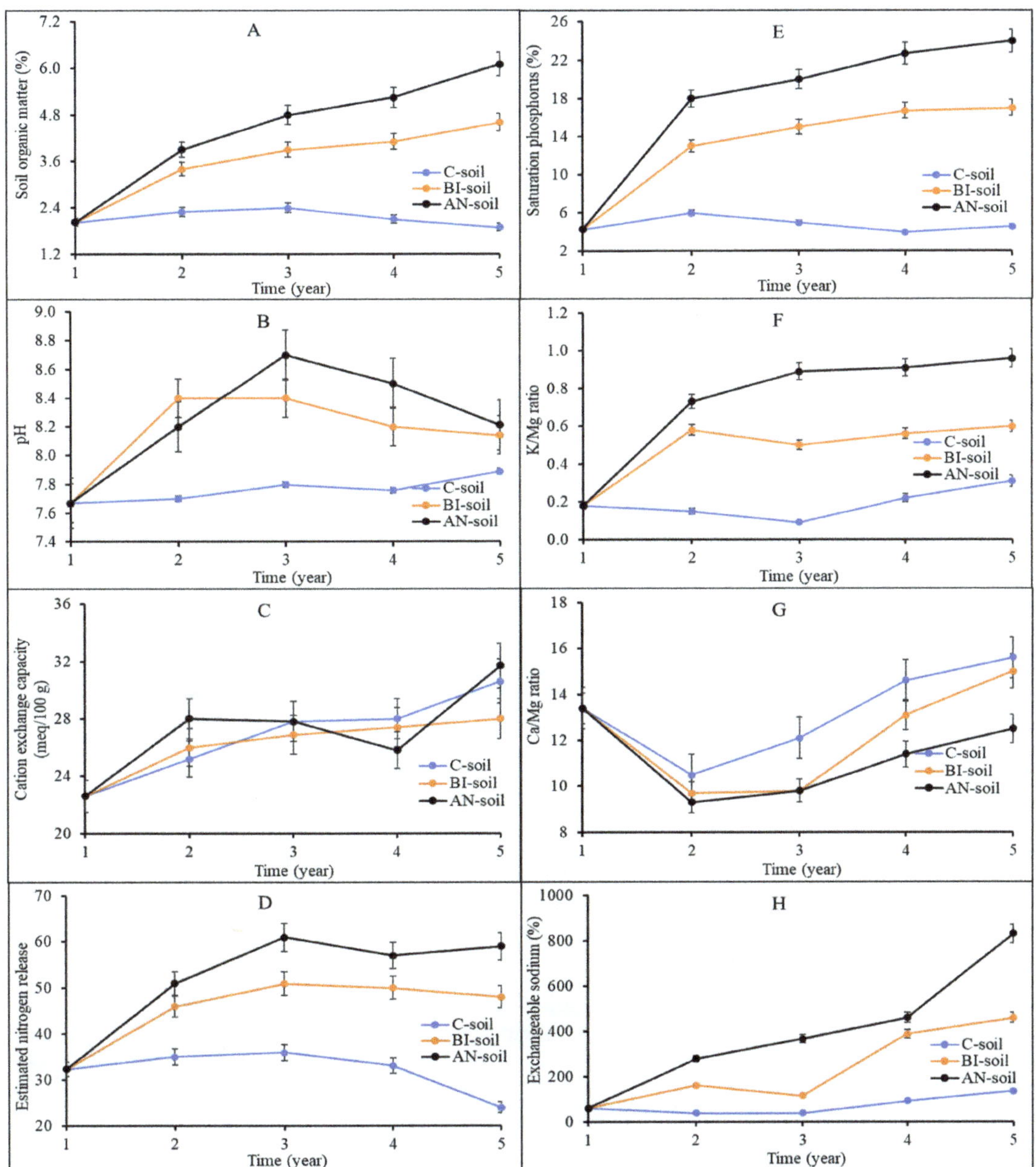

Figure 1. Changes in soil fertility status over five years of application of municipal solid waste compost at varying application frequency. C-soil, BI-soil and AN-soil represent no compost (control), biennial and annual application of compost, respectively. (**A–H**) is panel label. Vertical bars represent standard error bars (N = 9). K/Mg, potassium to magnesium ratio and Ca/Mg, calcium to magnesium ratio.

Table 2. Five-year accumulation of essential and beneficial chemical elements in soils as affected by frequency of application of municipal solid waste compost.

Compost Application	Soil Chemical Elements (mg/kg)							
	Total N	Phosphorus	Potassium	Magnesium	Calcium	Sulphur	Boron	Iron
Annual	50.0 ± 3.2 a	274.0 ± 4.1 a	3610.2 ± 6.1 a	5920.1 ± 23.1 a	22,000.8 ± 21.1 a	29.7 ± 4.5 a	13.3 ± 1.2 a	9330.4 ± 9.9 a
Biennial	20.1 ± 2.1 b	195.7 ± 2.8 b	1950.0 ± 6.5 b	4380.1 ± 22.9 b	17,401.3 ± 19.0 bc	28.0 ± 3.9 a	10.0 ± 1.8 ab	8740.6 ± 11.2 b
Control	10.2 ± 1.0 c	95.3 ± 2.3 c	940.4 ± 4.3 c	3220.0 ± 16.1 c	16,100.2 ± 19.0 c	12.7 ± 1.7 b	6.7 ± 1.1 b	7670.0 ± 8.7 c
Compost application	Soil chemical elements (mg/kg)							
	Manganese	Molybdenum	Cobalt	Sodium	Chromium	Copper	Nickel	Zinc
Annual	605.3 ± 8.7 a	0.8 ± 0.0 a	3.4 ± 0.0 a	119.3 ± 2.4 a	9.3 ± 0.0 a	13.3 ± 1.5 a	9.2 ± 0.0 a	59.1 ± 3.1 a
Biennial	511.1 ± 8.7 ab	0.51 ± 0.0 ab	3.1 ± 0.0 a	95.0 ± 1.1 a	8.1 ± 0.4 a	9.7 ± 1.3 a	9.1 ± 0.0 a	40.0 ± 2.3 b
Control	400.4 ± 7.9 b	0.24 ± 0.1 b	2.9 ± 0.2 a	22.3 ± 1.6 b	6.3 ± 0.09 a	4.0 ± 0.2 b	8.2 ± 0.3 a	28.2 ± 1.7 c

Means sharing the same alphabetical letters within the same column are not significantly different at the 5% level.

The other major plant required nutrient elements, i.e., P, K, Mg and S were significantly ($p < 0.05$) highest in AN-soil followed by BI-soil and the lowest in the C-soil (Table 2). This is expected due to the variations in frequency of MSW compost addition and the resultant differences in soil organic matter content and chemical indices as shown in Table 1. Besides, the trends in percentage P saturation (Psat%) (Figure 1E) and K/Mg ratio (Figure 1F) were similar. That is, AN-soil > BI-soil > C-soil. The Psat% ranged from 4–24%, 4–17% and 4–6% for the AN-soil, BI-soil and C-soil, respectively. Rheault [35] found a threshold range of ca. 6–18% for Manitoba soils with the different types of soil. Therefore, the addition of MSW compost increased Psat% to a maximum (BI) or exceeded the maximum (AN) threshold. Psat% is a function of soil Ca, Fe and Al contents, and an Indicator for environmental risk assessment [36]. The Psat% seemed to level off after Year 3, which suggested less environmental risk, particularly with the BI-soil compared to the AN-soil. This will require further investigation to ensure safe level of soil P for such an AN-soils. Rheault [35] also explained that stabilization of soil P occurs over time leading to a significant reduction in extractable P. We found that exchangeable K and Mg did not significantly ($p > 0.05$) change in the soils after Year 2 (data not presented). Exchangeable K and Mg in the C-soil changed from 0.41–0.37 meq/100 g and 2.76–2.12 meq/100 g in Years 1 and 5, respectively. This can be attributed to the lack of soil amendment and continuous soil nutrient depletion. For the AN- and BI-soils, exchangeable K was increased from 0.41 meq/100 g in Year 1 to 2.77 and 2.09 meq/100 g in Year 2, after which it did not change significantly ($p > 0.05$). Similarly, exchangeable Mg remained the same throughout the study after increasing from 2.76 meq/100 g in Year 1 to 3.78 meq/100 g for the AN-soil and 3.41 meq/100 g for the BI-soil, respectively, in Year 2. These results suggested that MSW compost amendment increased K and Mg availability to plants compared to the control. However, these cations remain the same due to consistency of crop species and MSW compost, i.e., compost type, amount, and time of application. The soil K/Mg ratio was less than one over the entire study period irrespective of the treatment (Figure 1F). This suggested inadequate K for plant use at time of soil sampling. A desirable soil K/Mg ratio is between 2–10. There was a sharp decline in Ca/Mg ratio in Year 2 but rose almost linearly with the highest increase recorded in the AN-soil followed by the BI-soil and then the C-soil (Figure 1G). The high Ca/Mg ratio in AN-soil suggested improved soil structure, and improved porosity and aeration. An exchangeable sodium percentage (ESP) of more than 10 suggests sodic soils. Overall, exchangeable Na was increased in the AN-soil and the BI-soil, which shows that the application of MSW compost increased soil sodicity (Figure 1H), especially in the AN-soil. The general trend for the soil macro-elements P, K, Mg and Ca was AN-soil > BI-soil > C-soil (Table 2). A similar trend was observed for the soil micro-elements sulphur (S), boron (B), iron (Fe), manganese (Mn), molybdenum (Mo), sodium (Na), copper (Cu) and zinc (Zn). However, soil content of cobalt (Co), chromium (Cr) and nickel (Ni) did not change (Table 2), irrespective of the differences in soil treatment. This suggested that soil amendment with this MSW compost will not have any negative environmental impact.

2.4. Soil Microbial Communities

Overall, Basidiomycota was the most relatively abundant fungal phylum found in the microbiome in all the soils, and it was represented by ca. 42% of all ITS reads (Supplementary Table S4) and contained 24%, 11% and 6% of the ITS reads annotated as *Tremellomycetes*, *Agaricomycetes*, and *Ustilaginomycetes*, respectively. *Mortierellomycota* were the other most abundant phyla represented by 42% of all the ITS reads. The ITS reads annotated as *Mortierellomycetes*. *Mortierellales*, *Filobasidiales*, *Cystofilobasidiales*, *Agaricales*, and *Ustilaginales* were the most abundant fungal orders represented by 41%, 15%, 9%, 9%, and 5%, respectively. On the other hand, *Actinobacteria*, *Proteobacteria*, *Acidobacteria*, *Chloroflexi*, and *Bacteroidetes* were the most abundant bacterial taxa and were represented by 35%, 23%, 11%, 11% and 5%, respectively (Supplementary Table S4). Based on total percentage reads, the most abundant bacterial taxa were *Rhizobiales* (10% of total reads), *Rubrobacterales* (7% of total reads), *Acidobacteria* Subgroup 6 (6% of total reads), *Propionibacteriales* (4% of total reads), and *Gaiellales* (4% of total reads).

In general, compost amendments significantly ($p < 0.05$) influenced soil bacterial community structure and diversity (Figure 2, Supplementary Table S5). Unlike bacteria, fungi communities were less affected by the MSW compost application (Figure 2A).

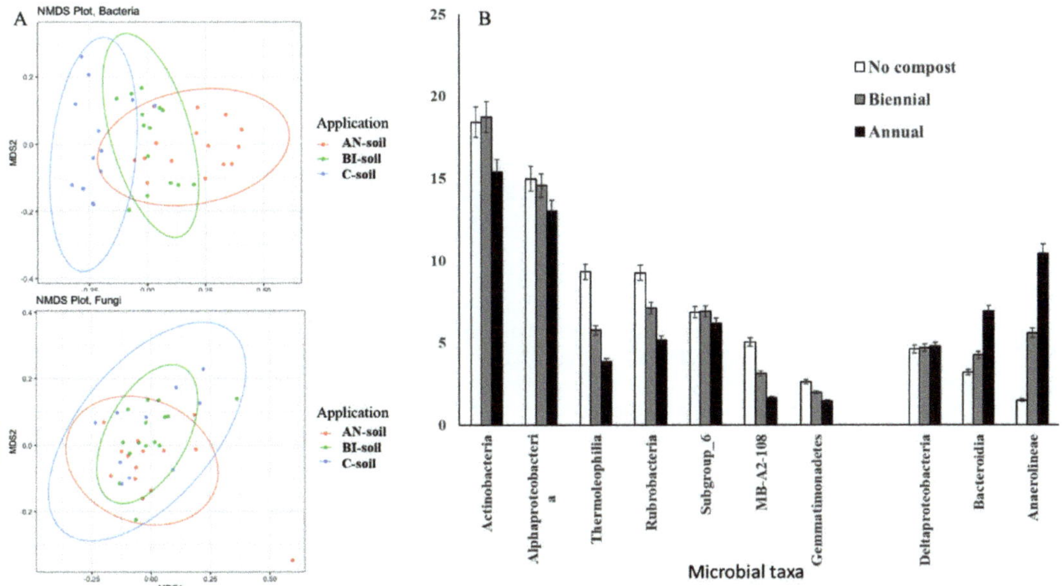

Figure 2. Diversity and structure of soils under different rates of compost application. (**A**): Non-metric multidimensional scaling plot of communities in soils under different rates of compost application. The analysis is based on Bray–Curtis distances of 16S rRNA (**top**) and ITS2 (**bottom**) samples. (**B**): Microbial taxa that were significantly overrepresented in comparison between soils with different rates of compost application. AN, BI, and C are annual, biennial and no compost (control) application, respectively. Corrected *p*-values (*q*-values) were calculated based on Benjamini-Hochberg FDR multiple test correction. Features with (Welc's *t*-test) *q* value < 0.01 were considered significant and were retained. Only microbial taxa represented by >2% total reads were shown at the class level.

Approximately 44% of bacterial community variation could be attributed to compost amendment. Moreover, compost significantly ($p < 0.001$) increased the relative number of observed bacterial features and Shannon diversity (Table 3). More specifically, several Actinobacteria classes such as Actinobacteria, Thermoleophilia, Rubrobacteria, Subgroup 6, and MB-A2-108 were less abundant in compost amended soils, i.e., AN- and BI-soils

(Figure 2B). The relative abundances of Alphaproteobacterial and Gemmatimonadetes were also reduced by compost application. On the other hand, Deltaproteobacteria, Bacteroidetes Bacteroidia, and Chloroflexi Anaerolineae were overrepresented in compost amended compared to the C-soil (Figure 2B).

Table 3. Microbial alpha-diversity as affected by frequency of application of municipal solid waste compost.

Treatment/Indexes	Observed Features	Evenness	Shannon Diversity
16S			
Control	1026 ± 6.1 b	0.892 ± 0.1 a	8.825 ± 1.2 b
Biennial compost application	1127 ± 5.1 a	0.902 ± 0.1 a	9.139 ± 1.1 a
Annual compost application	1138 ± 5.0 a	0.889 ± 0.1 a	9.023 ± 1.2 ab
ITS			
Control	43 ± 1.3 b	0.793 ± 0.2 a	4.280 ± 0.1 a
Biennial compost application	76 ± 2.2 a	0.734 ± 0.1 a	4.574 ± 0.2 a
Annual compost application	67 ± 1.8 a	0.753 ± 0.1 a	4.521 ± 0.2 a

For each variable, data followed by different letters are significantly different according to Kruskal–Wallis pairwise comparison ($q < 0.05$).

The relative abundances of *Alphaproteobacterial* and *Gemmatimonadetes* were also reduced by compost application. On the other hand, *Deltaproteobacteria*, *Bacteroidetes Bacteroidia* and *Chloroflexi Anaerolineae* were overrepresented in compost amended compared to the C-soil (Figure 2B). Our previous study showed that *Bacteroidetes* were among the 10 most abundant taxa in MSW compost sampled across Nova Scotia composting facilities [37]. Therefore, *Bacteroidetes* class *Bacteroidia* (Figure 2B) might be enriched in the MSW compost before adding to the soil resulting in the increase in relative abundance of this class in compost treated soils. The relative abundances of the two most abundant microbial taxa in our previous compost study *Alphaproteobacterial* and *Actinobacteria* [37], were reduced with frequent compost application in the present study (Figure 2B).

As such, the observed changes in microbial abundances could be due to promotion or repression of soil microbial growth due to differences in compost application frequency. Furthermore, the present results showed that bacterial metabolic pathways were influenced by frequency of compost application, which will be discussed later in this report. Specifically, compost application significantly ($p < 0.01$) increased fungal observed features, but not Shannon diversity (Table 3). Approximately, 9% of variations in fungal community can be attributed to the MSW compost application (Supplementary Table S5). We also did not detect any fungal classes differentially represented between treated and untreated soils. This agreed with our previous report that prokaryotes and eukaryotes differ in their responses to environmental factors [38].

To understand the effect of compost application on microbiome function, we extrapolated functional profile of bacterial community based on 16S rRNA marker gene using PICRUS2 software. Non-metric multidimensional scaling identified visual differences in functional profiles between bacterial community from soils with different compost application frequency. Additionally, ca. 38% KO and ca. 40% pathway of functional variation was attributed to differences in the frequency of compost application (Table 4).

When functional profiles of bacterial communities from the AN-soil and the BI-soil were combined and compared to the C-soil, ca. 23% KO and 26% pathway of functional variations were attributed to the former. With reference to C-soil, the BI-soil had lower effect on functional variation., i.e., ca. 15% KO and ca. 18% pathway as compared to those for the AN-soil., i.e., ca. 43% KO and ca. 46% pathway. Interestingly, functional profiles of bacterial communities from the AN-soil and the BI-soil also differed significantly ($p < 0.001$). Obviously, ca. 28% KO and ca. 27% pathway of functional variation was explained by the annual or the biennial compost application compared to the control.

Table 4. Variation in groupings of functions prediction by PICRUSt2 explained by Bray–Curtis distances as affected by frequency of application of municipal solid waste compost.

Grouping (Subset)	KO	Pathways
Treatments	0.383 ± 0.00 ***	0.400 ± 0.01 ***
Control vs. Biennial and Annual compost applications	0.226 ± 0.00 ***	0.264 ± 0.01 ***
Control vs. Biennial compost application	0.151 ± 0.00 ***	0.182 ± 0.00 ***
Control vs. Annual compost application	0.434 ± 0.01 ***	0.460 ± 0.02 ***
Annual vs. Biennial compost application	0.283 ± 0.00 ***	0.266 ± 0.00 ***

Adonis tests were used to assess whether beta-diversity is related to sample groupings, 999 permutations, R2, *** $p < 0.001$.

2.5. Soil Microbiome Function

In total, 205 pathways were differentially represented in microbiomes from the AN-soil, BI-soil, and the C-soil out of which 18 were overrepresented in the AN-soil and the BI-soil compared to the C-soil (Figure 3). The B-soil had medium level of functional profile of bacteria community compared to the AN-soil (high) and the C-soil (low) on the NMDS plot (Figure 3A). These 18 metabolic pathways were menaquinones (MK) and demethylmenaquinones (DMK) biosynthesis, i.e., superpathway of menaquinol-9 biosynthesis, superpathway of menaquinol-10 biosynthesis; superpathway of menaquinol-6 biosynthesis I, superpathway of demethylmenaquinol-6 biosynthesis I and superpathway of demethylmenaquinol-9 biosynthesis; fatty acid biosynthesis, i.e., oleate biosynthesis IV (anaerobic), stearate biosynthesis II (bacteria and plants), palmitoleate biosynthesis I (from (5Z)-dodec-5-enoate), (5Z)-dodec-5-enoate biosynthesis and superpathway of fatty acid biosynthesis initiation; DNA and RNA structures and enzymes cofactors, i.e., superpathway of purine nucleotides de novo biosynthesis II and pyrimidine deoxyribonucleotides de novo biosynthesis II; energy production, i.e., superpathway of thiamin diphosphate biosynthesis I, NAD biosynthesis II (from tryptophan), superpathway of thiamin diphosphate biosynthesis II, superpathway of pyridoxa–phosphate biosynthesis and salvage and carbon fixation, i.e., reductive acetyl coenzyme A pathway (Figure 3B).

The extent to which plants form mutually beneficial partnership with rhizosphere microbiome is dependent on the genotypic characteristics of the plant as explained by Kelly et al. [18]. This manifested in the plant growth and yield components of the different species in Tables 5 and 6. According to Zhou et al. [22], plants use their root exudates to mobilize specific microbial communities to combat disease pathogens, facilitate nutrient acquisition and crosstalk amongst various plant growth regulators, and to modulate signaling pathways and increase productivity. This can explain the results of the present study as shown below but will need to be validated in future studies.

2.6. Crop Morpho-Physiology

Application of MSW compost significantly ($p < 0.05$) increased SPAD (soil plant analysis development) value of leaf greenness (Table 5), which can be used to estimate leaf chlorophyll content because of the high positive regression coefficient ($R^2 > 0.93$) between the two [39]. F_o was comparatively high in all the control but did not differ between AN and BI treated crops. Fm was not significantly ($p > 0.05$) altered by MSW compost irrespective of the crop. F_v of lettuce and beets were not altered by compost, but it was significantly ($p < 0.05$) higher in the AN-green beans compared to the BI-green beans and the C-green beans. F_v, F_v/F_m and F_v/F_o were similarly high ($p < 0.05$) in AN- and BI-green beans and lettuce compared to the control plants (Table 5). Chlorophyll fluorescence indices were not altered in beets except for F_o (Table 5). Compost alteration of fundamental plant structural, biochemical, and physiological functions [14] translated into increased leaf chlorophyll content and photosynthetic activities.

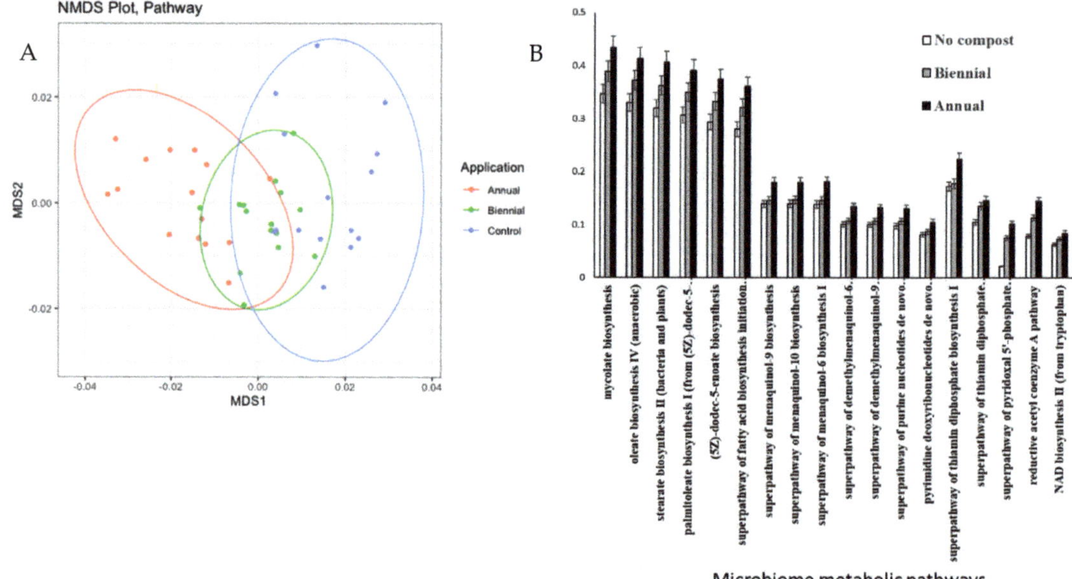

Figure 3. Functional profile of bacterial community based on 16S rRNA marker gene. (**A**)–Non-metric multidimensional scaling plot of community functions in soils under different rates of compost application. The analysis is based on Bray–Curtis distances of functional composition of bacterial community evaluated using PICRUSt. (**B**)–Pathways those were significantly overrepresented in Biennially and Annually treated soil compared to the untreated soil. Corrected *p*-values (*q*-values) were calculated based on Benjamini–Hochberg FDR multiple test correction. Features with (Welch's *t*-test) *q*-value <0.01 were considered significant and were thus retained.

Table 5. Leaf pigmentation and plant photosynthetic efficiency indices as affected by frequency of application of municipal solid waste compost.

Crop	Compost Application	SPAD [1] Value	ACI [2]	Chlorophyll Fluorescence Index				
				F_o [3]	F_v [4]	F_m [5]	F_v/F_m [6]	F_v/F_o [7]
Beans	Annual	42.9 ± 1.2 ab	7.34 ± 0.0 a	151.73 ± 3.2 b	407.37 ± 2.9 ab	559.33 ± 3.2 a	0.72 ± 0.0 a	2.82 ± 0.0 a
	Biennial	46.4 ± 1.0 a	7.97 ± 0.2 a	188.47 ± 3.1 b	491.10 ± 1.9 a	682.23 ± 2.8 a	0.72 ± 0.1 a	2.93 ± 0.2 a
	Control	36.3 ± 0.8 c	6.12 ± 0.0 a	225.13 ± 4.4 a	349.20 ± 3.2 b	570.33 ± 2.2 a	0.59 ± 0.0 b	1.72 ± 0.2 b
Lettuce	Annual	42.9 ± 1.2 a	6.32 ± 0.2 a	159.00 ± 3.6 b	574.93 ± 2.1 a	733.93 ± 2.1 a	0.78 ± 0.4 a	3.73 ± 0.1 ab
	Biennial	36.7 ± 0.9 b	5.44 ± 0.3 a	142.70 ± 2.3 b	558.23 ± 2.2 a	706.60 ± 2.3 a	0.79 ± 0.1 ab	3.96 ± 0.0 a
	Control	27.8 ± 1.0 c	3.53 ± 0.3 b	171.67 ± 1.9 a	528.30 ± 2.0 a	700.97 ± 3.1 a	0.75 ± 0.7 b	3.13 ± 0.3 b
Beet	Annual	50.2 ± 1.8 a	7.95 ± 0.0 b	139.30 ± 0.1 b	435.10 ± 2.0 a	573.73 ± 1.0 a	0.75 ± 0.7 a	3.16 ± 0.1 a
	Biennial	38.5 ± 1.0 b	9.45 ± 0.2 b	134.43 ± 1.9 b	444.47 ± 2.3 a	578.70 ± 1.7 a	0.75 ± 0.7 a	3.42 ± 0.1 a
	Control	30.1 ± 0.4 c	18.97 ± 0.4 a	177.13 ± 2.2 a	483.37 ± 2.1 a	660.53 ± 0.9 a	0.72 ± 0.6 a	2.85 ± 0.0 a

[1] SPAD, soil plant analysis development; [2] ACI, anthocyanin content index; [3] F_o, minimum, [4] F_m, maximum and [5] F_v, variable chlorophyll fluorescence indices; [6] F_v/F_m, maximum quantum yield of photosystem II; and [7] F_v/F_o, potential photosynthetic capacity. means sharing the same alphabetical letters within the same column are not significantly different at the 5% level.

Table 6. Plant growth components and yield of green beans (*Phaseolus vulgaris* cv. Golden Wax), beets (*Beta vulgaris* cv. Detroit Supreme) and lettuce (*Latuca sativa* cv. Grand Rapids) as affected by frequency of application of municipal solid waste compost.

Plant Growth Component			Compost Application		
			Annual	Biennial	Control
Shoot	Plant height (cm)	Green beans	79.84 ± 3.2 a	67.33 ± 2.3 b	55.6 ± 1.5 c
		Lettuce	30.10 ± 1.6 a	32.00 ± 1.7 a	24.16 ± 1.3 b
		Beets	40.80 ± 1.4 a	34.98 ± 1.6 b	28.05 ± 1.4 c
Leaf	Area (mm^2)	Green beans	8525 ± 13.9 a	5398 ± 12.2 b	2782 ± 9.9 c
		Lettuce	23,484 ± 10.7 a	20,442 ± 9.9 b	14,389 ± 8.3 c
		Beets	20,521 ± 20.0 a	14,219 ± 19.0 b	6481 ± 10.8 c
	LDMC (mg/g)	Green beans	121.98 ± 2.2 b	155.70 ± 2.6 b	223.24 ± 2.9 a
		Lettuce	61.06 ± 2.1 a	71.62 ± 2.4 a	75.17 ± 1.9 a
		Beets	137.92 ± 2.9 a	122.08 ± 1.6 a	140.52 ± 1.7 a
Stem	SSD (mg/mm^3)	Green beans	0.12 ± 0.0 a	0.30 ± 0.1 a	0.08 ± 0.0 b
		Lettuce	0.07 ± 0.3 b	0.07 ± 0.2 b	0.11 ± 0.0 a
		Beets	0.12 ± 0.0 b	0.13 ± 0.0 b	0.19 ± 0.0 a
Root	SRL (mg/g)	Green beans	0.12 ± 0.2 b	0.12 ± 0.0 b	0.27 ± 0.1 a
		Lettuce	0.15 ± 0.0 a	0.14 ± 0.0 a	0.08 ± 0.0 b
		Beets	0.36 ± 0.1 b	0.76 ± 0.1 a	0.77 ± 0.2 a
Edible portion	Yield (kg/m^2)	Green beans	434.16 ± 4.3 a	411.42 ± 4.2 ab	321.31 ± 8.1 c
		Lettuce	85.00 ± 2.7 a	72.09 ± 1.4 a	10.40 ± 1.3 c
		Beets	311.60 ± 3.0 a	260.00 ± 3.8 b	64.08 ± 3.3 c

LDMC, leaf dry matter content; SSD, specific stem density; SRL, specific root length; means sharing the same alphabetical letters within the same row are not significantly different at the 5% level.

Although the MSW compost only increased F_v/F_m and F_v/F_o of green beans grown in soils applied annually with compost, the comparatively low values of these indices in all the control plants suggest stressful conditions (Table 5). It is well established that nutrients imbalance can reduce photosynthetic efficiency and plant metabolism [40,41] as found in plants grown in the control plots. Furthermore, P deficiency reduced F_v/F_m while increasing F_o [42] with photosystem II being the most sensitive and vulnerable [43]. The comparatively low C-soil nutrients (Table 2) might have caused the nutrient-deficient control plants to switch to a survival mode by growing extensive root system (i.e., SRL) and biomass by way of increased leaf dry matter content and specific stem density (Table 6). According to Lohmus et al. [44], different plant species adopt different adaptation strategies to sustain and improve nutrition, which may include an increase in above-ground biomass or fine root length.

The positive impact of compost on crop Productivity is well established [2,12]. Plant heights of the AN-green beans and the AN-beets were significantly ($p < 0.01$) increased by ca. 19% and ca. 17% compared to their BI- counterparts; and by ca. 44% and ca. 46% compared to their C- counterparts, respectively (Table 6). Plant heights for the AN-lettuce and the BI-lettuce were similar ($p > 0.05$) but higher than that of the C-lettuce. Plant leaf area for the AN-green beans, AN-lettuce and AN-beets were increased significantly ($p < 0.001$) by ca. 58%, 15% and 44%, respectively, compared to their BI- counterparts; and by ca. 206%, 63% and 217% compared to their C- counterparts, respectively (Table 6). Compost did not affect leaf dry matter content (LDMC) of the lettuce and beets (Table 6). However, compost significantly ($p < 0.01$) reduced LDMC of the AN-green beans and the BI-green beans by an average of 38% compared to the C-green beans. Specific stem density (SSD) of the plants did not change with compost application (Table 6). However, SSD was reduced in the C-green beans by ca. 74% and increased in C-lettuce and C-beets by ca. 57% and 52%, respectively, compared to the average for their AN- and BI- counterparts.

Specific root length (SRL) was higher in the C-green beans and the C-beets than their AN- and BI- counterparts (Table 6). In contrast, SRL was significantly ($p < 0.01$) reduced in the C-lettuce by ca. 45% compared to the average for the AN-lettuce and the BI-lettuce. There was no significant ($p > 0.05$) difference between the AN- and BI-green bean fresh immature pod yield (Table 6). The fresh pod yield of C-green bean was reduced significant ($p < 0.01$) by ca. 35% compared to the average for the AN- and the BI-green beans. The yield difference between the AN- and the BI-lettuce was ca. 18% (Table 6). The average yield for the AN- and the BI-lettuce was ca. 717% more than the C-lettuce. The yield of the AN-beets was significantly ($p < 0.0001$) increased by ca. 20% compared to the BI-beets; and by ca. 386% compared to the C-beets.

The results proved that the different crop species responded differently to the MSW compost treatment. Additionally, the results of the plant growth indices demonstrated increased plant growth and productivity with the application of MSW compost compared with the control. According to Cornelissen et al. [45], plants with high LDMC (e.g., control plants) have high physical strength for survival under stress conditions and can be associated with long leaf life-span but may be less productive compared to plants with low LDMC (e.g., plants grown in AN-soil and BI-soil). Similarly, a high SSD demonstrates a dense stem that provides structural strength to the plant and an indication of carbon storage [45], which suggest a switch to survival mode of plants in the C-soil compared to plants that were grown in the AN- and BI-soils. Plants with high SRL develop longer roots per dry mass for water and nutrients uptake [45]. It seemed compost applied lettuce had higher SRL but lower in compost applied green beans and only AN-beets compared to the control. It can be suggested that C-soil had low soil nutrient content and as such, plants develop higher SRL to be able to reach available nutrients and water. Lettuce is shallow-rooted and was probably, stressed in the C-soil to the point where SRL was reduced.

2.7. Elemental Accumulation in Harvested Crops

Chemical elements accumulation in the edible portions of the crops was mostly increased by the annual application of MSW compost (Supplementary Table S6), which was previously reported in Abbey et al. [19]. Overall, lettuce accumulated most of the analyzed macro- and micro-elements (i.e., type and quantity) compared to the green beans (intermediate) and beets (lowest). On average, lettuce N (ca. 2100 mg/kg) > green bean N (ca. 967 mg/kg) > beets N (ca. 100 mg/kg); lettuce K (ca. 67,333 mg/kg) > green beans K (ca. 27,850 mg/kg) > beets K (ca. 27,200 mg/kg); lettuce Mg (ca. 5170 mg/kg) > green beans (ca. 2468 mg/kg) > beets (ca. 2183 mg/kg); lettuce Ca (ca. 13,033 mg/kg) > green beans Ca (ca. 4133 mg/kg) > beets (ca. 1890 mg/kg); and lettuce Fe (ca. 294 mg/kg) > green beans Fe (ca. 70 mg/kg) > beets Ca (ca. 39 mg/kg). Continuous application of MSW compost did affect how much micro-elements like Cu, Co and Ni accumulated in the tissues of the different crop species. Clearly, beetroots accumulated the highest Na (i.e., ca. 2530–7830 mg/kg) compared to green beans pod (i.e., ca. 30–40 mg/kg) and lettuce leaf (i.e., ca. 2020–3490 mg/kg). Abbey et al. [19] demonstrated that at a consumption rate of 400 g according to recommendations by WHO [46] and Statistics Canada [47], the crops are safe for human consumption and health based on previous analyses of bioaccumulation factor, estimated daily intake and health hazard quotient.

2.8. TCA Cycle Intermediate Metabolites

Plant metabolites involved in the TCA cycle were highly ($p < 0.001$) influenced by MSW compost application frequency (Figure 4A–F). Overall, glucose content was similarly high in lettuce and green beans but low in beets. We found that as part of the survival and adaptation modes, lettuce plants grown in the C-soil accumulated more glucose than their counterparts in the AN-soil and the BI-soil (Figure 4A).

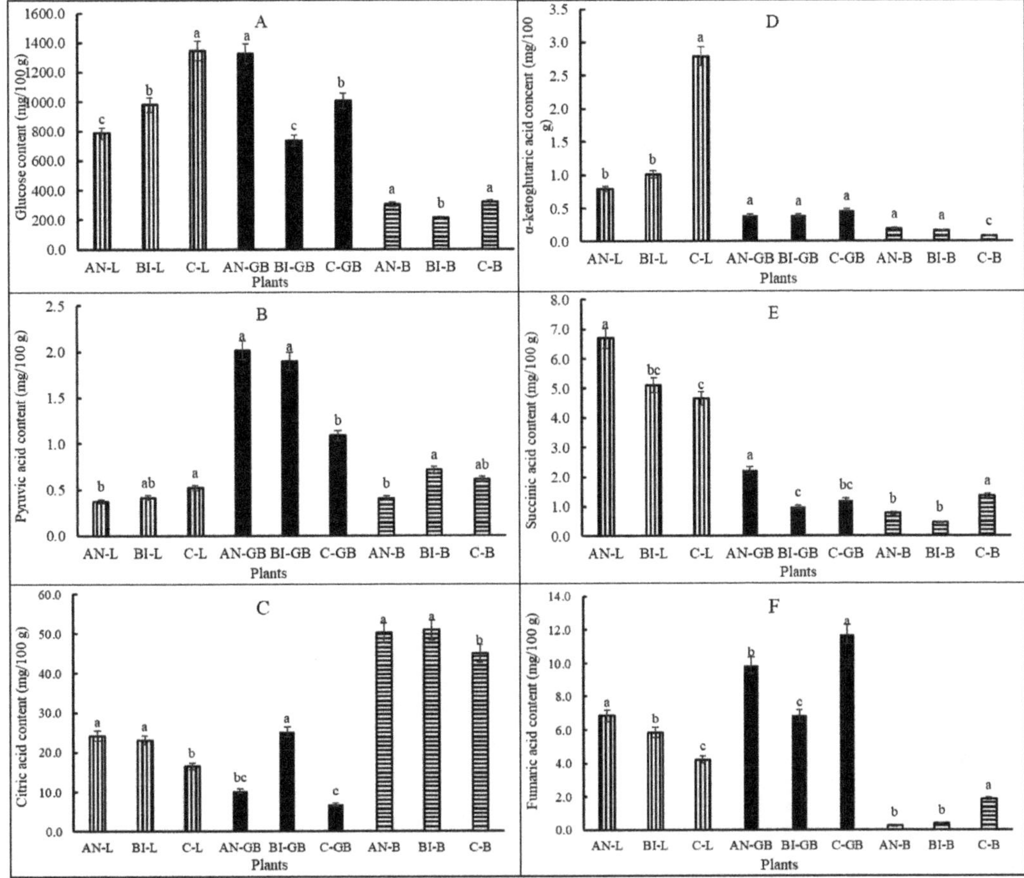

Figure 4. Plant metabolites of the tricarboxylic acid cycle as influenced by application of municipal solid waste compost at varying application frequency. Bars with vertical lines represent lettuce, solid black bars represent green beans and bars with horizontal lines represent beets; C-L, C-GB and C-B represent lettuce, green bean and beets grown with no compost (control), respectively; BI-L, BI-GB and BI-B represent lettuce, green bean and beets grown with biennial compost application, respectively; and AN-L, AN-GB and AN-B represent lettuce, green bean and beets grown with annual compost application, respectively; (**A–F**) is panel label. Differences in the letters on bars indicate significant differences at $p < 0.05$ (N = 9).

This conforms to previous observation on glucose accumulation in N-deficient plants [26,44]. AN-green beans glucose was ca. 80% and 37% more than that of the BI-green beans and the C-green beans, respectively. The AN-beets and the C-beets glucose were not significantly ($p > 0.05$) different, and their average was over 43% more than that of the BI-beets. The AN-green beans and the BI-green beans had similar and high pyruvic acid content than that of the C-green beans (Figure 4B). The general trend for pyruvic acid was green beans > beets > lettuce. For lettuce and beets citric acid contents, AN = BI > C (Figure 4C). In contrast, citric acid was significantly ($p < 0.001$) increased by ca. 147% and 270% in the BI-green beans compared to the AN-green beans and the C-green beans, respectively.

The AN-lettuce and BI-lettuce α-ketoglutaric acid contents were not significantly ($p > 0.05$) different and was less than that of the C-lettuce (Figure 4D). The C-lettuce α-ketoglutaric acid content was ca. 208% more than the average for the AN-lettuce and the BI-lettuce. There was no significant ($p > 0.05$) difference in α-ketoglutaric acid content between the AN-green beans and the BI-green beans or the AN-beets and the BI-beets. Succinic acid was significantly ($p < 0.05$) increased in the AN-lettuce by ca. 38% compared to the average for the BI-lettuce and the C-lettuce (Figure 4E). Likewise, succinic acid was significantly ($p < 0.001$) increased in the AN-green beans by ca. 103% compared to the average for the BI-green beans and the C-green beans. Succinic acid content was not significantly ($p > 0.05$) different between the BI-lettuce and the C-lettuce or between the BI-green beans and the C-green beans (Figure 4E). Moreover, the AN-lettuce fumaric acid was more than 17% and 63% compared to the BI-lettuce and the C-lettuce, respectively (Figure 4F). The trend for fumaric acid contents in the green beans and the beets was different from that for the lettuce. Fumaric acid content was highest in the C-green beans, and it was ca. 19% and 71% more than those for the AN-green beans and the BI-green beans, respectively. However, the AN-green beans had ca. 44% more fumaric acid than its BI- counterpart. In general, the application of MSW compost reduced fumaric acid content in the beets. This was confirmed by the over 509% increase in the C-beets fumaric acid content compared to the average for the AN-beets and the BI-beets.

The biennially treated plants consistently had high pyruvic acid and citric acid contents (Figure 4). α-ketoglutaric acid is formed from oxidation of isocitrate and found to be similar in plants grown with annual and biennial MSW compost application. Additionally, plant tissue contents of succinic acid (Figure 4E) and fumaric acid (Figure 4F) suggested that much of the high α-ketoglutaric acid in C-lettuce was not converted or its synthesis exceeded the rate at which it was converted to succinic acid. The oxidized succinic acid, i.e., succinate is converted to fumarate and then to malate and oxaloacetate to close the TCA cycle. Fumaric acid was high in AN-lettuce, C-green beans, and C-beets. The interconversion of these intermediate metabolites in the TCA cycle starting from photosynthesis to respiration is tightly regulated by enzymes [23]. The conversion is also dependent on plant species and type of tissue as demonstrated in the present study and in previous report by Fernie and Martinoia [48]. It is obvious from the present study that the composition of these metabolites can vary with plant species, plant developmental stage and type of plant tissue. However, further work will be required to investigate impact on respiratory products such as ATP and NADPH. According to van der Merwe [20], a reduction in TCA cycle activity in a given tissue (e.g., beet roots) reduces respiration and energy generation in that tissue. However, green tissues (e.g., lettuce and green beans) can compensate for TCA cycle deficiencies through photosynthesis and photorespiration as explained by Kromer [49].

2.9. Multivariate Assessment

To further examine the relationship among key variables—soil nutrients, microbiome function and citric acid intermediate metabolites—we first constructed a multivariate 2-D PCA biplot using an extrapolated functional profile of bacterial communities based on the 16S rRNA marker gene and the soil nutrient profile (Figure 5).

The two-dimension PCA plot showed that the AN-soil and the BI-soil can be highly and moderately associated with high content of soil elements and the 18 metabolic pathways, respectively, and negatively associated ($r = -0.76$) with the C-soil (Figure 5A). This is expected because compost increase of soil organic matter, chemical elements and microbiome composition is well established in the literature [2,3,50]. Moreover, DNA and RNA structures and enzymes cofactors (DENOVOPURINE2-PWY and PWY-7187) and MK and DMK pathways had positively strong association ($r = 0.8$) with soil N and Ca. Several studies confirmed that N source influenced bacteria growth and abundance [51–53], which is dependent on their ability to synthesize new cells, i.e., mainly new cellular structures [54]. This suggests that the high N and other elements in the AN-soil (Table 2) met the requirement for bacteria growth more than that of the BI-soil and the C-soil.

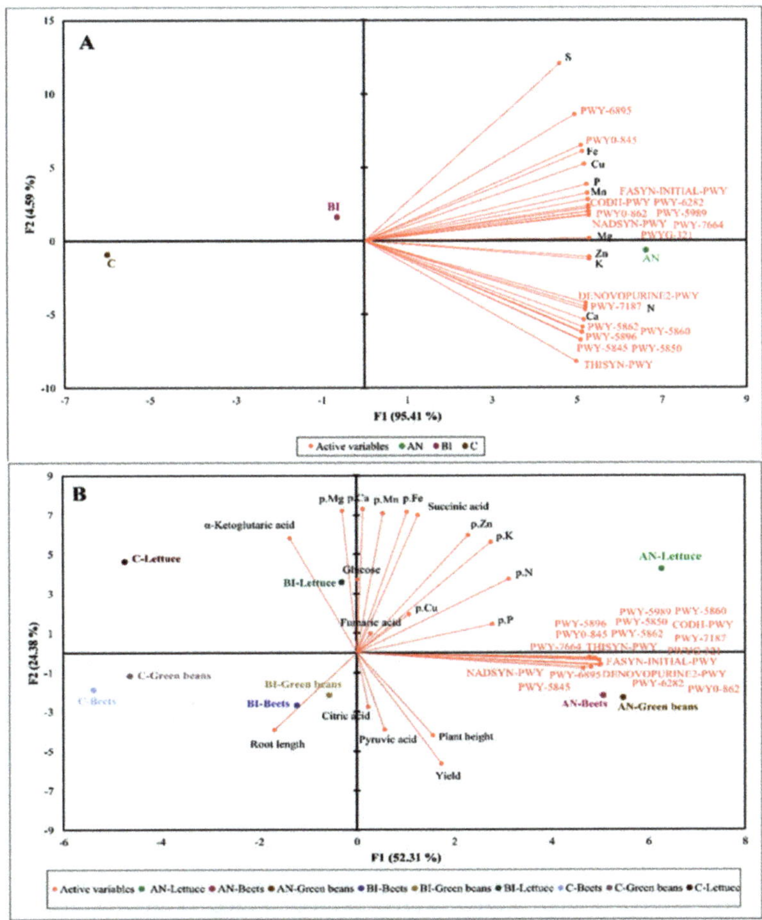

Figure 5. Multivariate analysis using principal component biplot. (**A**): shows association amongst municipal solid waste compost application frequency, soil elements and soil microbial function. (**B**): shows association amongst crop species, growth and yield components, plant tissue elements, tricarboxylic acid cycle metabolites and soil microbial function. AN, BI, and C are annual, biennial and no compost application, respectively. S, sulphur; Fe, iron; Cu, copper; P, phosphorus; Mn, manganese; Ca, calcium; N, nitrogen; Zn, zinc; K, potassium; Mg, magnesium. p represent plant tissue.

The green beans, lettuce and beets tissue N, P, K, Zn, and Cu were positively influenced (r = 0.9) by the 18 microbiome metabolic pathways in the AN-soil (Figure 5B). These pathways showed a moderate influence on citric acid and pyruvic acid accumulation as well as plant height and yield in the AN-green beans and the AN-beets, but less for the AN-lettuce. Under BI-soil, both BI-green beans and BI-beets were associated with high accumulation of pyruvic acid and citric acid, and increased plant height, root length and yield. In contrast, BI-lettuce exhibited high contents of glucose, fumaric acid, α-ketoglutaric acid and succinic acid, which were positively associated ($r \geq 0.78$) with all the determined plant tissue elements. On the other hand, BI-lettuce was negatively associated ($r \geq -0.92$) with the plant growth components and the metabolic pathways (Figure 5B). The C-lettuce showed a moderate positive association (r = 0.56) with only α-ketoglutaric acid accumulation and root length; and a strong negative association ($r = -0.89$) with plant tissue elements, plant growth and the remaining TCA cycle intermediate metabolites

(Figure 5B). Hence, the low accumulation of metabolites and nutrient elements in edible portions of the C-lettuce, the C-green beans and the C-beets beside the stunted growth and low yield. Overall, MSW compost increased organic acids in the plant tissues that can participate in the TCA cycle pathway for energy generation, especially in plants grown in soils that received compost annually.

A healthy and thriving soil food web is largely dependent on beneficial soil microbial diversity and abundance as influenced by soil organic matter and soil nutrient status [2,55]. Overall, the 2-D PCA biplot showed a strong association between MSW compost enhancement of soil health, particularly plots that received the annual compost application; and high functional microbial metabolic activities that was consistent with previous studies [2,55,56]. Soil microbes utilize carbon and other nutrients in organic matter for their metabolic activities, which facilitate nutrient mineralization and bioavailability [2,56]. Besides, the availability of soil N, P and S are crucial for enhancing microbial amino acids, proteins and nucleotides biosynthetic pathways, and C for structural building and energy production through glycolysis and peptidoglycan biosynthesis. These might have culminated in the high crop performance in the AN-soil followed by the BI-soil and the lowest in the C-soil.

3. Materials and Methods

3.1. Location and Materials

A 5-year organic field research was performed in Agaard Farms, Brandon, MB, Canada (longitude 99°56′59.9892″ W; latitude 49°50′53.9916″ N; altitude: 409 m above sea level) between fall 2015 and winter 2020. Brandon has dominant to moderate cool, boreal, subhumid continental climate [57]. The climatic conditions throughout the 5 years of the study were consistent with a 30-year average as presented in Supplementary Table S1. The soil of Agaard Farm is Orthic Black Chernozem solum on moderate to strong calcareous, loamy morainal till of limestone, granitic and shale origin under the classification of Newdale series [57]. The City of Brandon waste management facility donated the CQA tested MSW compost, and the 5-year average chemical elements is presented in Supplementary Table S2. The seeds for the green beans cv. Golden Wax, lettuce cv. Grand Rapids, and beets cv. Detroit Supreme were purchased from The Green Spot, Brandon, MB.

3.2. Field Preparation and Planting

Field preparation, MSW compost application and planting were previously described in Abbey et al. [7,19]. The experimental field measured 80 m × 50 m, and was further divided into three blocks of 20 m × 10 m. Each block was subdivided into three 6 m × 3 m plots, i.e., annual (AN), biennial (BI), and no compost (C, control) plots per block. Separations between blocks was 2 m and plots within blocks was 1 m. MSW compost application rate was ca. 15 t/ha at a bulk density of 650 kg/m^3. Compost was applied in the fall to AN plot every year or BI plots every two years. Planting begun after the last frost day between 20 and 30 May of each year when the soil temperature rose above 10 °C. Each of the three crops were planted in strips in each plot. The order of planting was rotated each year for five years. Green bean seeds were sown in rows 20 cm apart, and beets and lettuce seeds were sown 10 cm apart. Drip irrigation was applied when the soil moisture content was low or daily in summer. No synthetic chemical fertilizer or pesticide was used. Immature lettuce leaves, beet roots and green bean pods were harvested at edible maturity stage between 45 to 60 days after sowing.

3.3. Soil Analysis

Soil samples were randomly collected (n = 5) every year for five years from each plot at the start of the growing season and at peak harvesting time in August using a portable soil auger from 20 cm depth where most of the root mass were located. 300 g of composite soil samples were taken from each treatment plot and the soil physical and chemical properties were analyzed. For microbial analysis, 10 individual soil samples were taken per plot in

Year 5. The soils were processed, and DNA was extracted as described in Yurgel et al. [38] and stored at −80 °C for sequencing.

3.3.1. Soil Physical Properties

Soil physical properties and water retention characteristics from the AN, BI and C plots were determined in triplicate in the 5th-year as described by Blake and Hartge [58], Cassel and Nielsen [59], and Danielson and Sutherland [60]. In brief, bulk density (D_b) was determined from the weight (M) and volume (V_1) of soil core using a graduated glass cylinder after continuous tapping until there was no observable change in soil volume. For particle density (D_p), soil lumps were carefully broken, and the volume (V_2) was recorded using the tapping method.

$$\text{Bulk density } (D_b) = \frac{M}{V_1} \tag{1}$$

$$\text{Particle density } (D_p) = \frac{M}{V_2} \tag{2}$$

Water saturation, field capacity, and wilting point were determined after soil was air-dried under ambient conditions (ca. 22 °C). A known mass of the fresh soil sample (M_s) was placed in a 15.24 cm plastic pot with drainage holes and weighed (M_{sp}). The potted soil was placed in a saucer and saturated with distilled water, and the saturated soil weight (M_{sat}) was recorded after 48 h. Then, the saucer was removed for free water to drain out under atmospheric pressure for 72 h and weighed ($M_{drained}$). The drained soil was spread evenly in a flat aluminum tray and air-dried under ambient conditions for 72 h and weighed (M_{dried}).

$$\text{Water saturation } (S_c) = \frac{M_{sat} - M_{sp}}{M_s} \times 100 \tag{3}$$

$$\text{Field capacity } (F_c) = \frac{M_{drained} - M_{sp}}{M_s} \times 100 \tag{4}$$

$$\text{Wilting point } (W_c) = \frac{M_{dried} - M_{sp}}{M_s} \times 100 \tag{5}$$

Water-holding capacity (WHC) was determined as the difference between field capacity and wilting point., i.e., $F_c - W_c$.

3.3.2. Soil Chemical Properties

Air-dried soil samples were screened through a 2 mm sieve (Fieldmaster, Guangzhou, China) before sending for elemental analysis at RPC Inorganic Analytical Chemistry Laboratory, Fredericton, NB, Canada every year for five years. A portion of each soil sample was digested according to the United States Environmental Protection Agency (USEPA) method 3050B (Standard Operation Procedures #4.M19). The resulting solutions were analyzed for elemental composition by inductively coupled plasma-mass spectrometry/inductively coupled plasma optical emission spectroscopy (ICP-MS/ICP-ES) using EPA method 200.8/EPA 200.7 (Standard Operation Procedures #4.M01/4.M29). Organic matter content, cation exchange capacity and estimated nitrogen release were also reported. Portions of the soil samples were leached in dilute potassium chlorate for 1 hr, and the leachate was filtered and analyzed calorimetrically for nitrate-nitrite [61]. In brief, soil from each treatment plot was separately added to deionized water at a ratio of 1:4 and stirred vigorously for 2 min. In Year 5, the potential hydrogen concentration (pH), electric conductivity, total dissolved solids, and salinity were determined from the supernatant using ExStik® EC500 instrument (Extech Instruments Corporation, Nashua, NH, USA), and turbidity using an Oakton Turbidimeter (T-100) (Oakton Instruments, Vernon Hills, IL, USA).

3.3.3. Soil Microbial Properties

DNA Extraction and Sequencing

In Year 5, DNA extraction was carried out using E.Z.N.A soil DNA isolation kit (Omega Bio-tek, Norcross, GA, USA) according to manufacturer's protocol and the modified method of Yurgel et al. [38]. DNA quality and concentration were measured using a NanoDrop 1000 spectrophotometer (Thermo Scientific, Waltham, MA, USA). At least 10 μL of DNA samples were sent to Dalhousie University IMR centre (http://imr.bio/; accessed on 5 November 2021) for V6-V8 16S rRNA gene (16S; forward: ACGCGHNRAACCTTACC; reverse: ACGGGCRGTGWGTRCAA), and ITS2 gene (ITS2; forward: GTGAATCATCGAATCTTTGAA; reverse: TCCTCCGCTTATTGATATGC) library preparation and sequencing. Samples were multiplexed using a dual-indexing approach and sequenced using an Illumina MiSeq with paired-end 300 + 300 bp reads. All PCR procedures and Illumina sequencing details were as previously described by Comeau et al. [62]. All sequences generated in this study are available in the NCBI sequence read archive under the accession numbers PRJNA883731 (16S rRNA) and PRJNA883747 (ITS).

Amplicon Variant Sequence (AVS) Picking and Statistical Analyses

A Microbiome Helper standard operating procedure [63] and QIIME2 wrapper scripts [64] were used to process and analyze the sequencing data. Briefly, the primers were trimmed and overlapping paired-end reads were stitched together followed by quality filtering and open-reference AVSs picking. We filtered out AVSs that contained fewer than 0.1% of the total sequences to compensate for MiSeq run-to-run bleed-through. The complete statistics of Illumina sequencing data analysis is presented in Supplementary Table S3. In brief, 892,582 and 188,044 high-quality non-chimeric reads were obtained from 16S rRNA and fungal internal transcribed spacer 2 (ITS2). These sequences were clustered into 55,483 (16S rRNA) and 1281 (ITS2) AVS. AVSs annotated as mitochondria and chloroplast were removed and the datasets were normalized to the depth of 6350 and 1186 frequencies resulting in the normalized datasets comprising 1768 (16S rRNA) and 721 (ITS2) AVSs, respectively.

Microbiome Functional Analysis

To understand the impact of MSW compost on microbiome function, we extrapolated functional profile of bacterial community based on 16S rRNA marker gene. Functional potentials of the bacterial community were predicted using the AVS tables and reference sequences generated by QIIME2, which were processed by PICRUSt2 software [65]. Abundance tables were generated both for complete MetaCyc functional pathways as well as individual enzymes categorized by KEGG Orthology (KO) numbers.

Bioinformatic Analysis of Amplicon Sequencing and PICRUSt2 Outputs

Alpha-diversity, i.e., Chao1 richness, Simpson evenness and Shannon diversity, and beta-diversity metrics were generated using QIIME2 [66]. Variations in sample groupings explained by beta-diversity distances (i.e., Adonis tests, 999 permutations) were run in QIIME2 to calculate how sample groupings are related to microbial community structure and function. Differential abundances bacterial taxa and pathways were determined using ALDEx2 with Benjamini-Hochberg corrected p value of Welch's t-test ($p < 0.01$) [67]. Nonmetric multi-dimensional scaling (NMDS) plots were developed based on Bray–Curtis distances using Vegan R package [68]. Heatmap and NMDS plots were build used gglpot2 R package [69].

3.4. *Plant Morpho-Physiology and Yield*

3.4.1. Leaf Pigmentation and Plant Photosynthetic Efficiency

Leaf chlorophyll and anthocyanin contents were estimated from four leaves per plant using SPAD 502DL Plus Chlorophyll Meter (Spectrum Technologies Inc., Aurora, IL, USA) and Anthocyanin content meter ACM-200 plus (Opti-Science Inc., Hudson, NH, USA), respectively. A portable OS30p+ Chlorophyll fluorometer (Opti-Science Inc., Hudson, NH, USA) was used to record leaf fluorescence indices between 08:00 and 11:00 am Central Standard Time (GMT-6). Briefly, to measure fluorescence indices, the middle portion excluding mid-vein of each selected leaf was clipped with light exclusion clips for 20 min prior to recording minimum (F_o), maximum (F_m), variable (F_v) fluorescence indices, potential photosynthetic capacity (F_v/F_o) and maximum quantum yield of photosystem II (F_v/F_m). All of these plant pigmentation and photosynthetic efficiency data were collected at eight weeks after planting in Year 5.

3.4.2. Plant Growth and Yield

Plant growth and yield components were taken from six plants in the middle of each plant row per species per plot in Year 5. Data collected were plant height using a 60 cm ruler; leaf area (A) was measured using LI-3000C portable leaf area meter (LI-COR Biosciences, Lincoln, NE, USA); and leaf dry matter content (LDMC) was determined from the ratio of oven-dry mass (M_d) to a water-saturated leaf mass (M_f). In brief, leaf samples were submerged in 250 mL distilled water for 9 h to obtain M_f before oven-dried at 65 °C for 48 h in a mechanical convection oven (Cole-Parmer Instrumental Co., Vernon Hills, IL, USA) to obtain LDMC.

$$\text{LDMC} = \frac{M_d}{M_f} \tag{6}$$

For specific stem density (SSD), diameter (d) of a 10 cm long (L) stem from plant with full green foliage was measured using a pair of calipers and the volume (V) of the stem was calculated using the oven-dry mass of the stem (M).

$$\text{SSD} = \frac{M}{V}; V = (0.5d)^2 \times \pi \times L \tag{7}$$

Specific root length (SRL) was determined from plant absorptive roots that were excavated gently from the soil using a spade and taking extra care not to lose more that 5% of the root mass, and to ensure that the total root length was intact before washing under running water until clean. A 30 cm ruler was used to measure the root length (R_L) and the air-dry mass (R_m) was recorded after drying under ambient laboratory conditions at ca. 22 °C.

$$\text{SRL} = \frac{R_L}{R_m} \tag{8}$$

Total fresh weight of the edible portions of each vegetable plant species was used to estimate crop yield from the green leaves of lettuce cv. Grand Rapids, roots of beets cv. Detroit Supreme, and immature pods of green bean cv. Golden Wax.

3.5. Plant Chemical Composition

Part of the elemental composition data were previously reported in Abbey et al. [7,19]. Briefly, edible portions of the lettuce, beets, and green beans were harvested and oven-dried at 65 °C for 48 h. The dried plant tissues were ground using a hammer mill and screened through a 53-µm sieve after which samples were analyzed as described below.

3.5.1. Plant Elemental Composition

A 30 g sample of each ground tissue was shipped on ice to Research and Productivity Council (RPC) Laboratory, New Brunswick, Canada for further preparation and analysis in Year 5. Portions of the samples were prepared by microwave-assisted digestion in nitric acid (SOP 4.M26). The resulting solutions were analyzed for chemical elements by ICP-MS/ICP-ES (SOP 4.M01). Kjeldahl nitrogen and nitrates and nitrites concentrations were analyzed using the digestion, phenate colourimetry (SOP IAS-M16) and hydrazine red, derivatization, colourimetry (SOP IAS-M48) methods, respectively.

3.5.2. TCA cycle Intermediate Metabolites

A 35 g sample of each ground plant tissue was put in a screw cap tube before shipping on dry ice to Canada's national metabolomics core facility at The Metabolomics Innovation Centre (TMIC), AB for analysis of targeted organic acids (i.e., α-ketoglutaric acid, succinic acid, fumaric acid, citric acid, and pyruvic acid). The samples were analyzed by the method described in Abbey et al. [19]. Briefly, 150 µL of ice-cold methanol and 10 µL of isotope-labeled internal standard mixture were added to 50 µL of plant extract for overnight protein precipitation. The mixture was centrifuged at 13,000× g for 20 min. Afterwards, 50 µL of the supernatant was loaded into 96-deep well plate followed by the addition of 3-nitrophenylhydrazine (NPH) reagent. After incubation for 2 h, BHT stabilizer and water were added before LC-MS injection. Mass spectrometric analysis was performed on an ABSciex 4000 Qtrap® tandem mass spectrometry instrument (Applied Biosystems/MDS Analytical Technologies, CA, USA) equipped with an Agilent 1260 series ultra-high performance liquid chromatography (UHPLC) system (Agilent Technologies, Santa Clara, CA, USA). The samples were delivered to mass spectrometer by LC method followed by direct injection (DI) method.

3.6. Experimental Design and Data Analysis

Soil Physiochemical Properties and Plant Growth Components

The AN, BI and C treatments and the three crop species; namely, green beans, lettuce and beets were arranged in a randomized complete block design with three replications. The three plant species were treated independently per block. Ten samples of soil were collected per plot and bulked before taking a composite sample. Soil chemical analyses were performed from the composite sample in triplicate. For plants, tissue samples were collected from 10 plants in the middle of the plant row and tissue chemical analyses were performed in duplicate, and the average was presented due to large sample size and limited resources. Data collected on soil physical and chemical properties and crop growth were subjected to one-way analysis of variance (ANOVA) using Proc Mixed in SAS version 9.4 (SAS Institute Inc., Cary, NC, USA). Fisher's least significant difference (LSD) test was used for mean separation at $\alpha = 5\%$ when ANOVA showed significant difference at $p \leq 0.05$. Data analysis of the metabolites was done using Analyst 1.6.2. and XLSTAT version 19.1 (Addinsoft, New York, NY, USA). Microsoft Excel was used to plot graphs.

4. Conclusions

The present study focuses on application frequency of MSW compost effect on soil health, soil microbiome function, and crop productivity and TCA cycle intermediate metabolites. There are clear indications of positive impact of annual application of MSW compost on plant tissue nutrient status, photosynthetic activities, and primary metabolites accumulation. Currently, farmers are faced with many challenges related to soil health, which is exacerbated by sustained global climate change, and the current global shortage and high cost of fertilizers. Therefore, the present work is urgently needed and timely. We concluded that MSW compost application is safe and effective for food production. We affirm that annual application of MSW compost promoted positive association between microbiome function and ecosystem services for the benefit of crop plants. Therefore, annual application of MSW compost can be classified as climate-smart and sustainable practice for the maintenance or rejuvenation of agricultural soils for nutrient-dense food production over the medium to long-term. This will benefit farmers and if implemented, will immensely benefit consumers with safe, nutritious and affordable food crops. Plants are selective in the composition of rhizosphere environment and microbial communities. As such, our next step is to investigate the rhizosphere composition of different plant species in compost amended soil.

Supplementary Materials: The following supporting information can be downloaded at: https://www.mdpi.com/article/10.3390/plants11223153/s1, Table S1: Climate conditions in Brandon, Manitoba during the growing season between May to August 2015 to 2019, and 30-year average (1981–2010) (Government of Canada, 2020); Table S2: Five-year average chemical elements concentration of Compost Quality Assurance tested municipal solid waste compost; Table S3: Summary of deblur output samples; Table S4: Relative abundances of major microbial taxa identified in the study; Table S5: Variation in sample groupings explained by weighted UniFrac and Bray–Curtis beta-diversity distances; Table S6: Essential and beneficial chemical elements concentrations in edible portions of green beans (*Phaseolus vulgaris* cv. Golden Wax), beets (*Beta vulgaris* cv. Detroit Supreme) and lettuce (*Latuca sativa* cv. Grand Rapids) as affected by frequency of application of CQA-tested municipal solid waste compost.

Author Contributions: Conceptualization: L.A.; Formal analysis: O.A.A., R.O. and S.N.Y.; Funding acquisition: L.A. and S.N.Y.; Investigation: L.A., O.A.A. and S.N.Y.; Methodology: O.A.A., S.N.Y. and L.A.; Project administration: L.A. Resources: L.A. and S.N.Y.; Supervision: L.A.; Validation: O.A.A., S.N.Y., R.O., L.R.G. and L.A.; Writing—original draft: O.A.A., S.N.Y., J.A. and R.O.; Writing—review and editing: J.A., R.O., L.R.G., N.A., S.N.Y. and L.A. All authors have read and agreed to the published version of the manuscript.

Funding: This work was financially supported by The Compost Council of Canada, Ontario and Manitoba Conservation and Water Stewardship award to L.A. Additional support was provided by USDA ARS Project 2090-21000-003-00D and Natural Sciences and Engineering Research Council of Canada Discovery Grant to S.N.Y.

Data Availability Statement: Not applicable.

Acknowledgments: The authors wish to thank the Compost Council of Canada, Ontario for securing the funds for the project, Manitoba Conservation and Water Stewardship for providing the funding, Alan Plett of Agaard Farms for providing the land for the 5-year project and assisting with land preparation, planting and management of the fields, City of Brandon, Manitoba for providing the CQA-tested municipal solid waste compost, and The Metabolomics and Innovation Centre (TMIC), Alberta for the metabolites analyses.

Conflicts of Interest: The authors declare no conflict of interest.

References

1. UN. Economic and Social Affairs. United Nations Department of Economic and Social Affairs and Population Division, World Population to 2300. 2004. Available online: https://www.un.org/development/desa/pd/sites/www.un.org.development.desa.pd/files/files/documents/2020/Jan/un_2002_world_population_to_2300.pdf (accessed on 2 January 2022).
2. Wright, J.; Kenner, S.; Lingwall, B. Utilization of compost as a soil amendment to increase soil health and to improve crop yields. *Open J. Soil Sci.* **2022**, *12*, 216–224. [CrossRef]
3. Feodorov, C.; Velcea, A.M.; Ungureanu, F.; Apostol, T.; Robescu, L.D.; Cocarta, D.M. Toward a circular bioeconomy within food waste valorization: A case study of an on-site composting system of restaurant organic waste. *Sustainability* **2022**, *14*, 8232. [CrossRef]
4. Kaza, S.; Yao, L.; Bhada-Tata, P.; Van Woerden, F. *What a Waste 2.0: A Global Snapshot of Solid Waste Management to 2050*; The World Bank Group: Washington, DC, USA, 2018; p. 295. [CrossRef]
5. Maalouf, A.; Mavropoulos, A.; El-Fadel, M. Global municipal solid waste infrastructure: Delivery and forecast of uncontrolled disposal. *Waste Manag. Res.* **2020**, *38*, 1028–1036. [CrossRef] [PubMed]
6. Environment Canada. Technical Document on Municipal Solid Waste Organics Processing. 2013. Available online: https://www.ec.gc.ca/gdd-mw/3E8CF6C7-F214-4BA2-A1A3-163978EE9D6E/13-047-ID-458-PDF_accessible_ANG_R2-reduced%20size.pdf (accessed on 30 March 2021).
7. Abbey, L.; Ijenyo, M.; Spence, B.; Asunni, A.O.; Ofoe, R.; Amo-Larbi, V. Bioaccumulation of chemical elements in vegetables as influenced by application frequency of municipal solid waste compost. *Can. J. Plant Sci.* **2021**, *101*, 967–983. [CrossRef]
8. Bustamante, M.A.; Nogués, I.; Jones, S.; Allison, G. The effect of anaerobic digestate derived composts on the metabolite composition and thermal behaviour of rosemary. *Sci. Rep.* **2019**, *9*, 6489. [CrossRef] [PubMed]
9. Hargreaves, J.; Adl, M.; Warman, P. A review of the use of composted municipal solid waste in agriculture. *Agric. Ecosyst. Environ.* **2008**, *123*, 1–14. [CrossRef]
10. Giannakis, G.V.; Kourgialas, N.N.; Paranychianakis, N.V.; Nikolaidis, N.P.; Kalogerakis, N. Effects of municipal solid waste compost on soil properties and vegetables growth. *Compos. Sci. Util.* **2014**, *22*, 116–131. [CrossRef]

11. Ney, L.; Franklin, D.; Mahmud, K.; Cabrera, M.; Hancock, D.; Habteselassie, M.; Newcomer, Q.; Fatzinger, B. Rebuilding soil ecosystems for improved productivity in biosolarized soils. *Int. J. Agron.* **2019**, *1245*, 5827585. [CrossRef]
12. Singh, Y.P.; Arora, S.; Mishra, V.K.; Singh, A.K. Synergizing microbial enriched municipal solid waste compost and mineral gypsum for optimizing rice-wheat productivity in sodic soils. *Sustainability* **2022**, *14*, 7809. [CrossRef]
13. Machado, R.M.A.; Alves-Pereira, I.; Robalo, M.; Ferreira, R. Effects of municipal solid waste compost supplemented with inorganic nitrogen on physicochemical soil characteristics, plant growth, nitrate content, and antioxidant activity in spinach. *Horticulturae* **2021**, *7*, 53. [CrossRef]
14. Vaccaro, S.; Muscolo, A.; Pizzeghello, D.; Spaccini, R.; Piccolo, A.; Nardi, S. Effect of a compost and its water-soluble fractions on key enzymes of nitrogen metabolism in maize seedlings. *J. Agric. Food Chem.* **2009**, *57*, 11267–11276. [CrossRef] [PubMed]
15. Beffa, T.; Blanc, M.; Marilley, L.; Fischer, J.L.; Lyon, P.F.; Aragno, M. Taxonomic and metabolic microbial diversity during composting. In *The Science of Composting*; de Bertoldi, M., Sequi, P., Lemmes, B., Papi, T., Eds.; Springer: Dordrecht, The Netherlands, 1996; pp. 149–161. [CrossRef]
16. Mehta, C.M.; Palni, U.; Franke-Whittle, I.H.; Sharma, A.K. Compost: Its role, mechanism and impact on reducing soil-borne plant diseases. *J. Waste Manag.* **2014**, *34*, 607–622. [CrossRef]
17. Steel, H.; Moens, T.; Vandecasteele, B.; Hendrickx, F.; De Neve, S.; Neher, D.A. Factors influencing the nematode community during composting and nematode-based criteria for compost maturity. *Ecol. Indic.* **2018**, *85*, 409–421. [CrossRef]
18. Kelly, C.; Haddix, M.L.; Byrne, P.F.; Cotrufo, M.F.; Schipanski, M.E.; Kallenbach, C.M.; Wallenstein, M.D.; Fonte, S.J. Long-term compost amendment modulates wheat genotype differences in belowground carbon allocation, microbial rhizosphere recruitment and nitrogen acquisition. *Soil Biol. Biochem.* **2022**, *172*, 108768. [CrossRef]
19. Abbey, L.; Ofoe, R.; Gunupuru, L.R.; Ijenyo, M. Variation in frequency of CQA-tested municipal solid waste compost can alter metabolites in vegetables. *Food Res. Int.* **2021**, *143*, 110225. [CrossRef] [PubMed]
20. van der Merwe, M.J.; Osorio, S.; Araujo, W.L.; Balbo, I.; Nunes-Nesi, A.; Maximova, E.; Carrari, F.; Bunik, V.I.; Persson, S.; Fernie, A.R. Tricarboxylic acid cycle activity regulates tomato root growth via effects on secondary cell wall production. *Plant Physiol.* **2010**, *153*, 611–621. [CrossRef] [PubMed]
21. Neugart, S.; Wiesner-Reinhold, M.; Frede, K.; Jander, E.; Homann, T.; Rawel, H.M.; Schreiner, M.; Baldermann, S. Effect of solid biological waste compost on the metabolite profile of *Brassica rapa* ssp. *chinensis*. *Front. Plant Sci.* **2008**, *16*, 305. [CrossRef]
22. Zhou, D.; Huang, X.F.; Chaparro, J.M.; Badri, D.v.; Manter, D.K.; Vivanco, J.M.; Guo, J. Root and bacterial secretions regulate the interaction between plants and PGPR leading to distinct plant growth promotion effects. *Plant Soil* **2016**, *401*, 259–272. [CrossRef]
23. Zhang, Y.; Fernie, R. On the role of the tricarboxylic acid cycle in plant productivity. *J. Integr. Plant Biol.* **2018**, *60*, 1199–1216. [CrossRef]
24. Moffett, J.R.; Puthillathu, N.; Vengilote, R.; Jaworski, D.M.; Namboodiri, A.M. Acetate revisited: A key biomolecule at the nexus of metabolism, epigenetics and oncogenesis Part 1: Acetyl-CoA, Acetogenesis and Acyl-CoA Short-Chain Synthetases. *Front. Physiol.* **2020**, *11*, 580167. [CrossRef]
25. Omena-Garcia, R.P.; Araújo, W.L.; Gibon, Y.; Fernie, A.R.; Nunes-Nesi, A. Measurement of tricarboxylic acid cycle enzyme activities in plants. *Methods Mol. Biol.* **2017**, *1670*, 167–182. [CrossRef] [PubMed]
26. Fataftah, N.; Mohr, C.; Hajirezaei, M.R.; von Wirén, N.; Humbeck, K. Changes in nitrogen availability leads to a reprogramming of pyruvate metabolism. *BMC Plant Biol.* **2018**, *18*, 77. [CrossRef] [PubMed]
27. Wang, L.; Cui, D.; Zhao, X.; He, M. The important role of the TCA cycle in plants. *Genom. Appl. Biol.* **2017**, *8*, 25–29. [CrossRef]
28. Rosa, S.D.; Silva, C.A.; Carletti, P.; Sawaya, A.C.H.F. Maize growth and root organic acid exudation in response to water extract of compost application. *J. Soil Sci. Plant Nutr.* **2021**, *21*, 2770–2780. [CrossRef]
29. Adugna, G. A review on impact of compost on soil properties, water use and crop productivity. *Res. J. Agric. Sci.* **2016**, *4*, 93–104. [CrossRef]
30. Rawls, W.J.; Pachepsky, Y.A.; Ritchie, J.C.; Sobecki, T.M.; Bloodworth, H. Effect of soil organic carbon on soil water retention. *Geoderma* **2003**, *116*, 61–76. [CrossRef]
31. Metting, F.B.; Smith, J.L.; Amthor, J.S. Science needs and new technology for soil carbon sequestration. In *Carbon Sequestration in Soils: Science, Monitoring and Beyond*; Rosenberg, N.J., Izaurralde, R.C., Malone, E.L., Eds.; Battelle Press: Columbus, OH, USA, 1999; pp. 1–34.
32. Akpa, S.I.C.; Odeh, I.O.A.; Bishop, T.F.A.; Hartemink, A.E.; Amapu, I.Y. Total soil organic carbon and carbon sequestration potential in Nigeria. *Geoderma* **2016**, *271*, 202–215. [CrossRef]
33. McCauley, A.; Jones, C.; Jacobsen, J. *Nutrient Management-Module #8. Soil pH and Organic Matter*; Montana State University: Bozeman, MT, USA, 2016; pp. 1–16.
34. Solly, E.F.; Weber, V.; Zimmermann, S.; Walthert, L.; Hagedorn, F.; Schmidt, M.W.I. A critical evaluation of the relationship between the effective cation exchange capacity and soil organic carbon content in Swiss forest soils. *Front. For. Glob. Chang.* **2020**, *3*, 98. [CrossRef]
35. Rheault, D.G. Relationship between soil phosphorus and runoff phosphorus losses from plot scale rainfall simulations on Manitoba soils. Master's Thesis, Department of Soil Science, University of Manitoba, Winnipeg, MB, Canada, 2012; p. 197.
36. Nair, K.P. Soil fertility and nutrient management. In *Intelligent Soil Management for Sustainable Agriculture*; Springer International Publishing: Cham, Switzerland, 2019; pp. 165–189. [CrossRef]

37. Esan, E.O.; Abbey, L.; Yurgel, S. Exploring the long-term effect of plastic on compost microbiome. *PLoS ONE* **2019**, *14*, e0214376. [CrossRef]
38. Yurgel, S.N.; Douglas, G.M.; Comeau, A.M.; Mammoliti, M.; Dusault, A.; Percival, D.; Langille, M.G. Variation in bacterial and eukaryotic communities associated with natural and managed wild blueberry habitats. *Phytobiomes J.* **2017**, *1*, 102–113. [CrossRef]
39. Monje, O.A.; Bugbee, B. Inherent limitations of non-destructive chlorophyll meters: A comparison of two types of meters. *HortScience* **1992**, *27*, 69–71. [CrossRef] [PubMed]
40. Nielsen, N.E.; Hansen, E.M. Macronutrient cation uptake by plants. Effects of plant species, nitrogen concentration in the plant, cation concentration and activity ratio in soil solution. *Plant Soil* **1984**, *77*, 347–365. [CrossRef]
41. Seufert, V.; Jonathan, N.; Foley, A. Comparing the yields of organic and conventional agriculture. *Nature* **2012**, *485*, 229–232. [CrossRef]
42. Wu, Y.Y.; Zhao, K. Root-exuded malic acid versus chlorophyll fluorescence parameters in four plant species under different phosphorus level. *J. Soil Sci. Plant Nutr.* **2013**, *13*, 604–610. [CrossRef]
43. Maxwell, K.; Johnson, G.N. Chlorophyll fluorescence—A practical guide. *J. Exp. Bot.* **2000**, *51*, 659–668. [CrossRef]
44. Lõhmus, K.; Truu, M.; Truu, J.; Ostonen, I.; Kaar, E.; Vares, A.; Uri, V.; Alama, S.; Kanal, A. Functional diversity of culturable bacterial communities in the rhizosphere in relation to fine-root and soil parameters in alder stands on forest, abandoned agricultural, and oil-shale mining areas. *Plant Soil* **2006**, *283*, 1–10. [CrossRef]
45. Cornelissen, J.H.C.; Lavorel, S.; Garnier, E.; Diaz, S.; Buchmann, N.; Gurvich, D.E.; Reich, P.B.; ter Steege, H.; Morgan, H.D.; van der Heijden, M.G.A.; et al. A handbook of protocols for standardized and esay measurement of plant functional traits worldwide. *Aust. J. Bot.* **2003**, *51*, 335–380. [CrossRef]
46. World Health Organization (WHO). *Diet, Nutrition, and the Prevention of Chronic Diseases: Report of a Joint WHO/FAO Expert Consultation;* WHO Technical Report Series 961; Food and Agriculture Organization of the United Nations; World Health Organization: Geneva, Switzerland, 2003; pp. 23–25.
47. Statistics Canada. Health Fact Sheets. Fruit and Vegetable Consumption. 2017. Available online: https://www150.statcan.gc.ca/n1/pub/82-625-x/2019001/article/00004-eng.htm (accessed on 30 December 2021).
48. Paul, M.J.; Pellny, T.K. Carbon metabolite feedback regulation of leaf photosynthesis and development. *J. Exp. Bot.* **2003**, *54*, 539–547. [CrossRef]
49. Fernie, A.R.; Martinoia, E. Malate. Jack of all trades or master of a few? *Phytochemistry* **2009**, *70*, 828–832. [CrossRef]
50. Kromer, S. Respiration during photosynthesis. *Annu. Rev. Plant Physiol. Plant Mol. Biol.* **1995**, *46*, 45–70. [CrossRef]
51. Pot, S.; De Tender, C.; Ommeslag, S.; Delcour, I.; Ceusters, J.; Gorrens, E.; Debode, J.; Vandecasteele, B.; Vancampenhout, K. Understanding the shift in the microbiome of composts that are optimized for a better fit-for-purpose in growing media. *Front. Microbiol.* **2021**, *12*, 643679. [CrossRef]
52. Wang, J.; Yan, D.; Dixon, R.; Wang, Y.P. Deciphering the principles of bacterial nitrogen dietary preferences: A strategy for nutrient containment. *Mol. Biol.* **2016**, *7*, e00792-16. [CrossRef] [PubMed]
53. Chai, J.M.; Adnan, A. Effect of different nitrogen source combinations on microbial cellulose production by *Pseudomonas aeruginosa* in batch fermentation. In *IOP Conference Series: Materials Science and Engineering;* IOP Publishing: Bristol, UK, 2018; Volume 440, p. 012044. [CrossRef]
54. Wang, Q.; Wang, C.; Yu, W.; Turak, A.; Chen, D.; Huang, Y.; Ao, J.; Jiang, Y.; Huang, Z. Effects of nitrogen and phosphorus inputs on soil bacterial abundance, diversity, and community composition in Chinese fir plantations. *Front. Microbiol.* **2018**, *9*, 1543. [CrossRef]
55. Mooshammer, M.; Wanek, W.; Hämmerle, I.; Fuchslueger, L.; Hofhansl, F.; Knoltsch, A.; Schnecker, J.; Takriti, M.; Watzka, M.; Wild, B.; et al. Adjustment of microbial nitrogen use efficiency to carbon: Nitrogen imbalances regulate soil nitrogen cycling. *Nat. Commun.* **2014**, *5*, 3694. [CrossRef] [PubMed]
56. Tiquia, S.M.; Lloyd, J.; Herms, D.A.; Hoitink, H.A.J.; Michel, F.C. Effects of mulching and fertilization on soil nutrients, microbial activity and rhizosphere bacterial community structure determined by analysis of TRFLPs of PCR-amplified 16S rRNA genes. *Appl. Soil Ecol.* **2002**, *21*, 31–48. [CrossRef]
57. Tang, J.; Zhang, L.; Zhang, J.; Ren, L.; Zhou, Y.; Zheng, Y.; Luo, L.; Yang, Y.; Huang, H.; Chen, A. Physicochemical features, metal availability and enzyme activity in heavy metal-polluted soil remediated by biochar and compost. *Sci. Total Environ.* **2020**, *701*, 134751. [CrossRef] [PubMed]
58. Blake, G.R.; Hartge, K.H. Bulk density. In *Methods of Soil Analysis: Part 1 Physical and Mineralogical Methods, 5.1*, 2nd ed.; Arnold, K., Ed.; SSSA Book Series; John Wiley & Sons, Inc.: Hoboken, NJ, USA, 1986; pp. 363–375. [CrossRef]
59. Cassel, D.K.; Nielsen, D.R. Field capacity and available water capacity. In *Methods of Soil Analysis: Part 1 Physical and Mineralogical Methods, 5.1*, 2nd ed.; Arnold, K., Ed.; SSSA Book Series; John Wiley & Sons, Inc.: Hoboken, NJ, USA, 1986; pp. 901–926. [CrossRef]
60. Danielson, R.E.; Sutherland, P.L. Porosity. In *Methods of Soil Analysis: Part 1 Physical and Mineralogical Methods, 5.1*, 2nd ed.; Arnold, K., Ed.; SSSA Book Series; John Wiley & Sons, Inc.: Hoboken, NJ, USA, 1986; pp. 443–461. [CrossRef]
61. Cataldo, D.A.; Schrader, L.E.; Youngs, V.L. Analysis by digestion and colorimetric assay of total nitrogen in plant tissues high in nitrate. *Crop Sci.* **1974**, *14*, 854–856. [CrossRef]
62. Comeau, A.M.; Vincent, W.F.; Bernier, L.; Lovejoy, C. Novel chytrid lineages dominate fungal sequences in diverse marine and freshwater habitats. *Sci. Rep.* **2016**, *6*, 30120. [CrossRef]

63. Comeau, A.M.; Douglas, G.M.; Langille, M.G. Microbiome Helper: A Custom and Streamlined Workflow for Microbiome Research. *mSystems* **2017**, *2*, e00127-16. [CrossRef] [PubMed]
64. Bolyen, E.; Rideout, J.R.; Dillon, M.R.; Bokulich, N.A.; Abnet, C.C.; Al-Ghalith, G.A.; Alexander, H.; Alm, E.J.; Arumugam, M.; Asnicar, F.; et al. Reproducible, interactive, scalable and extensible microbiome data science using QIIME 2. *Nat. Biotechnol.* **2019**, *37*, 852–857. [CrossRef]
65. Douglas, G.M.; Maffei, V.J.; Zaneveld, J.R.; Yurgel, S.N.; Brown, J.R.; Taylor, C.M.; Huttenhower, C.; Langille, M.G.I. PICRUSt2 for prediction of metagenome functions. *Nat. Biotechnol.* **2020**, *38*, 685–688. [CrossRef] [PubMed]
66. Kim, B.-R.; Shin, J.; Guevarra, R.B.; Lee, J.H.; Kim, D.W.; Seol, K.-H.; Kim, H.B.; Isaacson, R.E. Deciphering diversity indices for a better understanding of microbial communities. *J. Microbiol. Biotechnol.* **2017**, *27*, 2089–2093. [CrossRef] [PubMed]
67. Fernandes, A.D.; Reid, J.N.; Macklaim, J.M.; McMurrough, T.A.; Edgell, D.R.; Gloor, G.R. Unifying the analysis of high-throughput sequencing datasets: Characterizing RNA-seq, 16S rRNA gene sequencing and selective growth experiments by compositional data analysis. *Microbiome* **2014**, *2*, 15. [CrossRef] [PubMed]
68. Oksanen, J.; Blanchet, F.G.; Friendly, M.; Kindt, R.; Legendre, P.; McGlinn, D.; Minchin, P.R.; O'Hara, R.B.; Simpson, G.L.; Solymos, P.; et al. Vegan. Community Ecology Package. R Package Version 2.5-6. 2019. Available online: https://CRAN.R-project.org/package=vegan (accessed on 3 April 2021).
69. Wickham, H. *ggplot2: Elegant Graphics for Data Analysis*; Springer: Berlin/Heidelberg, Germany, 2016.

MDPI

Article

Changes in Biogenic Amines of Two Table Grapes (cv. Bronx Seedless and Italia) during Berry Development and Ripening

Melek Incesu [1], Sinem Karakus [2,3], Hanifeh Seyed Hajizadeh [4], Fadime Ates [5], Metin Turan [6], Milan Skalicky [7,*] and Ozkan Kaya [8,*]

1 Department of Food Engineering, Faculty of Agriculture, Ataturk University, Erzurum 25100, Turkey
2 Çölemerik Vocational School, Hakkari University, Hakkari 30000, Turkey
3 Department of Biology, Faculty of Science and Art, Erzincan Binali Yıldırım University, Erzincan 24002, Turkey
4 Department of Horticulture, Faculty of Agriculture, University of Maragheh, Maragheh 55136-553, Iran
5 Manisa Viticulture Research Institute, Republic of Turkey Ministry of Agriculture and Forestry, Manisa 45125, Turkey
6 Department of Genetics and Bioengineering, Faculty of Engineering, Yeditepe University, Istanbul 34755, Turkey
7 Department of Botany and Plant Physiology, Faculty of Agrobiology, Food and Natural Resources, Czech University of Life Sciences Prague, Kamycka 129, 16500 Prague, Czech Republic
8 Erzincan Horticultural Research Institute, Republic of Turkey Ministry of Agriculture and Forestry, Erzincan 24060, Turkey
* Correspondence: skalicky@af.czu.cz (M.S.); kayaozkan25@hotmail.com (O.K.); Tel.: +90-553-4701308 (O.K.)

Citation: Incesu, M.; Karakus, S.; Seyed Hajizadeh, H.; Ates, F.; Turan, M.; Skalicky, M.; Kaya, O. Changes in Biogenic Amines of Two Table Grapes (cv. Bronx Seedless and Italia) during Berry Development and Ripening. *Plants* **2022**, *11*, 2845. https://doi.org/10.3390/plants11212845

Academic Editors: Lord Abbey, Josephine Ampofo and Mason MacDonald

Received: 13 October 2022
Accepted: 21 October 2022
Published: 26 October 2022

Abstract: Bronx Seedless and Italia (*Vitis vinifera* L.) are a variety preferred by consumers owing to their exciting flavour and widely cultivated in Aegean Region in Turkey. The aim was to identify the biogenic amines of these table grapes during berry ripeness. The biogenic amines were analyzed by HPLC in six different berry phenological stages. Italia grapes presented lower biogenic amine content than Bronx Seedless table grapes. The concentration of most of the biogenic amines analyzed linearly raised from the beginning of berry touch to when berries ripen for harvest stages. The most common biogenic amines in grape varieties were putrescine, followed by histamine, agmatine, and tyramine. There was also a positive correlation between all biogenic amines of the two grape varieties. The weakest correlation was found between spermine and cadaverine, whereas the strongest correlation was found among dopamine, trimethylamine, norepinephrine, tyramine, and histamine amines. The present study is the first report of a synthesis study regarding the effect of B.A.s on quality characteristics throughout berry ripeness in grape varieties containing foxy and muscat tastes. The concentration and composition of biogenic amines identified for both varieties might provide helpful information regarding human health and the vintage.

Keywords: grape; histamine; putrescine; agmatine; dopamine; spermine

1. Introduction

Global grape production reaches roughly 75 million tons annually, making grapes (*Vitis* sp.) one of the most cultivated fruit crops worldwide [1–3]. It has significant amounts of organic compounds, minerals, and carbohydrates, as well as its production, which increases day by day due to its derivatives, such as raisins, wine, fruit juice, jam, and flour consumption, which are known to be beneficial for human health [4]. The monitoring of these organic compounds is very important to producers because efficient quality control is needed, and their profile could influence the acceptability of consumers. It has been, indeed, noted that grape derivatives have a positive role in the prevention of various diseases such as liver diseases, anaemia, cardiovascular disease, cancer, and Alzheimer's owing to the inhibition of oxidation of low-density lipoproteins [5–7]. In addition, it has fruit acids and a fibrous structure that help clean the blood and regulate the functioning of the kidney and intestinal system [8]. On the other hand, biogenic amines (B.A.s) among bioactive compounds conducted on different grape varieties have been intensively investigated recently due to their effects on the quality, safety, and nutraceutical properties of fruit juices and wines [4]. Due to the fact that many authors have reported that B.A.s can have many adverse effects on human health, especially on susceptible individuals, as well as positive effects [9]. It has, indeed, been highlighted that the consumption of high concentrations of B.A.s, which are low molecular weight compounds produced by the decarboxylation of amino acids, may cause various health problems such as skin irritation, headache, blushing, itching, tachycardia, hypertension, impaired breathing, hypotension and vomit, kidney failure, anaphylactic shock [10–12].

Some amines such as agmatine, spermidine, serotonin, histamine, tryptamine, spermine, dopamine, norepinephrine, ethanolamine, phenylethylamine, putrescine, and tyramine in grapes have been reported by some authors [13,14]. It has, indeed, been noted that naturally occurring BAs [1,4-butanediamine = putrescine; 4-aminobutil guanidine = agmatine; 4-(2-aminoethyl) phenol = tyramine; 1,5-pentanediamine = cadaverine; 2-phenylethylamine = phenylethylamine; 4-aminomethylimidazole = histamine; (4-aminobutil) (4-[4-aminobutylamino] butyl) amine = spermine; 3-(2-aminomethyl) indole = tryptamine; N-(3-aminopropyl)-1,4-butanediamine = spermidine] are among the factors contributing to the quality of grapes and wines [15,16]. It has been detected in previous reports that these amines are both generally found in musts, and their values are affected by growing conditions, grape variety, degree of ripening, soil type, drying conditions of grape, and composition [17–19]. It was, indeed, highlighted that there was a relationship between varieties and B.A. content in seven different grape varieties under sterile conditions, except for agmatine [20,21]. It is well documented that the concentration and nature of B.A.s present found in wines, in general, depend mainly on the composition of the grape juice in literature by many authors [22–24]; however, we observed the lack of a synthesis study regarding the effect of B.A.s on quality characteristics throughout berry ripeness in grape varieties with foxy and muscat taste. As mentioned above, although there are many lines of research examining BAs in grapes and grape-derived products, there are still many possibilities to explore from microbiological, technological, analytical, and toxicological perspectives. Therefore, information on the variations or amounts of B.A.s in grapes and their derived products is mainly used as grape quality indicators and can assist in providing many insights into their potential impact on consumer health. Considering limited information on BA variation throughout berry ripeness in grape varieties with foxy and muscat taste, controlling these compounds can be of great importance to understanding the formation, reduction, and monitoring of B.A.s throughout berry ripeness, and even of the effects of B.A.s in consumers after the digestion of foods containing different levels of these compounds.

Grape varieties containing foxy (*Vitis vinifera* L. cv. Bronx Seedless) and muscat (*Vitis vinifera* L. cv. Italia) tastes are cultivated in different regions of Turkey, such as Central Anatolia Aegean, Southeastern Anatolia, and Marmara [25]. Italia is a seed variety characterized by a slight muscat flavor of fruits, a vigorous vegetative behavior, and mid-late maturity, whereas Bronx Seedless is a variety that is attracted and preferred by consumers because of its pink berries characterized by strawberry flavor. Recently, the cultivation of these new grapes that can be grown in different climatic conditions instead of traditional varieties and fresh consumption has increased in Turkey. These grape varieties have a distinct flavor from other *Vitis vinifera* grape cultivars and knowledge of the variation of BAs content in these varieties may be key for consumers. However, our knowledge of the dynamics of changes in the B.A.s profiles of Bronx Seedless and Italia table grapes throughout berry ripeness is limited. Monitoring B.A.s levels in these grape varieties can be an important marketing advantage and allow the establishment of B.A profiles for the safety and quality control of fresh consumption or by-products of these grapes. Therefore, this paper aims to present the changes in biogenic amines in Bronx Seedless and Italia table grapes throughout berry ripeness.

2. Results

Variety significantly affected the content of B.A.s except for serotonin, and the season factor affected the content of most B.A.s. There was no significant interaction between the variety and the phenological stage. The Italia grape variety showed a higher B.A. content than the Bronx Seedless grape variety for all B.A.s analyzed, except for spermine (Spn) and serotonin (Ser). For both grape cultivars, the most common B.A.s in grape varieties was putrescine (Put), followed by histamine (His), agmatine (Agm), and tyramine (Tyr). There was wide variation in the B.A profiles, with values ranging from 0.83 (norepinephrine for Bronx Seedless) to 12.24 $\mu g \cdot L^{-1}$ (putrescine for Italia). Both putrescine (Put) and histamine (His) were mostly amines, whereas less common BAs were spermine (Spn) and norepinephrine (Nor). Generally, berries collected at BBCH-89 and BBCH-85 stages showed higher B.A. content than berries collected at other phenological stages for both grape cultivars. Put and his content reached the highest level at BBCH-89, whereas lower rates of spermine, norepinephrine, and spermidine were found at BBCH-77 when compared to BBCH-89, BBCH-85, and BBCH-83 for both grape cultivars. Considering the phenological stages, there was a linear increase in B.A.s contents of both grape varieties (Table 1). On the other hand, regarding each phenological stage, there were significant differences ($p \leq 0.05$) between the two grape varieties. Total biogenic amine contents were higher in Italy grapes than in Bronx Seedless grapes (Figure 1).

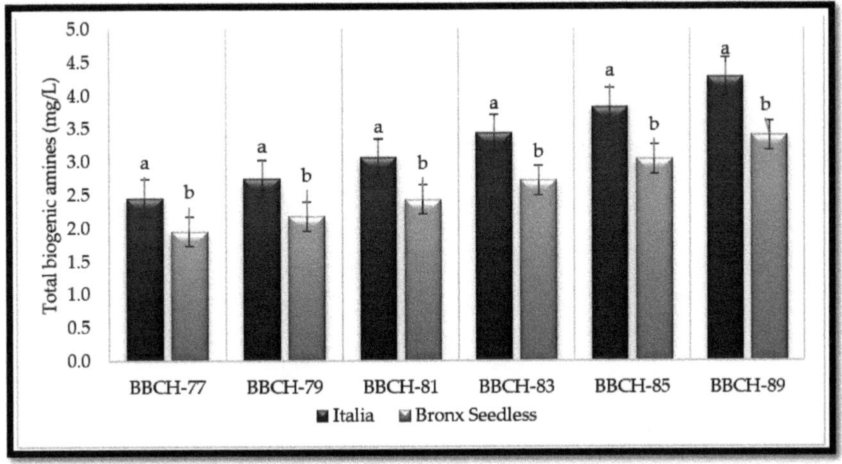

Figure 1. Total content of biogenic amines in Italia (black) and Bronx Seedless(grey) table grapes harvested in six different phenological stages (BBCH-77, BBCH-79, BBCH-81, BBCH-83, BBCH-85, and BBCH-89). Data are expressed as the mean of the data with their corresponding standard deviation. For a given variable and phenological stage, different letters over the bars represent significant differences (Duncan test, $p < 0.05$).

Pearson correlation analyses for the biogenic amine contents (i.e., Agm, Spd, Ser, His, Try, Dop, Nor, Cad, Tma, Put, Tyr) appear in Figure 2 for all data sets. Findings showed a positive correlation between all B.A.s of the two grape varieties. There was also a positive but strong correlation between grape varieties with B.A.s, except for Spn, Dop, His, Agm, Tyr, and Nor. Put, and Agm showed a strong correlation with other B.A.s ($p \leq 0.01$), whereas there was no correlation between Spn and Cad (Table 2). Apart from this, a P.C.A. was performed on all of the B.A.s identified in both grape varieties to detect individual B.A. associated with the grapes (Figure 3). PC1 (reveals 82.8% of the total data) was related to the B.A.s contents, such as Agm, Spd, Ser, His, Try, Dop, Nor, Cad, Tma, Put, Tyr. PC2 (which reveals 11.4% of total data) was related to B.A.s contents (Figure 3).

Table 1. Biogenic amines content (µg · L^{-1}) of Italia and Bronx Seedless table grapes harvested in six different phenological stages (BBCH-77, BBCH-79, BBCH-81, BBCH-83, BBCH-85, and BBCH-89).

Variety (V)	Put	Cad	Agm	Spn	Spd	His	Try	Ser	Tyr	Tma	Dop	Nor
Italia	12.24 ± 0.12 a	1.76 ± 0.03 a	4.62 ± 0.26 a	0.99 ± 0.08 b	1.23 ± 0.12 a	6.90 ± 1.26 a	3.45 ± 1.18 a	1.64 ± 0.68 a	1.84 ± 0.87 a	1.78 ± 0.43 a	1.93 ± 0.52 a	1.15 ± 0.12 a
Bronx Seedless	10.71 ± 0.56 b	1.21 ± 0.48 b	3.87 ± 0.26 b	1.16 ± 0.62 a	0.92 ± 0.08 b	4.47 ± 0.23 b	2.33 ± 0.15 b	1.59 ± 0.19 a	1.66 ± 0.05 b	1.30 ± 0.08 b	1.30 ± 0.11 b	0.83 ± 0.06 b
Phenological stage (S)												
BBCH-77	8.66 ± 1.13 e	1.22 ± 0.06 b	3.17 ± 1.01 e	0.82 ± 0.01 e	0.79 ± 0.05 d	4.08 ± 0.21 d	2.02 ± 0.10 d	1.22 ± 0.13 d	1.30 ± 0.18 f	1.16 ± 0.23 c	1.22 ± 0.10 c	0.73 ± 0.11 d
BBCH-79	9.62 ± 1.18 de	1.31 ± 0.11 b	3.54 ± 0.16 de	0.91 ± 0.02 dc	0.89 ± 0.07 d	4.61 ± 0.18 cd	2.31 ± 1.03 cd	1.36 ± 0.03 cd	1.46 ± 0.26 e	1.29 ± 0.16 cd	1.36 ± 0.18 c	0.82 ± 0.16 cd
BBCH-81	10.70 ± 0.23 cd	1.42 ± 0.62 ab	3.95 ± 0.18 cd	1.01 ± 016 cd	1.00 ± 0.05 cd	5.22 ± 1.12 bcd	2.63 ± 0.62 bcd	1.51 ± 0.05 c	1.63 ± 0.07 d	1.44 ± 0.19 bcd	1.51 ± 0.12 bc	0.92 ± 0.08 bcd
BBCH-83	11.90 ± 2.08 c	1.53 ± 0.03 ab	4.41 ± 0.09 bc	1.12 ± 0.09 bc	1.12 ± 0.06 bc	5.91 ± 1.07 abc	3.01 ± 0.15 bc	1.68 ± 0.28 c	1.82 ± 0.19 c	1.60 ± 0.25 bc	1.68 ± 0.08 abc	1.03 ± 0.01 abc
BBCH-85	13.23 ± 2.16 b	1.66 ± 1.02 ab	4.92 ± 0.83 ab	1.24 ± 0.02 b	1.26 ± 0.24 ab	6.69 ± 1.09 ab	3.44 ± 0.98 ab	1.87 ± 0.62 b	2.04 ± 0.69 b	1.78 ± 0.41 ab	1.87 ± 0.65 ab	1.16 ± 0.16 ab
BBCH-89	14.72 ± 2.61 a	1.79 ± 0.066 a	5.49 ± 0.98 a	1.38 ± 0.07 a	1.41 ± 0.31 a	7.57 ± 2.16 a	3.92 ± 0.86 a	2.07 ± 0.75 a	2.28 ± 0.53 a	1.98 ± 0.26 a	2.07 ± 0.45 a	1.30 ± 0.29 a
Significance												
V	0.0000	0.0000	0.0000	0.0000	0.0000	0.0000	0.0000	0.0000	0.0000	0.2724	0.0000	0.0000
S	0.0000	0.0000	0.0000	0.0000	0.0000	0.0000	0.0000	0.0000	0.0000	0.0000	0.0000	0.0000
V × S	0.8800	0.9291	0.8272	0.7403	0.0813	0.0738	0.4716	0.9992	0.9091	0.7115	0.3302	0.8841

Data are expressed as mean of the data. aSignificance (*p*-value) of variety (V), season (S), and V–S interactions. For a given factor and significance ($p < 0.05$), different letters within a column represent significant differences (Duncan test, $p < 0.05$). Put; Putrescine, Cad; Cadaverine, Agm; Agmatine, Spn; Spermine, Spd; Spermidine, His; Histamine, Try; Tryptamine, Ser; Serotonin, Tyr; Tyramine, Tma; Trimethylamine, Dop; Dopamine, Nor; Norepinephrine.

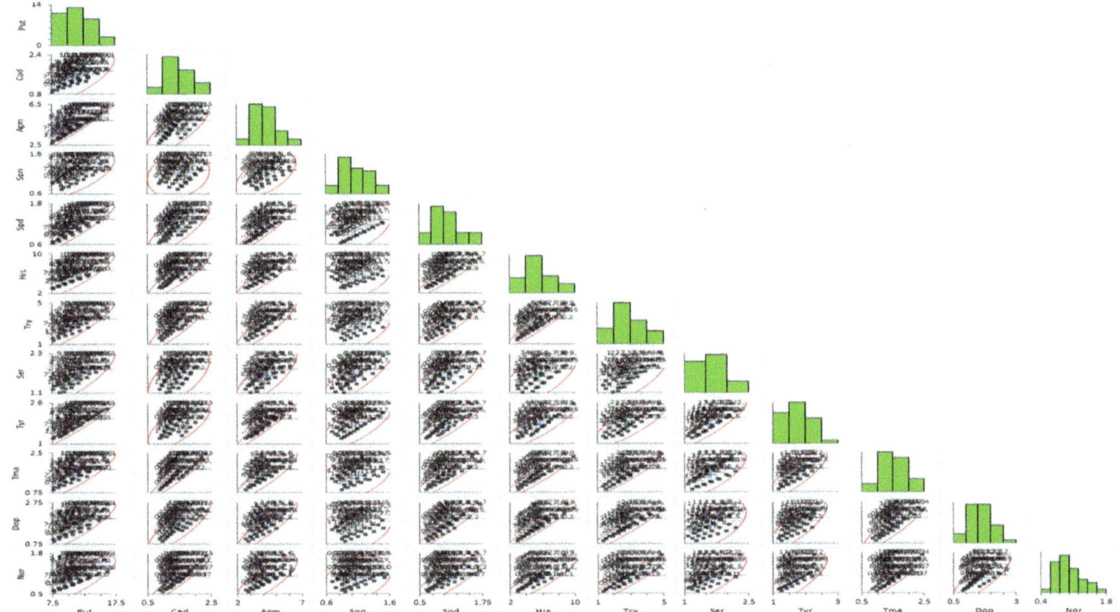

Figure 2. Grape varieties data correlation analysis. Scatterplot matrix representation for the entire data set belonging to biogenic amines values of grape varieties.

Table 2. Pearson correlation between biogenic amines in grape varieties. Significant correlations are reported for $0.01 < p < 0.05$ (**) and $p < 0.01$ (*). The color intensity is proportioned to the Pearson Index.

Pearson's r	Put	Cad	Agm	Spn	Spd	His	Try	Ser	Tyr	Tma	Dop	Nor
Put	1											
Cad	0.739 **	1										
Agm	0.984 **	0.757 **	1									
Spn	0.684 **	0.193 ns	0.593 **	1								
Spd	0.948 **	0.814 **	0.958 **	0.487 **	1							
His	0.891 **	0.897 **	0.928 **	0.302 *	0.952 **	1						
Try	0.912 **	0.915 **	0.908 **	0.390 **	0.933 **	0.968 **	1					
Ser	0.809 **	0.613 **	0.776 **	0.819 **	0.744 **	0.642 **	0.666 **	1				
Tyr	0.963 **	0.689 **	0.971 **	0.721 **	0.902 **	0.851 **	0.832 **	0.886 **	1			
Tma	0.891 **	0.961 **	0.895 **	0.416 **	0.922 **	0.943 **	0.967 **	0.740 **	0.842 **	1		
Dop	0.889 **	0.838 **	0.910 **	0.295 *	0.966 **	0.975 **	0.956 **	0.577 **	0.806 **	0.905 **	1	
Nor	0.818 **	0.952 **	0.811 **	0.400 **	0.841 **	0.895 **	0.933 **	0.773 **	0.798 **	0.955 **	0.827 **	1

**. Correlation is significant at the 0.01 level; *. Correlation is significant at the 0.05 level; ns; not significant.

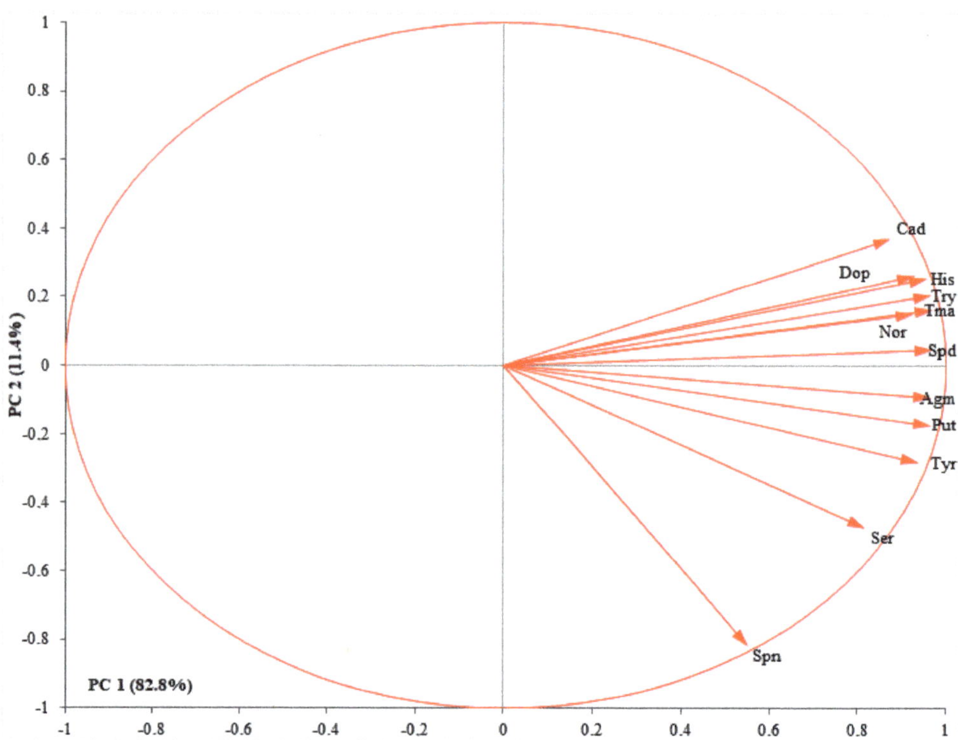

Figure 3. P.C.A. biplot (score and loadings plots) of berries colored by varieties. All biogenic amines are displayed. The size of the arrows indicates the contribution strength of the compound. Each point is the average of quaternaryplicate of each organic acid.

3. Discussion

Although there have been new approaches and reports on a wide range of functions of B.A.s in wine and table grapes in the last few decades [4,22,26], to our knowledge, information on the reactions of B.A. both in grape varieties containing foxy and muscat taste and in berry ripeness stages has yet to be revealed. Previous results on the functions of B.A.s in wine and table grapes have allowed us to understand positive and negative effects on human health in greater and greater detail; however, the picture's complexity has dramatically increased, and negative effects on human health appear in greater and greater detail; however, the complexity of the picture has greatly increased. Much previous research has focused on the effects of different applications on grape and wine chemical composition in various grape varieties [4,22,26]; however, less study has been conducted with B.A.s of grape varieties containing foxy and muscat taste. In this context, the results of our study are similar to the previous reports by researchers who analyzed the B.A.s of wine, and pomace from different varieties grown in various viticultural regions of the world. In our study, Put, His, Agm, and Try were the most abundant amino acids in raisin varieties, confirming previous reports that these B.A.s occur at high concentrations in grapes [27]. However, when the general literature is examined, the effect of grape evaluation forms or grape by-products (i.e., grape seeds, skins, wines, pomace, bagasse, stalks) on the concentration of B.A.s of raisin is not clear and consistent among the different varieties and even for the seeded and seedless grape varieties.

Experimental observations on highly complex systems for B.A.s in grape varieties are mainly based on measurements performed in wine and must be rationalized in terms of elementary article science. The present study is, in this context, the first report of a synthesis study regarding the effect of B.A.s on quality characteristics throughout berry ripeness in grape varieties containing foxy and muscat tastes. Based on our results, we can state that there was wide variation in the total B.A.s between grape varieties, with values ranging from 12.24 (Put) to 0.83 $\mu g \cdot L^{-1}$) (Nor). In addition, Put, His, Agm, and Try were the majority amine, coinciding with the results of Gomez et al. [27]. Moreover, some authors who agree with our findings reported on the presence of primary amines being generally histamine, tyramine, putrescine, cadaverine, tryptamine, agmatine, phenylethylamine, methylamine, isoamylamine, and ethylamine in wine products obtained from different grape [10,28].

On the other hand, the Bronx Seedless grape variety showed lower content of all B.A.s than the Italia grape variety, except for Spn and Spd (Table 1). This was not surprising given the reports that the content of B.A. in grapes varies by variety [24,29]. The Spn and Spd contents were relatively low for both varieties, and this could be explained by the fact that probably by the metabolism of Agm, which is the most abundant amino acid in grape or Spn and Spd originate from Put, the latter being produced by decarboxylation of ornithine [30]. Although Put has also been found that this amine may be present in berries without external microbial contamination, it has been related to poor sanitary conditions of grapes [28]. Some studies have highlighted that the quantity of amines in food depends not only on the amount of microorganisms present but on the activity of the decarboxylase enzyme on specific amino acids and the favourability of the enzymatic conditions, like, pH, temperature [28–30].

As far as phenological stages are concerned, as the berry development progresses, individual B.A.s concentrations also rise, and there was a linear increase in all B.A.s contents of both grape varieties. The berries of BBCH-89 and BBCH-85 stages reached higher B.A. content than the berries of other phenological stages. The most common B.A.s in grape varieties were Put, His, Agm, and Try, whereas among these amines, Spn, Nor, and Spd were the lowest (Table 1). It has been reported that some bio-components exhibit patterns of decline and subsequent accumulation during ripening, suggesting their covalent association with other cellular compounds or utilization and degradation for the biosynthesis of other compounds [31]. However, there is not much information in the literature about the anabolism or the catabolism of B.A.s during berry development and ripening. Logically here, we hypothesize that the smaller berry size in very early berry development contains low concentrations of B.A. compared to larger berry size. On the other hand, we think that B.A.s could have increased in musts of berry due to the hydrolysis of hydroxycinnamic amide compounds throughout berry ripeness. It is, indeed, known that amino acid precursors could accumulate in berries as well during berry ripeness, supporting a further B.A.s increase. It has also been noted that amino acids vary depending on the berry development stages [32], which confirms our hypothesis. Although it is known that there is no consensus on the correlation between amino acids in the environment and total B.A.s [21], this point should not be neglected by researchers.

Regarding the Pearson correlation values calculated to determine the relationships between individual B.A.s analyzed in all data, a significant positive correlation ($p \leq 0.05$) was obtained among B.A.s (Figure 2). This is consistent with the results that there is a significant correlation between some B.A.s identified in raisins [14]. Most significant correlations for data obtained from both grape varieties show a high confidence level ($p \leq 0.05$), except for His, Dop, Tyr, Spn, Agm, and Nor (Table 2). A marked correlation was determined between Put and other B.A.s (r values greater than 0.73, with $p \leq 0.05$); between Cad and other B.A.s, except for Spn, Ser, and Tyr (r values greater than 0.61, with $p \leq 0.05$). The strongest correlation was found among Dop, Tma, Nor, Try, and His amines, whereas the weakest correlation was found between Spn and Cad (Table 2). It has been noted that there is a high correlation between serotonin, spermidine, spermine, and tryptophan [4], which is consistent with our results. In addition, the relationship among cadaverine, tryptamine, and phenylethylamine [33], between histamine-tyramine [34], putrescine-tyramine [35], or some amines [36] have been reported in different works for wines. On the other hand, we opted to compare the detected findings by P.C.A. in terms of creating a descriptive model for grouping B.A.s as a function of grape varieties throughout berry ripeness. A PCA was obtained that detected for (PC1 + PC2) 94.2 % of the variance in the twelve B.A.s for grape varieties in our findings. Cad, Dop, His, Try, Tma, Nor, and Spd were close to each other in the second quadrant, whereas the Agm, Put, Try, and Ser was close to each other (except for Spn) and was located in the third quadrant (Figure 3). This correlation among B.A.s concentration during berry ripening in our study supports the results of Ates et al. [14], who found a high correlation between spermine, tryptamine, putrescine, agmatine, and dopamine.

Given the dual importance (quality and health effects) of B.A.s, on the other hand, efforts must be made to control B.A.s in grape products and regulatory limits for B.A.s have not yet been established by the OIV (Organization International de la Vigne et du Vin) for fresh grape, raisins, and wines; however, Lehtonen [37], stated that for France 8 mg/L, Belgium 5–6 mg/L, Austria 10 mg/L, Holland 3.5 mg/LGermany 2 mg/L, and Switzerland 10 mg/L as the maximum limit for histamine in wine are recommended. A legal limit for histamine in wine does not exist in either the European Union or Austria [37]. Although our findings for histamine are lower than the above-mentioned limit values, the maximum BA limit that is generally considered safe for consumers is not yet known for fresh grapes. It has been reported that histamine and tyramine are generally assumed to be the most toxic amines among B.A.s European Food Safety Authority [38], (2011), but to the best of our knowledge, there is no scientific data to confirm this yet. Despite the increase in B.A.s content as the berry development progresses in our study, these values are well below the values considered harmful to human health and do not cause any problems related to berry quality.

4. Materials and Methods

4.1. Plant Material and Sample Preparation

The study was conducted in 2021 on twenty-year-old Bronx Seedless (New York 8536 × Sultanina) and Italia (Bicane × muscat Hamburg) vines grafted onto 5 B.B. rootstock in the Manisa Province (Manisa Viticulture Research Institute), Aegean Region, Turkey (27°23′57.36″ East Longitude and 38°37′57.14″ North Latitude at 3.3 m above sea level. Bronx Seedless and Italia vines were planted at 2.0 m between vines and 3.0 m between rows. Vines have a high trunk (about 1 m) cordon trellis system with 12–15 shoots per vine and one cluster per shoot at a northwest orientation, and vines were spur-pruned. Healthy berries (450 per cultivar) were randomly harvested in triplicate from the cluster's bottom, middle, and top parts. Samplings were carried out on July 27 (the first week before veraison; stage BBCH-77), and the latest sampling was August 28 (harvest time; BBCH-89), for a total of six selections. Based on BBCH scale published by Lorenz et al. [39], clusters were harvested in six different stages as follows: BBCH-79, BBCH-81, BBCH-83, BBCH-85 (these stages follow in order; begin berry touch, berry touch complete, berries begin to brighten in color, berries brightening in color, softening of berries, berries ripe for harvest). Clusters were collected, put into plastic bags, and transported at 4 °C to the laboratory, where they were stored at −80 °C until analysis.

4.2. Chemicals

Standard solutions of spermine (in the 1 to 30 mg/L range), agmatine, spermidine, serotonin, histamine, tryptamine, dopamine, norepinephrine, cadaverine, trimethylamine, putrescine, and tyramine were obtained from Sigma-Aldrich Chemie, Steinheim, Germany.

4.3. Isolation of Amines from Grape Varieties

Berries (5 g) were homogenized using an Ultra-turax homogenizer, including 0.5 mL of 70% perchloric acid. Homogenate (X g) was centrifuged at 10,000 rpm for 10 min. The supernatant was recovered, filtered over a 0.22 mm membrane, and diluted with 10%

perchloric acid to the initial homogenate weight. The sample was then filtered over 0.45 µm and injected into the HPLC.

4.4. Identification of Amines from Grape Varieties by HPLC

Biogenic amines (B.A.) were separated and quantified according to the method of Nagy et al. [40], with modifications. Samples were injected on a reverse phase column (Bondapak C18, 300 × 3.9 mm, 10 mm; Waters, Milford, MA, USA) mounted on a Waters Alliance Liquid Chromatograph attached to a Waters 474 fluorescence detector (Milford, MA, USA). Post-column derivatization (2-mercaptoethanol, o-phtalaldehyde) was used to improve detection. Peaks were identified using authentic standards. Calibration curves in the range of 1 to 30 mg/L (spermine) and 0.1 to 10 mg/L (agmatine, spermidine, serotonin, histamine, tryptamine, dopamine, norepinephrine, cadaverine, trimethylamine, putrescine, and tyramine) were used for quantitation. The biogenic amine content of samples was expressed in $\mu g \cdot L^{-1}$ fresh weight.

4.5. Statistical Analysis

Descriptive statistics for the studied variables were analyzed considering a completely randomized design with the factorial arrangement, accounting for two grape varieties involving six phenological stages. The variables were subjected to a variety analysis (ANOVA) that was performed using SPSS 21.0 (SPSS Inc., Chicago, IL, USA). Duncan's test detected the significance of the differences in data ($p \leq 0.05$). The Pearson correlation coefficients were evaluated as scatterplot matrices using Analyse-it statistical software, and they were significant at the 0.01 (**) and 0.05 (*) levels.

5. Conclusions

Although scientific advances have been achieved in understanding the biochemical, molecular, and physiological aspects of berry ripening in grapes, there is no previous research on the change of B.A.s throughout berry ripeness. Therefore, this study is the first report on B.A.s. The findings showed that Bronx Seedless table grapes presented higher contents of Put, Cad, Agm, Spd, His, Try, Ser, Tyr, Tma, Dop, Nor, and lower contents of Spn than Italia table grapes. The analyzed B.A.s concentration increased linearly as the berry ripened, and there reached the highest level in BBCH-77 and BBCH-89 stages. In addition, a significant positive correlation was detected between B.A.s and a strong correlation between Dop, Tma, Nor, Try, and His amines. Given the therapeutic value and importance of some amines of these grapes for humans, the pharmaceutical industry may capitalize on the potential human health benefits of pharmaceutical preparations for plants, animals, and humans. Indeed, it is likely that soon, we will be able to see the processing of pre-and post-harvest food crops and the production of fruits and vegetables with higher amine levels by combining traditional and modern growing approaches.

Author Contributions: Conceptualization, visualization, data curation, methodology, resources, software, O.K. and F.A.; validation, formal analysis, investigation, M.I., S.K., H.S.H., F.A., M.T. and M.S.; writing—original draft preparation O.K.; writing—review and editing, O.K. and M.S. All authors have read and agreed to the published version of the manuscript.

Funding: This research received no external funding.

Institutional Review Board Statement: Not applicable.

Informed Consent Statement: Not applicable.

Data Availability Statement: Not applicable.

Acknowledgments: The authors would like to thank Manisa Viticulture Research Institute for support in providing raisin varieties samples for the assessment. All individuals included have consented to the acknowledgement.

Conflicts of Interest: The authors declare no conflict of interest.

References

1. FAO. Grapevine Production. Available online: http://www.fao.org/faostat (accessed on 20 July 2022).
2. Keskin, N.; Bilir Ekbic, H.; Kaya, O.; Keskin, S. Antioxidant Activity and Biochemical Compounds of *Vitis vinifera* L. (cv. 'Katıkara') and *Vitis labrusca* L. (cv. 'Isabella') Grown in Black Sea Coast of Turkey. *Erwerbs-Obstbau* **2021**, *63*, 115–122. [CrossRef]
3. Keskin, N.; Kunter, B.; Çelik, H.; Kaya, O.; Keskin, S. ANOM approach for the statistical evaluation of organic acid contents of clones of the grape variety'Kalecik Karasi'. *Mitt. Klosterneubg.* **2021**, *71*, 126–138.
4. Gomes, E.P.; Borges, C.V.; Monteiro, G.C.; Belin, M.A.F.; Minatel, I.O.; Junior, A.P.; Tecchio, M.A.; Lima, G.P.P. Preharvest salicylic acid treatments improve phenolic compounds and biogenic amines in 'Niagara Rosada' table grape. *Postharvest Biol. Technol.* **2021**, *176*, 111–505. [CrossRef]
5. Nandakumar, V.; Singh, T.; Katiyar, S.K. Multi-targeted prevention and therapy of cancer by proanthocyanidins. *Cancer Lett.* **2008**, *269*, 378–387. [CrossRef] [PubMed]
6. Anastasiadi, M.; Pratsinis, H.; Kletsas, D.; Skaltsounis, A.L.; Haroutounian, S.A. Bioactive non-coloured polyphenols content of grapes, wines and vinification by-products: Evaluation of the antioxidant activities of their extracts. *Food Res. Int.* **2010**, *43*, 805–813. [CrossRef]
7. Frankel, E.N.; Bosanek, C.A.; Meyer, A.S.; Silliman, K.; Kirk, L.L. Commercial grape juices inhibit the in vitro oxidation of human low-density lipoproteins. *J. Agric. Food Chem.* **1998**, *46*, 834–838. [CrossRef]
8. Çelik, H.; Ağaoğlu, Y.S.; Fidan, Y.; Marasalı, B.; Söylemezoğlu, G. *Genel Bağcılık. Sun Fidan AŞ Mesleki Kitaplar Serisi: I*; Fersa Matbaacılık San. Tic. Ltd.: Ankara, Turkey, 1998; Volume 1, pp. 178–190.
9. Basile, T.; Alba, V.; Suriano, S.; Savino, M.; Tarricone, L. Effects of ageing on stilbenes and biogenic amines in red grape winemaking with stem contact maceration. *J. Food Process. Preserv.* **2018**, *42*, e13378. [CrossRef]
10. Anli, E.; Bayram, M. Ochratoxin A in wines. *Food Rev. Int.* **2009**, *25*, 214–232. [CrossRef]
11. Ancin-Azpilicueta, C.; Gonzalez-Marco, A.; Jimenez-Moreno, N. Current knowledge about the presence of amines in wine. *Crit. Rev. Food Sci. Nutr.* **2008**, *48*, 257–275. [CrossRef]
12. Ladero, V.; Calles-Enríquez, M.; Fernández, M.A.; Alvarez, M. Toxicological effects of dietary biogenic amines. *Curr. Nutr. Food Sci.* **2010**, *6*, 145–156. [CrossRef]
13. Wang, Y.Q.; Ye, D.Q.; Zhu, B.Q.; Wu, G.F.; Duan, C.Q. Rapid HPLC analysis of amino acids and biogenic amines in wines during fermentation and evaluation of matrix effect. *Food Chem.* **2014**, *163*, 6–15. [CrossRef] [PubMed]
14. Ates, F.; Kaya, O.; Keskin, N.; Turan, M. Biogenic amines in raisins of one vintage year: Influence of two chemical pre-treatments (dipping in oak ash solution or potassium carbonate solution). *Mitt. Klosterneubg.* **2022**, *72*, 51–59.
15. Hajós, G.; Sass-Kiss, A.; Szerdahelyi, E.; Bardocz, S. Changes in Biogenic Amine Content of Tokaj Grapes, Wines, and Aszu-wines. *J. Food Sci.* **2000**, *65*, 1142–1144. [CrossRef]
16. Sass-Kiss, A.; Hajós, G. Characteristic biogenic amine composition of Tokaj aszú-wines. *Acta Aliment.* **2005**, *34*, 227–235. [CrossRef]
17. Guo, D.L.; Zhang, G.H. A new early-ripening grape cultivar—'Fengzao'. *Acta Hortic.* **2015**, *1082*, 153–156.
18. Mah, J.H.; Park, Y.K.; Jin, Y.H.; Lee, J.H.; Hwang, H.J. Bacterial production and control of biogenic amines in Asian fermented soybean foods. *Foods* **2019**, *8*, 85. [CrossRef]
19. Keskin, N.; Kaya, O.; Ates, F.; Turan, M.; Gutiérrez-Gamboa, G. Drying Grapes after the Application of Different Dipping Solutions: Effects on Hormones, Minerals, Vitamins, and Antioxidant Enzymes in Gök Üzüm (*Vitis vinifera* L.) Raisins. *Plants* **2022**, *11*, 529. [CrossRef]
20. Del Prete, V.; Costantini, A.; Cecchini, F.; Morassut, M.; Garcia-Moruno, E. Occurrence of biogenic amines in wine: The role of grapes. *Food Chem.* **2009**, *112*, 474–481. [CrossRef]
21. Martínez-Pinilla, O.; Guadalupe, Z.; Hernández, Z.; Ayestarán, B. Amino acids and biogenic amines in red varietal wines: The role of grape variety, malolactic fermentation and vintage. *Eur. Food Res. Technol.* **2013**, *237*, 887–895. [CrossRef]
22. Emer, C.D.; Marques, S.; Colla, L.M.; Reinehr, C.O. Biogenic amines and the importance of starter cultures for malolactic fermentation. *Aust. J. Grape Wine Res.* **2021**, *27*, 26–33. [CrossRef]
23. Khiari, B.; Jeguirim, M. Pyrolysis of grape marc from Tunisian wine industry: Feedstock characterization, thermal degradation and kinetic analysis. *Energies* **2018**, *11*, 730. [CrossRef]
24. Yue, X.; Zhao, Y.; Ma, X.; Jiao, X.; Fang, Y.; Zhang, Z.; Ju, Y. Effects of leaf removal on the accumulation of anthocyanins and the expression of anthocyanin biosynthetic genes in Cabernet Sauvignon (*Vitis vinifera* L.) grapes. *J. Sci. Food Agric.* **2021**, *101*, 3214–3224. [CrossRef]
25. Kaya, O.; Incesu, M.; Ates, F.; Keskin, N.; Verdugo-Vásquez, N.; Gutiérrez-Gamboa, G. Study of Volatile Organic Compounds of Two Table Grapes (cv. Italia and Bronx Seedless) along Ripening in Vines Established in the Aegean Region (Turkey). *Plants* **2022**, *11*, 1935. [CrossRef] [PubMed]
26. Marques, A.P.; Leitão, M.C.; San Romão, M.V. Biogenic amines in wines: Influence of oenological factors. *Food Chem.* **2008**, *107*, 853–860. [CrossRef]
27. Gomez, H.A.G.; Marques, M.O.M.; Borges, C.V.; Minatel, I.O.; Monteiro, G.C.; Ritschel, P.S.; Lima, G.P.P. Biogenic amines and the antioxidant capacity of juice and wine from brazilian hybrid grapevines. *Plant Foods Hum. Nutr.* **2020**, *75*, 258–264. [CrossRef]
28. Restuccia, D.; Loizzo, M.R.; Spizzirri, U.G. Accumulation of biogenic amines in wine: Role of alcoholic and malolactic fermentation. *Fermentation* **2018**, *4*, 6. [CrossRef]

29. Cecchini, F.; Morassut, M. Effect of grape storage time on biogenic amines content in must. *Food Chem.* **2010**, *123*, 263–268. [CrossRef]
30. Mangani, S.; Guerrini, S.; Granchi, L.; Vincenzini, M. Putrescine accumulation in wine: Role of Oenococcus oeni. *Curr. Microbiol.* **2005**, *51*, 6–10. [CrossRef]
31. Conde, C.; Silva, P.; Fontes, N.; Dias, A.C.P.; Tavares, R.M.; Sousa, M.J.; Gerós, H. Biochemical changes throughout grape berry development and fruit and wine quality. *Foods* **2007**, *1*, 1–22.
32. Ali, M.A.; Poortvliet, E.; Strömberg, R.; Yngve, A. Polyamines in foods: Development of a food database. *Food Nutr. Res.* **2011**, *55*, 5572–5586.
33. García-Villar, N.; Hernández-Cassou, S.; Saurina, J. Characterization of wines through the biogenic amine contents using chromatographic techniques and chemometric data analysis. *J. Agric. Food Chem.* **2007**, *55*, 7453–7461. [CrossRef] [PubMed]
34. Soufleros, E.; Barrios, M.L.; Bertrand, A. Correlation between the content of biogenic amines and other wine compounds. *Am. J. Enol. Vitic.* **1998**, *49*, 266–278.
35. Herbert, P.; Cabrita, M.J.; Ratola, N.; Laureano, O.; Alves, A. Free amino acids and biogenic amines in wines and musts from the Alentejo region. Evolution of amines during alcoholic fermentation and relationship with variety, sub-region and vintage. *J. Food Eng.* **2005**, *66*, 315–322. [CrossRef]
36. Romero, R.; Jönsson, J.Å.; Gázquez, D.; Bagur, M.G.; Sánchez-Viñas, M. Multivariate optimization of supported liquid membrane extraction of biogenic amines from wine samples prior to liquid chromatography determination as dabsyl derivatives. *J. Sep. Sci.* **2002**, *25*, 584–592. [CrossRef]
37. Lehtonen, P. Determination of amines and amino acids in wine—A review. *Am. J. Enol. Vitic.* **1996**, *47*, 127–133.
38. European Food Safety Authority (EFSA). Scientific opinion on risk based control of biogenic amine formation in fermented foods. Panel on Biological Hazards (BIOHAZ). *EFSA J.* **2011**, *9*, 2393–2486. [CrossRef]
39. Lorenz, D.H.; Eichhorn, K.W.; Bleiholder, H.; Klose, R.; Meier, U.; Weber, E. Growth Stages of the Grapevine: Phenological Growth Stages of the Grapevine (*Vitis vinifera* L. Ssp. Vinifera)—Codes and Descriptions According to the Extended BBCH Scale. *Aust. J. Grape Wine Res.* **1995**, *1*, 100–103. [CrossRef]
40. Nagy, A.; Bálo, B.; Ladányi, M.; Fazekas, I.; Kellner, N.; Nagy, B.; Nyitrainé Sárdy, D. Examination of biogenic amines in grapevine musts originating from vineyards treated with different viticultural practices. *J. Wine Res.* **2018**, *29*, 151–158. [CrossRef]

MDPI

Article

Field Application of a Vis/NIR Hyperspectral Imaging System for Nondestructive Evaluation of Physicochemical Properties in 'Madoka' Peaches

Kyeong Eun Jang [1,†], Geonwoo Kim [2,3,†], Mi Hee Shin [3], Jung Gun Cho [4], Jae Hoon Jeong [4], Seul Ki Lee [4], Dongyoung Kang [2] and Jin Gook Kim [3,5,*]

1 Division of Applied Life Science, Graduate School of Gyeongsang National University, 501, Jinju-daero, Jinju-si, Gyeongsangnam-do 52828, Korea
2 Department of Bio-industrial Machinery Engineering, College of Agriculture and Life Science, Gyeongsang National University, 501, Jinju-daero, Jinju-si, Gyeongsangnam-do 52828, Korea
3 Institute of Agriculture and Life Sciences, Gyeongsang National University, 501, Jinju-daero, Jinju-si, Gyeongsangnam-do 52828, Korea
4 Fruit Research Division, National Institute of Horticultural and Herbal Science, Wanju 55365, Korea
5 Department of Horticulture, College of Agriculture and Life Science, Gyeongsang National University, 501, Jinju-daero, Jinju-si, Gyeongsangnam-do 52828, Korea
* Correspondence: jgkim119@gnu.ac.kr
† These authors contributed equally to this work.

Abstract: Extensive research has been performed on the in-field nondestructive evaluation (NDE) of the physicochemical properties of 'Madoka' peaches, such as chromaticity (a*), soluble solids content (SSC), firmness, and titratable acidity (TA) content. To accomplish this, a snapshot-based hyperspectral imaging (HSI) approach for filed application was conducted in the visible and near-infrared (Vis/NIR) region. The hyperspectral images of 'Madoka' samples were captured and combined with commercial HSI analysis software, and then the physicochemical properties of the 'Madoka' samples were predicted. To verify the performance of the field-based HSI application, a lab-based HSI application was also conducted, and their coefficient of determination values (R^2) were compared. Finally, pixel-based chemical images were produced to interpret the dynamic changes of the physicochemical properties in 'Madoka' peach. Consequently, the a* values and SSC content shows statistically significant R^2 values (0.84). On the other hand, the firmness and TA content shows relatively lower accuracy (R^2 = 0.6 to 0.7). Then, the resultant chemical images of the a* values and SSC content were created and could represent their different levels using grey scale gradation. This indicates that the HSI system with integrated HSI software used in this work has promising potential as an in-field NDE for analyzing the physicochemical properties in 'Madoka' peaches.

Keywords: fruit quality; quality prediction; plant phenotyping; orchard management

Citation: Jang, K.E.; Kim, G.; Shin, M.H.; Cho, J.G.; Jeong, J.H.; Lee, S.K.; Kang, D.; Kim, J.G. Field Application of a Vis/NIR Hyperspectral Imaging System for Nondestructive Evaluation of Physicochemical Properties in 'Madoka' Peaches. *Plants* **2022**, *11*, 2327. https://doi.org/10.3390/plants11172327

Academic Editors: Lord Abbey, Josephine Ampofo and Mason MacDonald

Received: 17 August 2022
Accepted: 2 September 2022
Published: 5 September 2022

1. Introduction

Hyperspectral imaging (HSI) has been intensively used as an effective tool for plant phenotyping in the nondestructive evaluation (NDE) fields [1], agricultural products [2], food safety, and quality [3] due to its rapid and accurate detection and classification of target materials. Moreover, HSI can simultaneously provide the spectral and spatial information of a target material by combining spectroscopic analysis and image processing [4–6]. It, therefore, has a high potential to be used for online sorting application systems of various foods [5]. Consequently, this allows users to detect or classify a target material through pixel-based chemical images from agricultural products [7–9].

Despite the great advantages of HSI technologies for agricultural NDE, the in-field-based HSI technologies for agricultural NDE still have some critical issues. The total cost of an in-field based HSI system is expensive, and it is also difficult for users to handle the HSI system due to its complexity, such as in the scanning method (line or snapshot scanning), long image acquisition time (more than regular machine vision), sensitive control for line scanning, low pixel resolution, large amount of sample information, etc. [10].

Moreover, to build robust prediction models based on various chemometric methods, spectral calibration is essential for establishing high accuracy [8,11]. However, because the spectral calibration is strongly affected by weather conditions such as brightness and the movement of clouds, the majority of detection and classification chemometric models have been developed under a precisely controlled environment [6,12,13].

To overcome these difficulties, commercial companies such as Perception Park (Graz, Austria), Prediktera (Umeå, Sweden), and perClass BV (Delft, Netherlands) have released their commercial hyperspectral imaging software, which allows researchers and agricultural entrepreneurs to analyze hyperspectral images more easily. They have been attempting to use this software in a wide variety of industrial [14,15], medical [16], and agricultural fields [17,18]. In the agricultural field, the HSI system has been widely applied to fruit ripeness, soluble solids content (SCC), firmness, bruising, fungal contamination, nutrients, and so on [10,19–23]. Therefore, in this study, we conducted an NDE of fruit quality using a commercial hand-held visible near-infrared (Vis/NIR) hyperspectral camera (Specim IQ, Specim, Oulu, Finland) and classification software (perClass Mira, Delft, Nederlands) which performs an automatic selection of machine learning models with various preprocessing methods for field applications.

Peaches (*Prunus persica* (L.) Batsch) are one of the most widely consumed fruits, which are cultivated in both hemispheres because of their unique flavor and high nutrition [24,25]. Peaches can be divided into melting (< 8 N firmness) and non-melting (>16 N firmness) types depending on how much of the flesh softens as they ripen [26,27]. Moreover, peach fruits are vulnerable to climate change because they quickly ripen and lose their quality even at room temperature [28]. Thus, ripeness is an important factor for fresh peaches during market distribution.

As a result of global warming, the average yearly temperature might rise by 0.3 to 4.8 degrees. The worldwide mean annual temperature will rise by 0.3 to 4.8 °C by 2100, and many agricultural crops will likely experience temperatures that are higher than the global average [29]. Hence, predicting the quality parameters of peaches has recently attracted substantial attention in the fields of agricultural NDE because of their temperature sensitivity [23,24,30].

Approximately 30 kinds of peach cultivars, including 'Madoka' peaches, have been cultivated in the Republic of Korea. 'Madoka' is the typical melting type that quickly softens during its ripening stage [25]; therefore, we have selected 'Madoka' peaches as our target in field applications due to its physiological response to temperature conditions.

In a lab-based environment, important physicochemical properties of agricultural products could be measured and predicted under light source, controlled scanning speed, and minimized vibration [3,19,20]. However, field-based HSI for agricultural products has some critical issues, such as irregular intensity of sunlight, the movement of clouds, dirt on the surface of fruits, and so on. They can affect the brightness of the obtained hyperspectral images and decrease the overall HSI performance. Accordingly, improving the performance of the field-based HSI technologies is an emerging issue for the NDE of agricultural products [20].

Therefore, the main objective of this study was to evaluate the physicochemical properties of 'Madoka' peaches. Their chromaticity, soluble solids content (SSC), firmness, and titratable acidity (TA) were evaluated using a commercial hyperspectral camera and software for field application of a commercial HSI system. To verify its accuracy, the predicted values produced by the commercial software were compared with their measured values by a colorimeter, refractometer, firmness meter, and pH meter. In addition, the accuracy of the predicted values was acquired from laboratory (indoor) and field (outdoor) studies to analyze the differences and accuracies between the two environments. Our detailed objectives are summarized as follows:

- Establish two Vis/NIR HSI systems (indoor and outdoor) for obtaining hyperspectral data of 'Madoka' peaches, with an increase in their growth period,
- Measure the physicochemical properties of 'Madoka' peaches in an indoor and outdoor environment and compare their predicted values with those of the commercial software,
- Analyze the predicted results from both the indoor and outdoor environment and determine a preprocessing method which produces the highest accuracy among various preprocessing methods,
- Demonstrate feasibility by providing pixel-based visualization of physicochemical distribution created by hyperspectral images taken from an indoor and outdoor environment.

2. Results

2.1. Spectra Extraction First Item

Figure 1 shows the average spectra acquired from the ROI regions in both lab- and field-based hyperspectral sample images, ranging from 400 to 1000 nm. Both spectra were divided into six groups according to each harvesting date. The obtained spectra show a high absorption valley appearing at the 680 nm and 970 nm regions in both spectra plots. It was found that there was no major difference between the two methods.

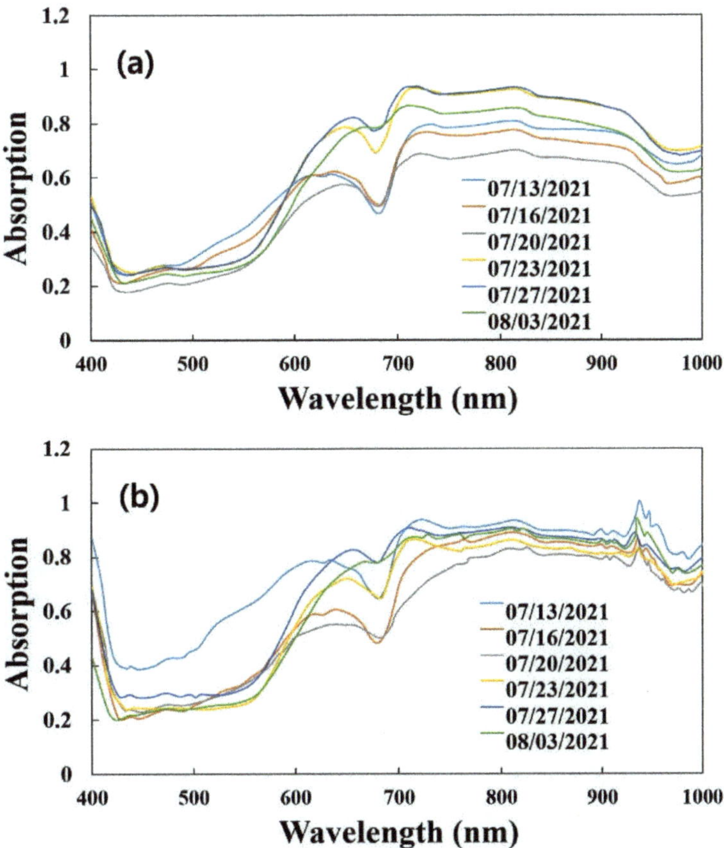

Figure 1. Average spectra acquired form lab- (**a**) and field-based hyperspectral images (**b**).

2.2. Measured Physicochemical Properties

Table 1 shows the measured physicochemical properties of peach 'Madoka'. The ranges of chromaticity (a*), SSC, firmness, and TA were shown, and their average, standard error (SE), standard deviation (STDEV) values were represented, respectively.

Table 1. Measured physicochemical properties of peach 'Madoka'.

Parameters	Range	Average	SE	STEDV
Chromaticity (a*)	−3 to 30	12.88	1.48	8.84
SSC (°Brix)	10 to 18	13.2	0.5	2.47
Firmness (N)	25 to 50	36.77	2.18	11.88
TA (pH)	0.27 to 0.36	0.36	0.02	0.02

2.3. Chromaticity (a*) Prediction

The result of predicting chromaticity (a*) in 'Madoka' is summarized in Table 2. As shown in the table, under lab-based HSI, SGD 1st derivative differentiation shows the highest R^2 value (R^2 = 0.87 at the validation set) in both the calibration and validation set with the lowest bias values. On the other hand, under field-based HSI, smoothing has the highest accuracy (R^2 = 0.85 at the validation set).

Table 2. Model outcome of predicting chromaticity (a*) values of 'Madoka' peaches.

Place	Preprocessing	Calibration Set				Validation Set				RMSECV
		R^2	SEP	RPD	Bias	R^2	SEP	RPD	Bias	
Lab	Raw	0.79	3.49	2.21	−0.30	0.84	4.63	1.64	−0.92	0.02
	Smoothing	0.80	3.46	2.24	0.07	0.85	4.61	1.65	−0.27	0.01
	Savitzky golay 1st	0.89	2.57	3.01	0.11	0.87	4.21	1.81	−0.72	0.01
	Savitzky golay 2nd	0.89	2.51	3.08	−0.02	0.79	5.32	1.43	−0.67	0.01
Field	Raw	0.52	4.83	1.59	−2.23	0.82	4.27	1.81	−2.62	0.02
	Smoothing	0.89	2.54	3.03	0.01	0.85	4.59	1.68	0.65	0.01
	Savitzky golay 1st	0.85	2.93	2.62	0.10	0.84	4.59	1.68	−0.08	0.03
	Savitzky golay 2nd	0.86	2.86	2.68	−0.09	0.78	5.35	1.44	−1.11	0.02

2.4. SSC Prediction

The results of the SSC content analysis of 'Madoka' peaches are summarized in Table 3. As shown in the table, SGD 1st derivative differentiation shows the highest (R^2 = 0.87) accuracy under lab-based HSI application. In addition, smoothing and SGD 2nd derivative differentiation show good performance (R^2 = 0.89 and 0.85) under field-based HSI.

Table 3. Model outcome of predicting SSC concentrations of 'Madoka' peaches.

Place	Preprocessing	Calibration Set				Validation Set				RMSECV
		R^2	SEP	RPD	Bias	R^2	SEP	RPD	Bias	
Lab	Raw	0.77	1.19	2.09	0.00	0.84	1.45	1.66	0.50	0.01
	Smoothing	0.90	0.77	3.23	0.00	0.86	1.39	1.73	0.21	0.01
	Savitzky golay 1st	0.91	0.73	3.40	0.00	0.87	1.37	1.76	0.20	0.01
	Savitzky golay 2nd	0.88	0.87	2.85	0.00	0.82	1.62	1.49	0.22	0.01
Field	Raw	0.63	1.54	1.64	−0.05	0.87	1.36	1.68	−0.11	0.01
	Smoothing	0.84	1.02	2.48	0.00	0.89	1.24	1.85	0.33	0.01
	Savitzky golay 1st	0.78	1.17	2.15	0.01	0.91	1.12	2.05	0.35	0.01
	Savitzky golay 2nd	0.86	0.95	2.66	0.00	0.85	1.40	1.63	0.57	0.01

2.5. Firmness Prediction

The firmness prediction results of 'Madoka' are represented in Table 4. As can be seen, it was observed that the accuracies of the calibration set were relatively much lower than those of the validation set. SGD 2nd derivative differentiation shows the best accuracies (R^2 = 66) under the lab-based HSI application. In the field-based HSI, SGD 1st derivative differentiation shows the highest accuracies (R^2 = 0.68) under field-based HSI.

Table 4. Model outcome of predicting firmness values of 'Madoka' peaches.

Place	Preprocessing	Calibration Set				Validation Set				RMSECV
		R^2	SEP	RPD	Bias	R^2	SEP	RPD	Bias	
Lab	Raw	0.54	8.35	1.48	0.08	0.81	7.56	1.36	−3.15	0.04
	Smoothing	0.53	8.46	1.46	0.08	0.82	7.53	1.37	−3.01	0.04
	Savitzky golay 1st	0.60	7.80	1.58	−0.08	0.82	7.72	1.33	−2.36	0.05
	Savitzky golay 2nd	0.66	7.20	1.72	0.00	0.79	8.00	1.29	−3.52	0.03
Field	Raw	0.50	8.59	1.41	0.11	0.75	9.12	1.18	−2.20	0.07
	Smoothing	0.67	7.00	1.73	0.00	0.51	11.87	0.91	−5.22	0.04
	Savitzky golay 1st	0.68	6.91	1.76	0.00	0.58	11.40	0.94	−3.72	0.06
	Savitzky golay 2nd	0.55	8.14	1.49	−0.32	0.78	8.56	1.26	−1.67	0.04

2.6. TA Prediction

Table 5 shows the results of the TA contents in 'Madoka'. In the lab-based HSI, R^2 values of the calibration and validation set were between 0.5 and 0.7. In the field-based HSI, the model accuracy was about 0.7, which is a slightly higher value than that of the lab-based HSI outcome.

Table 5. Model outcome of predicting TA values in 'Madoka' peaches.

Place	Preprocessing	Calibration Set				Validation Set				RMSECV
		R^2	SEP	RPD	Bias	R^2	SEP	RPD	Bias	
Lab	Raw	0.56	0.07	1.51	0.00	0.68	0.09	1.18	−0.04	0.00
	Smoothing	0.56	0.07	1.50	0.00	0.69	0.09	1.19	−0.03	0.00
	Savitzky golay 1st	0.60	0.07	1.59	0.00	0.68	0.09	1.15	−0.03	0.00
	Savitzky golay 2nd	0.71	0.06	1.86	0.00	0.67	0.09	1.13	−0.03	0.00
Field	Raw	0.73	0.06	1.92	0.00	0.63	0.10	0.90	−0.01	0.00
	Smoothing	0.56	0.07	1.51	0.00	0.75	0.08	1.11	0.01	0.00
	Savitzky golay 1st	0.78	0.05	2.14	0.00	0.74	0.08	1.07	0.00	0.00
	Savitzky golay 2nd	0.70	0.06	1.81	0.00	0.77	0.08	1.14	0.00	0.00

2.7. Visualization of Physicochemical Properties

The pixel-based chemical imaging of the physicochemical component distribution in 'Madoka' was created by the prediction model chosen by perClass Mira software. Figure 2 shows the representative resultant hyperspectral images with an increase in SSC contents and chromaticity (a*) values. The measured values of a* and SSC contents were acquired from the 'Madoka' samples harvested between 13 July and 3 August. The resultant images for firmness and TA content distribution were not shown due to their low R^2 values. In Figure 2, the resultant hyperspectral images were described by a linear gray scale with different intensities being used for each pixel and compared with their RGB images, facilitating improved understanding of the spatial variation in chromaticity (a*) and SSC in the 'Madoka' samples.

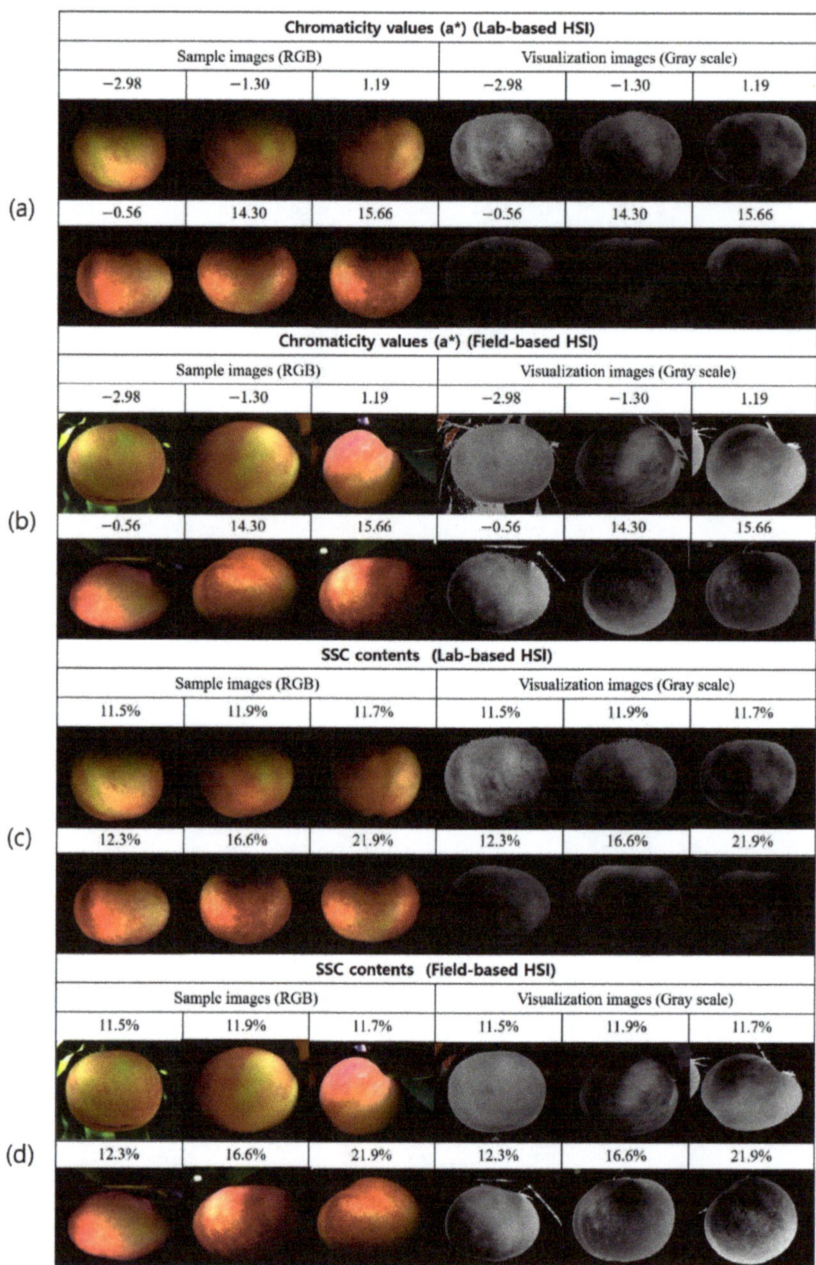

Figure 2. RGB and HSI images for the evaluation of chromaticity (a*) values and SSC in 'Madoka' peaches.

3. Discussion

3.1. Spectral Analysis

Spectral analysis of a target material can provide material information about its chemical composition and physical properties. Every agricultural product includes nearly the same constituents contributing to the absorption (or reflection) spectra, such as chlorophyll, carotenoids, water, proteins, waxes, starches, structural biochemical molecules, and so on [31]. Therefore, these components respond to electromagnetic radiation as a function of wavelengths to their composition and physical properties.

The typical shape and absorption peaks (675 nm and 970 nm) of the 'Madoka' spectra were acquired, as shown in Figure 1 [19,23,32]. The acquired sample spectra presented similar spectral trends with different intensities. The absorption peaks at 675 nm and 970 nm are related to chlorophyll a and the second overtone stretching of O-H vibrations of water and sugar components [19,23]. Increasing the time until harvest leads to the maturity of yellow 'Madoka'. This decreases the chlorophyll content and can be used to evaluate the maturity of fresh peach fruits [33]. From the spectral data in both lab- and field-based HSI, it was difficult to determinate the criteria for high quality 'Madoka' as well as distinguish the critical differences between the two methods.

3.2. Model Evaluation Factors

The model accuracy was evaluated by conducting a quantitative analysis based on four preprocessing methods with evaluation factors such as the coefficient of determination (R^2), standard error of prediction (SEP), residual prediction deviation (RPD), root mean square error of cross validation (RMSECV), and bias values. These evaluation factors have been commonly used for cross-validation and robust model establishment [4,6,34].

SEP indicates the prediction ability of the used model, and RPD is the ratio of the standard deviation of the validation set divided by the root mean square errors of prediction (RMSEP), describing by which component its prediction accuracy has been affected compared to the mean composition for the total samples [34]. RPD can also assess NIR calibration performance. In general, an excellent prediction model can be evaluated by the following criteria: high R^2, high RPD, low bias, low SEP, and low RMSECV [35,36].

These factors are intensively used to evaluate the performance of chemometric methods such as partial least squares-discriminant analysis (PLS-DA), principal component analysis (PCA), least-squares support-vector machine (LS-SVM), and so on [5,36,37]. Among them, the PLS based chemometric technique with the four preprocessing methods may be automatically selected by perClass Mira software because they have been intensively applied to NDE applications for assessing fresh peach quality with high accuracy [1,19–23,38,39]. Therefore, an appropriate model was selected for the NDE of the physicochemical properties of 'Madoka' peaches, and their pixel-based chemical images were created.

3.3. Analysis of Model Prediction

The model prediction accuracies of the physicochemical properties in 'Madoka' are summarized in Tables 2–5. Each preprocessing method includes the calculated values of the above evaluation factors under lab- and field-based HSI applications. Besides the preprocessing methods used here, other methods, such as normalization, multiplicative scatter correction, and standard normal variate, can be utilized to develop chemometric models to improve model performance. However, only four preprocessing methods could be selected by the software used.

The overall values of the firmness and TA indices gradually decrease as the time until harvesting increases; however, chromaticity (a*) and SSC content values show an opposite trend [33,40]. This color change on the peaches' surface is consistent with their decreasing chlorophyll a content and increasing anthocyanin components [41]. Therefore, the a* component can describe the dramatic maturity stage change in red colored fruits changing from green (negative) to red (positive) color [42]. As shown in Table 2, the accuracies of the a* values of lab- and field-based HSI results were nearly identical and statistically significant (R^2 = 0.84 to 0.87).

Conversely, the firmness and TA prediction exhibited a major difference in R^2 values between the calibration (R^2 = 0.6 to 0.7) and validation set (R^2 = 0.7 to 0.85). Moreover, their prediction accuracies were much lower than chromaticity (a*) and SSC prediction. This indicates that the performance of firmness and TA contents, predicted by the automatically selected model, was statistically non-significant. In the current study, a single model was developed and applied to simultaneous identification of the physicochemical qualities of 'Madoka' for rapid and real-time field application using a commercial HSI camera and software. Based on the above analysis, we concluded that the developed model cannot perform multiple quality factors; thus, a different approach for developing optimal models for each quality factor is needed for the simultaneous prediction of 'Madoka' qualities.

3.4. Visualization of Physicochemical Properties

The conventional spectroscopy technique can create one set of spectral data per individual target. However, hyperspectral images can produce full spectral data from every pixel of the target image. This is the critical benefit of HSI and can provide an interpretation of the particular chemical components in a specific pixel of the target image. Therefore, the sample images obtained under two different environments can show different spatial and spectral information from their pixel data.

As shown in Figure 2, the RGB images were used for comparison with their hyperspectral images, and the resultant hyperspectral images present different a* values and SSC on a linear grey scale with different intensities being used for each pixel. This facilitates improved interpretation of the spatial variation in a* values and SSC in 'Madoka' samples. The differences of the a* values and SSC are hardly distinguished by the naked eye, except for yellow 'Madoka' samples. However, the developed model could clearly classify the different a* and SSC values based on the number of black and white pixels, which increase with their levels. As seen in Figure 2a, it is particularly difficult to distinguish the peach's maturity in an RGB image with the naked eye, except for an a* value of −2.98, because red and pink colors are irregularly mixed with the yellow regions of the RGB sample images. However, the developed model detects and differentiates the increasing a* values with a grey scale. This can be also observed in the SSC (see Figure 2c and d) from 12.3 to 21.9%.

Although the a* values and SSC contents and their different levels could be detected, regular intervals of a* values and SSC could not be achieved in this study due to their fast ripening characteristics. According to the measured SSC values, it seems that the 'Madoka' samples used ripened between 23 and 27 July. Therefore, the hyperspectral images for gradual variation in peach maturity could not be obtained with the 'Madoka' samples used. Moreover, the dramatic change in the grey scale can be observed in Figure 2c; thus, color difference between the two groups reveals the SSC level. Consequently, the field application of the commercial HSI camera and software used in this study has considerable potential for the NDE of chromaticity and SSC in 'Madoka' peaches.

4. Materials and Methods

4.1. Sample Preparation

The 'Madoka' peaches utilized in this study were grown at an orchard located in Jiphyeon-myeon, Jinju-si, Gyeongsangnam-do, Republic of Korea. Hyperspectral images were taken from 13 July to 3 August 2021 between 10:00 am and 12:00 during the growing season of peach plants on sunny days. Detailed weather information can be found at Korea Meteorological Administration (https://data.kma.go.kr/). In all, 180 'Madoka' samples were collected and divided into 2 subsets: a calibration set (70 % of the total samples) for model development and a validation set (30 % of total samples). After taking field-based images of the samples, they were then taken to an indoor laboratory for the performance comparison between the two different methods.

4.2. HSI System

Both a lab- and field-based HSI system were utilized for taking indoor and outdoor sample images. Their conceptual diagrams are shown in Figure 3. The main components of the laboratory-based system consist of a Vis/NIR hyperspectral camera (Specim IQ, Specim Ltd., Oulu, Finland), 2 650 W halogen lamps (H1000, FOMEX, Seoul, Republic of Korea), a calibration plate for relative reflectance correction (99% barium sulfate [$BaSO_4$]

reference plate), and a darkened room for sample image acquisition. For the field-based HSI application, the same hyperspectral camera was used only under sun light conditions. Then, these sample images were analyzed.

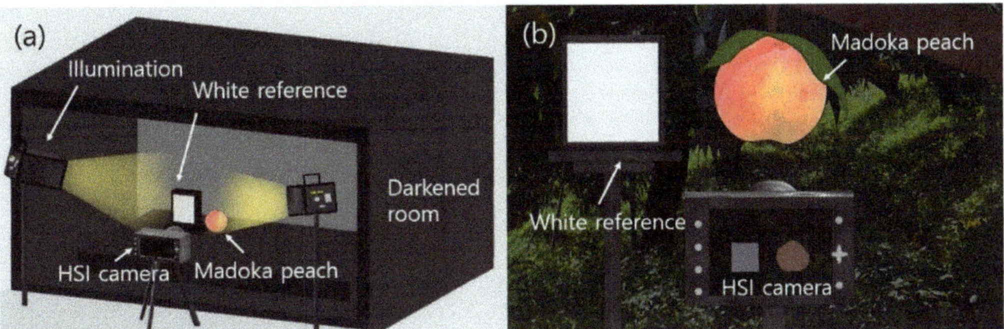

Figure 3. Conceptual diagram of the lab- (**a**) and field-based hyperspectral imaging system (**b**).

In general, the line-scan method is used for agricultural applications because food products linearly move along a production line. Moreover, online application is well suited to line scanning. However, in the current study, a hyperspectral camera for area scanning, which is able to obtain two dimensional hyperspectral images (x and y plane) with full spatial data of a target at once, was selected due to its rapid image acquisition ability for field application [43,44].

Halogen lamps are a typical broadband light source, which can perform reflectance and transmittable imaging because they produce a stable and continuous spectrum from the visible to infrared wavelength region without sudden peaks [3]. The Vis/NIR region has been widely used for HSI applications in the fields of agricultural NDE owing to its strong response to the major chemical bonds of C–H, N–H, and O–H functional groups at specific frequencies. Therefore, halogen lamps were selected as the illumination source for our analysis. In addition, as shown in Figure 3a, a dark curtain was used to prevent external light noise.

4.3. Spectral Calibration and Image Acquisition

The spectral range of the hyperspectral camera used was 400 nm to 1000 nm with a 2.9 nm interval (total of 204 wavebands). The region of interest (ROI) was composed of 512 × 512 pixels per single sample image. The background regions of the acquired sample images were removed by perClass Mira software because the background pixels include totally irrelevant spectral information from the sample pixels. The field of view of the camera was 200 × 200 mm at a distance of 300 mm. The mean spectra within the ROI region were then acquired to analyze both sample types.

For every hyperspectral image, the sample and white reference images were simultaneously taken for spectral (reflectance) calibration. The reflectance calibration is an important process to minimize the quantum effect of the imaging sensor mounted in the hyperspectral camera [45]. It produces an unwanted irregular radiance even when taking the same sample images under the same conditions [7]. To prevent this, a white reference and dark current images are typically captured, and the relative reflectance of a target is then calculated. In this study, white reference images were obtained without current images, because the dark current image was automatically obtained by the camera used. Spectral calibration was performed for both lab- and field-based imaging.

4.4. Measurement of Physicochemical Properties

After obtaining the sample images, the physicochemical properties of 'Madoka' peaches such as chromaticity, SSC, firmness, and TA were measured for use as a calibration set for model development. According to USDA peach grade and standards, the best peach grade should have no less than one-third of its surface showing a pink or red color [46]. To apply a red color to the digital imaging system for peach grading, CIELAB color space has been widely used [41,47,48]. Among various color parameters, parameter a* is closely related to peach grading, with the axis indicating negative values for immature green and positive values for mature red [40,48]. Therefore, in this study, the a* parameter was selected as a major indicator for peach grading and was measured using a colorimeter (CR-400, Minolta, Osaka, Japan).

The firmness and SSC are also critical quality factors which can directly influence customers when purchasing fresh peaches [49]. Firmness was measured 3 times using a Sun Rheometer with an 8 mm probe (CR-100, Sun Scientific Inc., Tokyo, Japan) at a loading rate of 2 mm/s after taking all of the lab- and field-based hyperspectral peach sample images. The measuring point of firmness was the same location where the hyperspectral images of the 'Madoka' samples were taken. The SSC was measured using a digital refractometer (PAL-1, Atago Co. Ltd., Tokyo, Japan) after all samples were ground to juice. The refractometer used had a refractive index accuracy of $\pm$ 0.2, and the °Brix (%) range was 0 to 53% with a 0.1 % Brix resolution at room temperature. The TA was measured using a pH meter (BP3001, Trans Instruments, Jalan Kilang Barat, Singapore). Titration was accomplished by a 1 mL pulp diluted into 80 mL of distilled water. Then, NaOH (0.1 N) was added at a rate of 1 mL/min to reach pH 8.3.

4.5. Model Accuracy Parameters

In this study, the model accuracy was evaluated by R^2, SEP, RPD, RMSECV, and bias values. R^2 values range from zero to one and can measure how well a statistical model predicts a result. A high R^2 value corresponds to a model that describes the variations in its data fit well [8,44]. SEP is the parameter commonly used in the HSI technique to describe the prediction ability of the developed model. Therefore, SEP can be compared to the expected accuracy to decide whether or not the developed method is acceptable. RPD value is the ratio of the standard deviation to the root mean square errors of prediction, explaining by which factor the model accuracy has been increased compared to utilizing the mean composition for total samples. In general, quantitative prediction can be achieved when the RPD value is higher than 2. RMSECV is calculated to evaluate the performance of the partial least square model. The optimal wavelength combinations for predicting physicochemical properties can be obtained by the lowest RMSECV values [36].

4.6. Visualization

Prior to model development, smoothing and Savitzky–Golay derivative (SGD) 1st and 2nd derivative differentiation preprocessing techniques were applied to the raw data to improve model forecasting performance by correcting and lowering the baseline shift and multiplicative scatter effects generated from electrical devices and irregular intensities of light sources [14]. As previously mentioned, after background removal of the 'Madoka' sample images, the spectra of the ROI were extracted, and their spectral data were used for model development. Then, an appropriate machine learning method was automatically selected by perClass Mira software for rapid and real-time image analysis [50]. The flow chart of the model development process is represented in Figure 4.

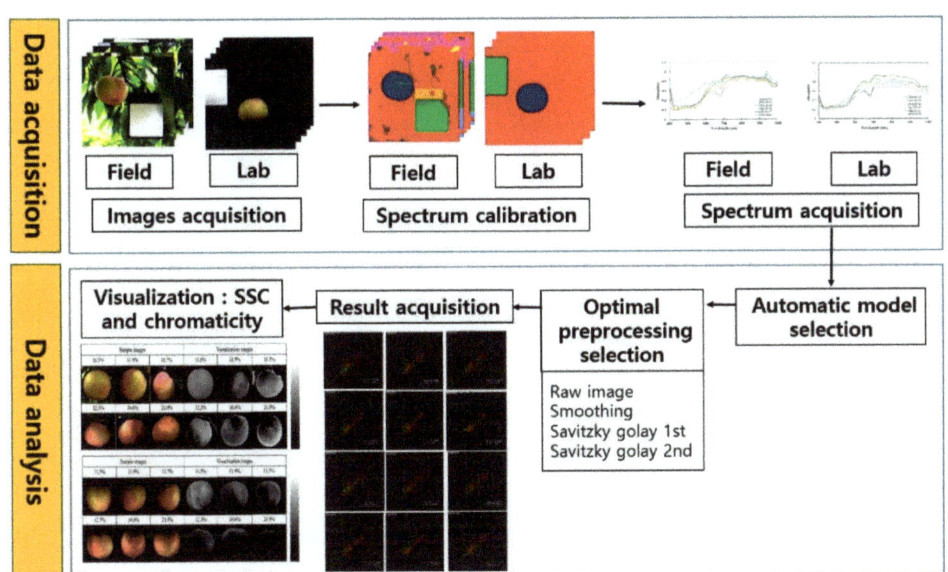

Figure 4. Flow chart of data processing procedures for prediction of physicochemical properties of 'Madoka' peaches based on lab- and field-based HSI applications.

The novel benefit of HSI is that chemical mapping of a component distribution in a target material can be accomplished by various chemometric models constructed by simultaneous measurement of spatial and spectral information [4,43,44]. Thus, in this study, pixel-based chemical images based on the spatial distribution of the physicochemical components in 'Madoka' samples were obtained from both lab- and field-based HSI information. Spectral data in each pixel of the sample images were extracted and utilized to analyze model prediction performance. Consequently, the prediction results were shown in color, indicating different physicochemical concentrations in the sample images, and the field-based application outcome was compared with the lab-based result to verify its prediction accuracy. All data analyses for model development, preprocessing, and visualization were conducted by perClass Mira software.

5. Conclusions

This study investigated the field application feasibility for the NDE of physicochemical properties in 'Madoka' peaches using a Vis/NIR HSI system integrated with commercial HSI analysis software. Before hyperspectral image acquisition, spectral calibration was conducted by white balance reference to prevent the quantum effect on the imaging sensor of the camera. Then, the absorption spectra of the 'Madoka' samples were extracted from the sample images and the spectral features of both groups were analyzed.

To analyze the physicochemical properties (chromaticity (a*), SSC, firmness, and TA), a machine learning model was automatically selected by perClass Mira software. Three preprocessing methods (smoothing, SGD 1st, and SGD 2nd) were also applied to the selected model to improve the model accuracy. The performance of the lab-based HSI analysis was conducted by using the same samples used in the field-based application.

Accordingly, statistically significant R^2 values (about 0.85) were found in the a* values and SSC content; however, the firmness and TA content exhibited much lower R^2 values (0.5 to 0.7). There was also a major difference in R^2 values between the calibration and validation set. Based on these results, resultant chemical distribution images of a* values and SSC were constructed by the selected model. The final outcome clearly demonstrates that the HSI camera and commercial software used have potential for the field NDE for analyzing physicochemical properties in 'Madoka' peaches.

Author Contributions: Conceptualization, J.G.K.; methodology, J.G.K., G.K., J.H.J., and S.K.L.; investigation, K.E.J. and M.H.S.; writing—original draft preparation, K.E.J., D.K., and G.K.; writing—review and editing, J.G.K. and G.K.; supervision, J.G.K.; funding acquisition, J.G.C. and J.G.K. All authors have read and agreed to the published version of the manuscript.

Funding: This research was funded by the Cooperative Research Program for Agriculture Science and Technology Development (Project No. PJ014988042022), Rural Development Administration, Republic of Korea.

Institutional Review Board Statement: Not applicable.

Informed Consent Statement: Not applicable.

Data Availability Statement: Not applicable.

Conflicts of Interest: The authors declare no conflict of interest.

References

1. Mangalraj, P.; Cho, B.K. Recent Trends and Advances in Hyperspectral Imaging Techniques to Estimate Solar Induced Fluorescence for Plant Phenotyping. *Ecol. Indic.* **2022**, *137*, 108721. [CrossRef]
2. Lu, B.; Dao, P.D.; Liu, J.; He, Y.; Shang, J. Recent Advances of Hyperspectral Imaging Technology and Applications in Agriculture. *Remote Sens.* **2020**, *12*, 2659. [CrossRef]
3. Qin, J.; Chao, K.; Kim, M.S.; Lu, R.; Burks, T.F. Hyperspectral and Multispectral Imaging for Evaluating Food Safety and Quality. *J. Food Eng.* **2013**, *118*, 157–171. [CrossRef]
4. Kim, G.; Lee, H.; Cho, B.-K.; Baek, I.; Kim, M.S. Quantitative Evaluation of Food-Waste Components in Organic Fertilizer Using Visible–Near-Infrared Hyperspectral Imaging. *Appl. Sci.* **2021**, *11*, 8201. [CrossRef]
5. Faqeerzada, M.A.; Perez, M.; Lohumi, S.; Lee, H.; Kim, G.; Wakholi, C.; Joshi, R.; Cho, B.-K. Online Application of a Hyperspectral Imaging System for the Sorting of Adulterated Almonds. *Appl. Sci.* **2020**, *10*, 6569. [CrossRef]
6. Kim, G.; Baek, I.; Stocker, M.D.; Smith, J.E.; Van Tassell, A.L.; Qin, J.; Chan, D.E.; Pachepsky, Y.; Kim, M.S. Hyperspectral Imaging from a Multipurpose Floating Platform to Estimate Chlorophyll-a Concentrations in Irrigation Pond Water. *Remote Sens.* **2020**, *12*, 2070. [CrossRef]
7. Faqeerzada, M.A.; Lohumi, S.; Kim, G.; Joshi, R.; Lee, H.; Kim, M.S.; Cho, B.-K. Hyperspectral Shortwave Infrared Image Analysis for Detection of Adulterants in Almond Powder with One-Class Classification Method. *Sensors* **2020**, *20*, 5855. [CrossRef]
8. Kim, G.; Lee, H.; Baek, I.; Cho, B.K.; Kim, M.S. Quantitative Detection of Benzoyl Peroxide in Wheat Flour Using Line-Scan Short-Wave Infrared Hyperspectral Imaging. *Sens. Actuators B Chem.* **2022**, *352*, 130997. [CrossRef]
9. Kim, G.; Lee, H.; Baek, I.; Cho, B.K.; Kim, M.S. Short-Wave Infrared Hyperspectral Imaging System for Nondestructive Evaluation of Powdered Food. *J. Biosyst. Eng.* **2022**, *47*, 223–232. [CrossRef]
10. Wang, N.N.; Sun, D.W.; Yang, Y.C.; Pu, H.; Zhu, Z. Recent Advances in the Application of Hyperspectral Imaging for Evaluating Fruit Quality. *Food Anal. Methods* **2016**, *9*, 178–191. [CrossRef]
11. Access, O. Prediction of Moisture Contents in Green Peppers Using Hyperspectral Imaging Based on a Polarized Lighting System. *Korean J. Agric. Sci.* **2020**, *47*, 995–1010.
12. Seo, Y.; Lee, H.; Bae, H.J.; Park, E.; Lim, H.S.; Kim, M.S.; Cho, B.K. Optimized Multivariate Analysis for the Discrimination of Cucumber Green Mosaic Mottle Virus-Infected Watermelon Seeds Based on Spectral Imaging. *J. Biosyst. Eng.* **2019**, *44*, 95–102. [CrossRef]
13. Wakholi, C.; Kandpal, L.M.; Lee, H.; Bae, H.; Park, E.; Kim, M.S.; Mo, C.; Lee, W.H.; Cho, B.K. Rapid Assessment of Corn Seed Viability Using Short Wave Infrared Line-Scan Hyperspectral Imaging and Chemometrics. *Sens. Actuators B Chem.* **2018**, *255*, 498–507. [CrossRef]
14. Munawar, A.A.; von Hörsten, D.; Wegener, J.K.; Pawelzik, E.; Mörlein, D. Rapid and Non-Destructive Prediction of Mango Quality Attributes Using Fourier Transform near Infrared Spectroscopy and Chemometrics. *Eng. Agric. Environ. Food* **2016**, *9*, 208–215. [CrossRef]
15. Kilgus, J.; Zimmerleiter, R.; Duswald, K.; Hinterleitner, F.; Langer, G.; Brandstetter, M. Application of a Novel Low-Cost Hyperspectral Imaging Setup Operating in the Mid-Infrared Region. In *Proceedings of the EUROSENSORS 2018, Graz, Austria, 9–12 September 2018*; MDPI: Basel, Switzerland, 2018; p. 800.
16. Holmer, A.; Tetschke, F.; Marotz, J.; Malberg, H.; Markgraf, W.; Thiele, C.; Kulcke, A. Oxygenation and Perfusion Monitoring with a Hyperspectral Camera System for Chemical Based Tissue Analysis of Skin and Organs. *Physiol. Meas.* **2016**, *37*, 2064–2078. [CrossRef]

17. Somaratne, G.; Reis, M.M.; Ferrua, M.J.; Ye, A.; Nau, F.; Floury, J.; Dupont, D.; Singh, R.P.; Singh, J. Mapping the Spatiotemporal Distribution of Acid and Moisture in Food Structures during Gastric Juice Diffusion Using Hyperspectral Imaging. *J. Agric. Food Chem.* **2019**, *67*, 9399–9410. [CrossRef]
18. Sricharoonratana, M.; Thompson, A.K.; Teerachaichayut, S. Use of near Infrared Hyperspectral Imaging as a Nondestructive Method of Determining and Classifying Shelf Life of Cakes. *LWT* **2021**, *136*, 110369. [CrossRef]
19. Liu, Q.; Zhou, D.; Tu, S.; Xiao, H.; Zhang, B.; Sun, Y.; Pan, L.; Tu, K. Quantitative Visualization of Fungal Contamination in Peach Fruit Using Hyperspectral Imaging. *Food Anal. Methods* **2020**, *13*, 1262–1270. [CrossRef]
20. Shao, Y.; Wang, Y.; Xuan, G. In-Field and Non-Invasive Determination of Internal Quality and Ripeness Stages of Feicheng Peach Using a Portable Hyperspectral Imager. *Biosyst. Eng.* **2021**, *212*, 115–125. [CrossRef]
21. Lu, R.; Peng, Y. Hyperspectral Scattering for Assessing Peach Fruit Firmness. *Biosyst. Eng.* **2006**, *93*, 161–171. [CrossRef]
22. Abenina, M.I.A.; Maja, J.M.; Cutulle, M.; Melgar, J.C.; Liu, H. Prediction of Potassium in Peach Leaves Using Hyperspectral Imaging and Multivariate Analysis. *AgriEngineering* **2022**, *4*, 27. [CrossRef]
23. Xuan, G.; Gao, C.; Shao, Y. Spectral and Image Analysis of Hyperspectral Data for Internal and External Quality Assessment of Peach Fruit. *Spectrochim. Acta-Part A Mol. Biomol. Spectrosc.* **2022**, *272*, 121016. [CrossRef] [PubMed]
24. Yang, B.; Gao, Y.; Yan, Q.; Qi, L.; Zhu, Y.; Wang, B. Estimation Method of Soluble Solid Content in Peach Based on Deep Features of Hyperspectral Imagery. *Sensors* **2020**, *20*, 5021. [CrossRef]
25. Tilahun, S.; Jeong, M.J.; Choi, H.R.; Baek, M.W.; Hong, J.S.; Jeong, C.S. Prestorage High CO2 and 1-MCP Treatment Reduce Chilling Injury, Prolong Storability, and Maintain Sensory Qualities and Antioxidant Activities of "Madoka" Peach Fruit. *Front. Nutr.* **2022**, *9*, 1–11. [CrossRef] [PubMed]
26. Lurie, S.; Crisosto, C.H. Chilling Injury in Peach and Nectarine. *Postharvest Biol. Technol.* **2005**, *37*, 195–208. [CrossRef]
27. Lee, E.J. Chilling Injury and Phytochemical Composition of Peach Fruits as Affected by High Carbon Dioxide Treatment before Cold Storage. *Hortic. Environ. Biotechnol.* **2014**, *55*, 190–195. [CrossRef]
28. Brizzolara, S.; Manganaris, G.A.; Fotopoulos, V.; Watkins, C.B.; Tonutti, P. Primary Metabolism in Fresh Fruits During Storage. *Front. Plant Sci.* **2020**, *11*, 1–16. [CrossRef]
29. Jagadish, S.V.K.; Way, D.A.; Sharkey, T.D. Plant Heat Stress: Concepts Directing Future Research. *Plant Cell Environ.* **2021**, *44*, 1992–2005. [CrossRef]
30. Pan, L.; Zhang, Q.; Zhang, W.; Sun, Y.; Hu, P.; Tu, K. Detection of Cold Injury in Peaches by Hyperspectral Reflectance Imaging and Artificial Neural Network. *Food Chem.* **2016**, *192*, 134–141. [CrossRef]
31. Kokaly, R.F.; Despain, D.G.; Clark, R.N.; Livo, K.E. Mapping Vegetation in Yellowstone National Park Using Spectral Feature Analysis of AVIRIS Data. *Remote Sens. Environ.* **2003**, *84*, 437–456. [CrossRef]
32. Lleó, L.; Roger, J.M.; Herrero-Langreo, A.; Diezma-Iglesias, B.; Barreiro, P. Comparison of Multispectral Indexes Extracted from Hyperspectral Images for the Assessment of Fruit Ripening. *J. Food Eng.* **2011**, *104*, 612–620. [CrossRef]
33. Li, B.; Yin, H.; Liu, Y.-d.; Zhang, F.; Yang, A.-k.; Su, C.-t.; Ou-yang, A.-g. Study on Qualitative Impact Damage of Yellow Peaches Using the Combined Hyperspectral and Physicochemical Indicators Method. *J. Mol. Struct.* **2022**, *1265*, 133407. [CrossRef]
34. Li, S.; Shao, Q.; Lu, Z.; Duan, C.; Yi, H.; Su, L. Rapid Determination of Crocins in Saffron by Near-Infrared Spectroscopy Combined with Chemometric Techniques. *Spectrochim. Acta-Part A Mol. Biomol. Spectrosc.* **2018**, *190*, 283–289. [CrossRef] [PubMed]
35. Liu, J.; Han, J.; Xie, J.; Wang, H.; Tong, W.; Ba, Y. Assessing Heavy Metal Concentrations in Earth-Cumulic-Orthic-Anthrosols Soils Using Vis-NIR Spectroscopy Transform Coupled with Chemometrics. *Spectrochim. Acta-Part A Mol. Biomol. Spectrosc.* **2020**, *226*, 117639. [CrossRef]
36. Li, H.; Chen, Q.; Zhao, J.; Wu, M. Nondestructive Detection of Total Volatile Basic Nitrogen (TVB-N) Content in Pork Meat by Integrating Hyperspectral Imaging and Colorimetric Sensor Combined with a Nonlinear Data Fusion. *LWT* **2015**, *63*, 268–274. [CrossRef]
37. Roggo, Y.; Chalus, P.; Maurer, L.; Lema-Martinez, C.; Edmond, A.; Jent, N. A Review of near Infrared Spectroscopy and Chemometrics in Pharmaceutical Technologies. *J. Pharm. Biomed. Anal.* **2007**, *44*, 683–700. [CrossRef]
38. Shao, Y.; Bao, Y.; He, Y. Visible/Near-Infrared Spectra for Linear and Nonlinear Calibrations: A Case to Predict Soluble Solids Contents and PH Value in Peach. *Food Bioprocess Technol.* **2011**, *4*, 1376–1383. [CrossRef]
39. Sun, Y.; Xiao, H.; Tu, S.; Sun, K.; Pan, L.; Tu, K. Detecting Decayed Peach Using a Rotating Hyperspectral Imaging Testbed. *LWT-Food Sci. Technol.* **2018**, *87*, 326–332. [CrossRef]
40. Miller, B.K.; Delwiche, M.J. Color Vision System for Peach Grading. *Trans. Am. Soc. Agric. Eng.* **1989**, *32*, 1484–1490. [CrossRef]
41. Bible, B.B.; Singha, S. Canopy Position Influences CIELAB Coordinates of Peach Color. *HortScience* **1993**, *28*, 992–993. [CrossRef]
42. Saad, A.; Ibrahim, A.; El-Bialee, N. Internal Quality Assessment of Tomato Fruits Using Image Color Analysis. *Agric. Eng. Int. CIGR J.* **2016**, *18*, 339–352.
43. Qin, J.; Kim, M.S.; Chao, K.; Chan, D.E.; Delwiche, S.R.; Cho, B.K. Line-Scan Hyperspectral Imaging Techniques for Food Safety and Quality Applications. *Appl. Sci.* **2017**, *7*, 125. [CrossRef]
44. Kim, M.S.; Chen, Y.R.; Mehl, P.M. Hyperspectral Reflectance and Fluorescence Imaging System for Food Quality and Safety. *Trans. Am. Soc. Agric. Eng.* **2001**, *44*, 721–729. [CrossRef]
45. Huang, M.; Kim, M.S.; Delwiche, S.R.; Chao, K.; Qin, J.; Mo, C.; Esquerre, C.; Zhu, Q. Quantitative Analysis of Melamine in Milk Powders Using Near-Infrared Hyperspectral Imaging and Band Ratio. *J. Food Eng.* **2016**, *181*, 10–19. [CrossRef]

46. D'Souza, M.C.; Singha, S.; Ingle, M. Lycopene Concentration of Tomato Fruit Can Be Estimated from Chromaticity Values. *HortScience* **1992**, *27*, 465–466. [CrossRef]
47. Cáceres, D.; Díaz, M.; Shinya, P.; Infante, R. Assessment of Peach Internal Flesh Browning through Colorimetric Measures. *Postharvest Biol. Technol.* **2016**, *111*, 48–52. [CrossRef]
48. Mrázová, M.; Rampáčková, E.; Šnurkovič, P.; Ondrášek, I.; Nečas, T.; Ercisli, S. Determination of Selected Beneficial Substances in Peach Fruits. *Sustainability* **2021**, *13*, 14028. [CrossRef]
49. Muhua, L.; Peng, F.; Renfa, C. Non-Destructive Estimation Peach SSC and Firmness by Mutispectral Reflectance Imaging. New Zeal. *J. Agric. Res.* **2007**, *50*, 601–608. [CrossRef]
50. Mishra, P.; Sytsma, M.; Chauhan, A.; Polder, G.; Pekkeriet, E. All-in-One: A Spectral Imaging Laboratory System for Standardised Automated Image Acquisition and Real-Time Spectral Model Deployment. *Anal. Chim. Acta* **2022**, *1190*, 339235. [CrossRef] [PubMed]

Article

Cultivar and Postharvest Storage Duration Influence Fruit Quality, Nutritional and Phytochemical Profiles of Soilless-Grown Cantaloupe and Honeydew Melons

Boitshepo L. Pulela [1,2], Martin M. Maboko [1], Puffy Soundy [1] and Stephen O. Amoo [2,3,4,*]

[1] Department of Crop Sciences, Tshwane University of Technology, Private Bag X680, Pretoria 0001, South Africa
[2] Agricultural Research Council, Vegetables, Industrial and Medicinal Plants, Private Bag X293, Pretoria 0001, South Africa
[3] Department of Botany and Plant Biotechnology, University of Johannesburg, P.O. Box 524, Auckland Park, Johannesburg 2006, South Africa
[4] Indigenous Knowledge Systems Centre, Faculty of Natural and Agricultural Sciences, North-West University, Private Bag X2046, Mmabatho 2790, South Africa
* Correspondence: amoos@arc.agric.za

Citation: Pulela, B.L.; Maboko, M.M.; Soundy, P.; Amoo, S.O. Cultivar and Postharvest Storage Duration Influence Fruit Quality, Nutritional and Phytochemical Profiles of Soilless-Grown Cantaloupe and Honeydew Melons. *Plants* **2022**, *11*, 2136. https://doi.org/10.3390/plants11162136

Academic Editors: Lord Abbey, Josephine Ampofo and Mason MacDonald

Received: 29 July 2022
Accepted: 11 August 2022
Published: 17 August 2022

Abstract: There is an increasing demand for sweet melon (*Cucumis melo* L.) fruit in fruit and vegetable markets due to its nutritional content, resulting in different cultivars being grown in different production systems. This study evaluated the nutritional and phytochemical contents of soilless-grown cantaloupe and honeydew sweet melon cultivars at harvest and postharvest. At harvest, vitamin C and β-carotene concentrations were higher in orange-fleshed (cantaloupe) cvs. Magritte, Divine, Majestic, Cyclone, MAB 79001, E25F.00185, E25F.00075 and Adore, compared to green-fleshed (honeydew) cvs. Honey Brew and Honey Star. The zinc (Zn), phosphorus (P), potassium (K), magnesium (Mg) and calcium (Ca) contents were higher in orange-fleshed compared to green-fleshed cultivars. Total phenolics content (TPC) in cv. E25F.00075 was the highest (2.87 mg GAE·g^{-1} dry weight). A significant, positive, correlation occurred between β-carotene and Zn, P, K, Ca and Mg contents. Postharvest storage duration affected TPC and total soluble solid content. The interaction of cultivar and postharvest storage duration affected flavonoid, vitamin C and β-carotene contents, free radical scavenging activity and fruit juice pH. Vitamin C and β-carotene contents decreased with increased postharvest storage duration while flavonoid content increased. The cantaloupe cultivars performed significantly better compared to the honeydew cultivars as evident in their high mineral element content, and vitamin C and β-carotene concentrations. Selection of appropriate cultivars in a production system should consider variation in nutritional traits of cultivars and postharvest storage duration. Soilless production of sweet melon cultivars in tunnels offers a viable alternative to open field to produce high-quality melons at harvest and postharvest.

Keywords: antioxidant; β-carotene *Cucumis melo*; flavonoids; mineral element; vitamin C

1. Introduction

Sweet melons (*Cucumis melo* L., family: Cucurbitaceae), commonly known as spanspek or muskmelons, have become a popularly consumed fresh fruit across the world. Although most melons are grown in open fields, commercial production of melons in soilless culture under greenhouse conditions is becoming a preferred option among farmers to improve yield [1–3]. Changes in weather conditions and availability of land and suitable soil type coupled with build-up of soil-borne pathogens and fluctuations in available nutrient supply are among the challenges associated with field-based monoculture systems [4]. Protected soilless culture alleviates some of these problems while extending the harvest period with the potential to increase yield compared to production under open field conditions [5].

Improved fruit yield per unit area may be achieved in soilless culture under high-tunnel protection than in a field because plants are arranged more uniformly, large gaps between plants and rows are avoided and light interception is optimized [6]. Soilless culture can produce fruit and vegetables with better taste, color, texture and higher nutritional value than those from soil cultivation [7]. Melon growers should consider soilless culture under protection or greenhouse conditions that can provide a consistent supply of high-quality fruit with low risk of crop failure. Systems such as non-temperature controlled (NTC) plastic tunnels can be employed as they use natural ventilation to reduce temperature [4,8]. However, the use of soilless culture under protection does not automatically result in production of high-quality melons. High yields and quality are dependent on cultivar choice, crop management and growing season [9]. Sweet melon cultivars vary in development and yield in NTC plastic tunnels [10], with no information on nutritional and phytochemical content at harvest and postharvest.

Orange-fleshed cantaloupe and green-fleshed honeydew melons are rich sources of β-carotene, vitamin C (ascorbic acid), vitamin E and folic acid [11–13]. Melons have low fat and sodium with no cholesterol [11], making them a good addition to a healthy diet. There is increased consumer awareness regarding melon fruit with good visual and eating quality, health-promoting compounds and properties such as antioxidant capacity. Unlike cooked vegetables in which nutritional values become altered during cooking, the nutritional profiles of both orange- and green-fleshed sweet melon fruit are retained due to being eaten raw.

The increase in demand for sweet melon fruit has resulted in production and release of new sweet melon varieties. Nutritional content can vary among varieties, and the production system used to grow them [14–16]. This study evaluated nutritional and phytochemical qualities at harvest and after postharvest storage of fruit from sweet melon cultivars grown in soilless culture.

2. Results and Discussion

2.1. *Effect of Cultivars on Fruit Nutritional and Phytochemical Quality at Harvest*

At harvest, fruit mineral element content varied among cultivars (Table 1). Cultivar Magritte had the highest fruit potassium and phosphorus contents. Cultivar Majestic had the highest magnesium and calcium contents, which were 2-fold that of the green-fleshed sweet melons, cvs. Honey Brew and Honey Star. All orange-fleshed cultivars had higher Zn content compared to the green-fleshed cultivars. Selection of orange-fleshed cultivars may be a particularly important consideration in sub-Saharan Africa and South Asia where zinc deficiency is prevalent [17]. Green-fleshed cvs. Honey Brew and Honey Star contained the lowest levels of all mineral elements. Variations in fruit mineral element content might thus be linked to cultivar differences.

The β-carotene content in harvested fruit, which is responsible for the orange color pigments in orange-fleshed sweet melon cultivars, was higher, as expected, in orange-fleshed cultivars (Figure 1). Honey Brew and Honey Star cvs. have a pale green color and lack orange color pigments. Thus, the concentration of β-carotene is dependent on the cultivar and fruit flesh color, which is linked to the pigments present in the fruit. Pigment (carotenoid and chlorophyll) compositions are largely different between cantaloupe and honeydew fruits [16]. With some limitations, color may be used as an indicator for β-carotene-rich sweet melon cultivars. Cantaloupe fruits do not contain chlorophyll, and β-carotene accounts for the majority of total carotenoids, while in honeydew melons, the major pigments are chlorophylls a and b [16]. β-ionone is a major norisoprenoid derived from β-carotene with a typical aroma, rarely found or in small quantities when present in green- and white-fleshed melon varieties [18]. Honeydew melons accumulate less β-carotene than cantaloupes [16]. In the current study, there were significant variations in the β-carotene content of the cantaloupe types. Cultivar type, fruit size, growing location and year may interact to influence β-carotene content of orange-fleshed muskmelon fruit [19]. The latter two factors can be excluded in the current study as the fruit were obtained from cultivars grown at the same time and in the same location/production system.

Table 1. Sweet melon fruit mineral element content (mg kg^{-1} dry weight) at harvest.

Cultivar	Fruit Flesh Color	Potassium	Phosphorus	Magnesium	Calcium	Zinc
Adore	Orange	23,627 ± 262 b,c	1502 ± 23 c,d	1120 ± 20 b,c	657 ± 20 b	22 ± 1 a
Cyclone	Orange	24,646 ± 398 a,b	1674 ± 83 b,c	1191 ± 139 b	521 ± 39 c	23 ± 1 a
Divine	Orange	23,428 ± 778 b,c	1636 ± 61 b,c	1088 ± 20 b,c	405 ± 34 d	20 ± 1 a
E25F.00075	Orange	19,783 ± 594 e,f	1428 ± 65 d	998 ± 81 b,c,d	305 ± 15 e,f	22 ± 1 a
E25F.00185	Orange	23,651 ± 726 b,c	1646 ± 101 b,c	1050 ± 63 b,c	556 ± 36 c	21 ± 1 a
Honey Brew	Green	20,882 ± 138 d,e,f	1121 ± 13 e	809 ± 42 d,e	250 ± 28 f	13 ± 0 c
Honey Star	Green	19,488 ± 781 f	1213 ± 23 e	717 ± 32 e	323 ± 23 e,f	17 ± 2 b
MAB 79001	Orange	22,227 ± 383 c,d	1703 ± 87 a,b	921 ± 10 c,d,e	352 ± 11 d,e	22 ± 2 a
Magritte	Orange	25,290 ± 182 a	1873 ± 25 a	1015 ± 60 b,c,d	309 ± 13 e,f	23 ± 1 a
Majestic	Orange	21,266 ± 307 d,e	1696 ± 50 a,b	1495 ± 110 a	733 ± 13 a	22 ± 1 a
$LSD_{0.05}$		1537	185	207	73	3

Values (±standard error of means) in the same column followed by the same letters are not significantly different at $p \leq 0.05$ (Fisher's protected LSD test, $n = 4$).

Cultivar also influenced fruit vitamin C content at harvest (Figure 1). All orange-fleshed cultivars had higher vitamin C content compared to the green-fleshed cultivars. The vitamin C contents of orange-fleshed cvs. Adore, E25F.00075 and E25F.00185 were approximately 2-fold that of the green-fleshed cvs. Honey Brew and Honey Star. Variation in fruit color influenced concentration of vitamin C.

Cultivar E25F.00075 exhibited the highest total phenolic content with 51% increased total phenolic content, compared to cv. Magritte with the lowest total phenolic content at harvest (Figure 2). Cultivars E25F.00075 and Divine (both cantaloupe type) exhibited higher total phenolic content than honeydew sweet melon cvs. Honey Brew and Honey Star. Cultivar E25F.00185 (cantaloupe type) had the highest flavonoid content; more than double the flavonoid content in all other cultivars (including the honeydew types) (Figure 2). Different cantaloupe type cultivars had the highest total phenolic and flavonoid contents whereas the honeydew type cultivars were among the cultivars with low total phenolic and flavonoid contents. Thus, fruit color seemed to be associated with β-carotene, vitamin C, flavonoid and total phenolic contents, with the cantaloupe cultivars in this study as more nutrient-rich sweet melons.

Higher free radical scavenging activity was recorded with 50% methanol extract for all cultivars compared to their respective water extracts (Figure 3). Water extracts from cvs. Majestic and E25F.00075 (cantaloupe type) had higher free radical scavenging activity than the honeydew melons, cvs. Honey Brew and Honey Star. The 50% methanol extracts of cvs. MAB 79001 and Adore (cantaloupe type) had higher free radical scavenging activity than honeydew melons. The antioxidant amount in muskmelon varied between stages and between cultivars [20]. Cultivar type had a significant influence on the antioxidant activity regardless of the solvent used as both extracts recorded higher activity in the orange-fleshed cantaloupes, therefore cultivar type remains the main determinant factor. Zeb [21] reported higher radical scavenging activity with the aqueous methanolic extracts than individual methanol and water extracts, however, these results were from seeds while the current study compared fruit extracts. The efficiency of extraction, as well as extract yield, depends on type of solvents used, solubility and polarity of the active compounds in the solvent, including time and temperature of extraction [22]. High total antioxidant activity occurs in the early stages of muskmelon fruit ripening with its maximum in the premature stage [23]. Though the antioxidant activity may be at its maximum at the premature stage, sweet melons are rarely harvested at this stage of growth, as this may affect eating quality for local and commercial markets. Sweet melons are harvested at the fully ripe stage when they have attained optimum quality.

The correlation between sweet melon nutritional and phytochemical properties at harvest varied (Table 2). A significant, positive, correlation was established between β-carotene content and all mineral elements, except calcium, with β-carotene and phosphorus having the highest positive correlation. Both phosphorus and zinc contents were individually significantly positively correlated with fruit vitamin C, β-carotene and all other mineral element contents. This relationship may explain why green-fleshed sweet melon cultivars recorded low vitamin C and β-carotene contents as these cultivars had low concentrations of all mineral elements. Nonetheless, a strong positive multivariate correlation is advantageous for direct and indirect selection in biofortification as well as nutritional and phytochemical quality improvement programs.

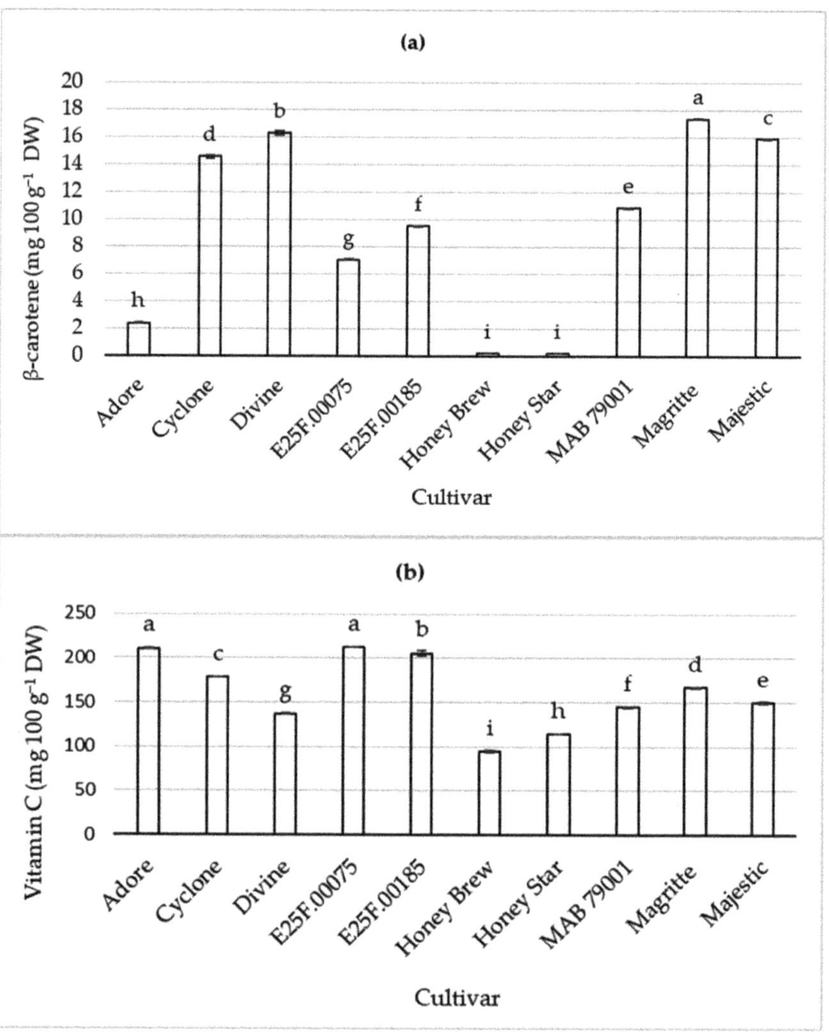

Figure 1. β-Carotene (**a**) and vitamin C (**b**) contents of fruit from sweet melon cultivars at harvest. Bars (with error bars indicating standard error of means) with different letters are significantly different at $p \leq 0.05$ (Fisher's protected LSD test, $n = 4$). DW = dry weight.

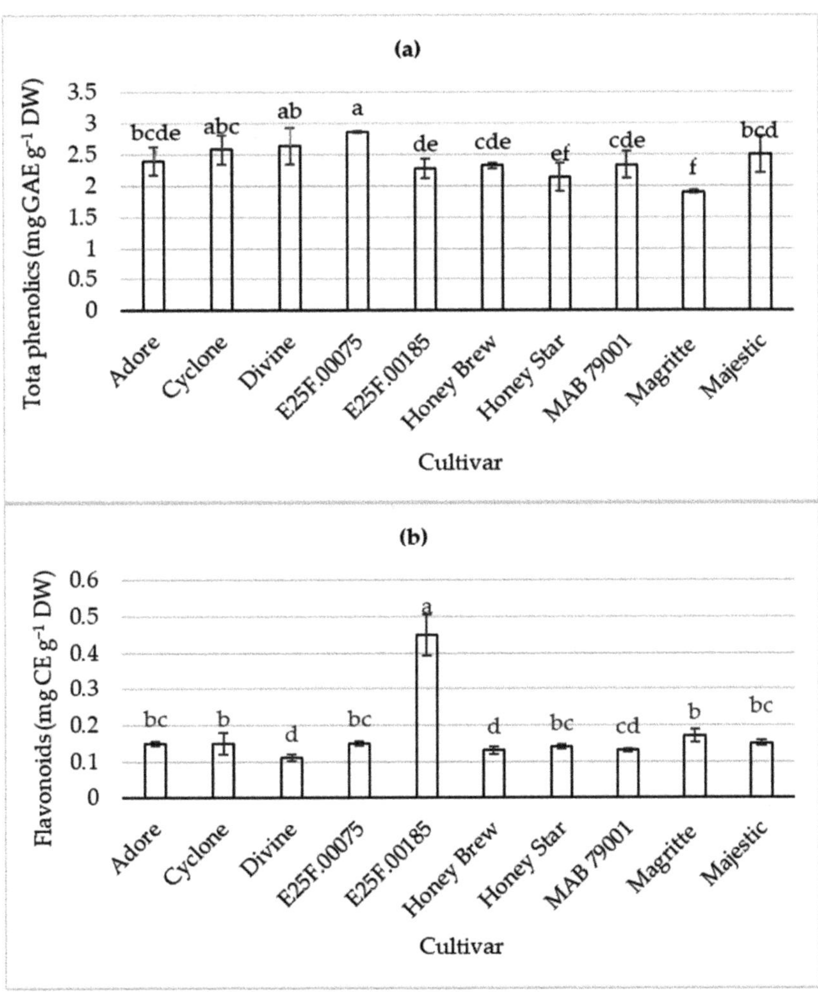

Figure 2. Total phenolics (**a**) and flavonoid (**b**) contents of fruit from sweet melon cultivars at harvest. Bars (with error bars indicating standard error of means) with different letters are significantly different at $p \leq 0.05$ (Fisher's protected LSD test, $n = 4$). GAE = gallic acid equivalent; CE = catechin equivalent; DW = dry weight.

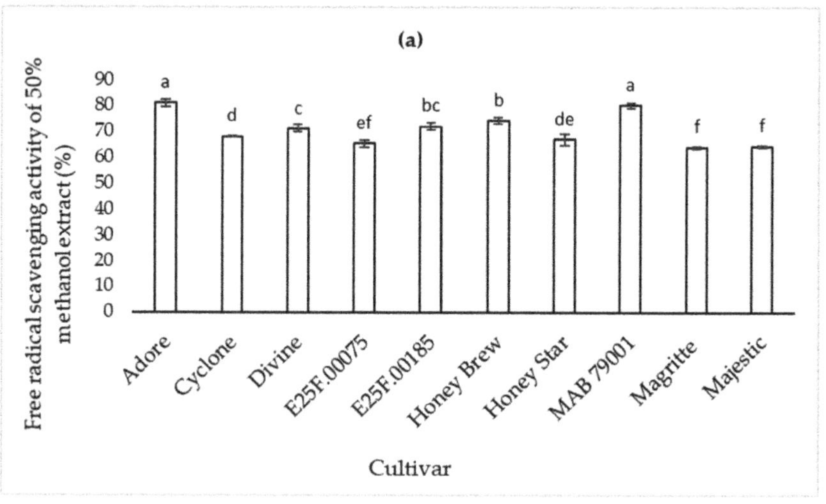

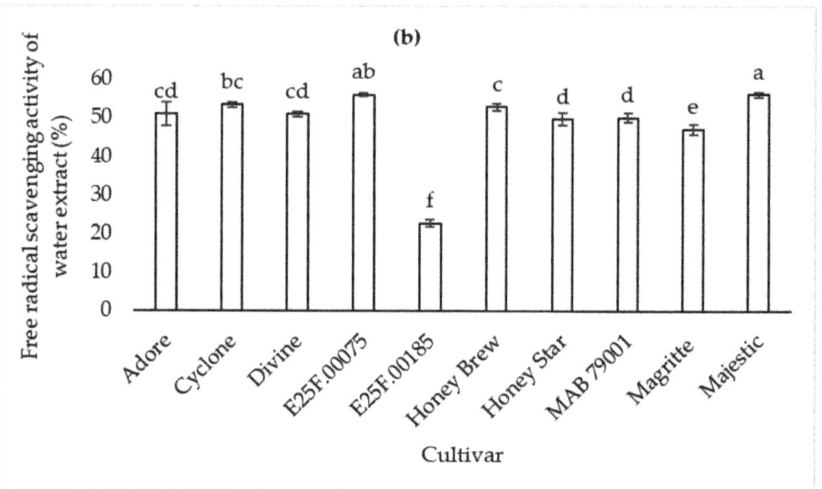

Figure 3. Free radical scavenging activity of 50% methanol (**a**) and water (**b**) extracts of fruit from sweet melon cultivars at harvest. Bars (with error bars indicating standard error of means) with different letters are significantly different at $p \leq 0.05$ (Fisher's protected LSD test, $n = 4$).

The cluster analysis based on the fruit nutritional and phytochemical qualities grouped the cultivars into two main clusters: the honeydew melons in group I and the cantaloupe melons in group II (Figure 4). Furthermore, the principal component analysis revealed four principal components (PCs), each with an eigenvalue $\geq$ 1.00 (Table S1). The first four PCs cumulatively accounted for 85.15% of the total variation. Vitamin C, β-carotene and all the mineral elements (except calcium) contributed above 10% each to PC1, which accounted for 40.39% of the total variation (Table S2). In the PC2 that accounted for 20.88% of the total variation, flavonoids and free radical scavenging activity of the water extract contributed 28.88% and 36.81%, respectively (Table S2). Thus, a principal component biplot, based on the first two PCs that cumulatively accounted for 61.27% of the total variation, revealed cultivars with high associations with the measured fruit quality at harvest (Figure 5). All the nutritional and phytochemical traits were closely associated with the cantaloupe cultivars as indicated by their proximity to the vector line. In particular, cvs. Divine, Majestic, Cyclone, Magritte and Adore were closely associated with all the mineral elements evaluated, β-carotene and vitamin C. Cultivar E25F.00185 was closely associated with flavonoid, while cvs. E25F.00075 and Divine were closely associated with total phenolics. The two honeydew melons cvs. Honey Brew and Honey Star were not associated with any of the evaluated nutritional and phytochemical qualities.

Table 2. Pearson correlation of nutritional and phytochemical traits of cantaloupe and honeydew melon cultivars at harvest.

Variable	TPC	Flav	Vit. C	β-Car	K	P	Mg	Ca	Zn	FSA (W)
Flavonoids	−0.162									
Vitamin C	0.236	**0.439** [a]								
β-Carotene	0.051	0.008	0.184							
K	−0.270	0.229	0.343	**0.524**						
P	−0.042	0.183	**0.387**	**0.825**	**0.679**					
Mg	0.249	0.055	0.260	**0.538**	0.238	**0.572**				
Ca	0.204	0.267	**0.417**	0.228	0.273	**0.370**	**0.761**			
Zn	0.175	0.129	**0.699**	**0.591**	**0.437**	**0.691**	**0.457**	**0.467**		
FSA (W)	0.229	**−0.919**	−0.309	−0.009	−0.240	−0.059	0.192	−0.090	−0.045	
FSA (AM)	0.043	0.011	0.022	**−0.391**	0.066	−0.184	−0.192	0.134	−0.035	−0.124

[a] Values in bold indicate significant ($p = 0.05$) correlation between 2 parameters. TPC = total phenolics content; Flav = flavonoid; Vit. C = vitamin C; β-Car = β-carotene; K = potassium; P = phosphorus; Mg = magnesium; Ca = calcium; Zn = zinc; FSA (W) = free radical scavenging activity of water extract; FSA (AM) = free radical scavenging activity of 50% methanol extract.

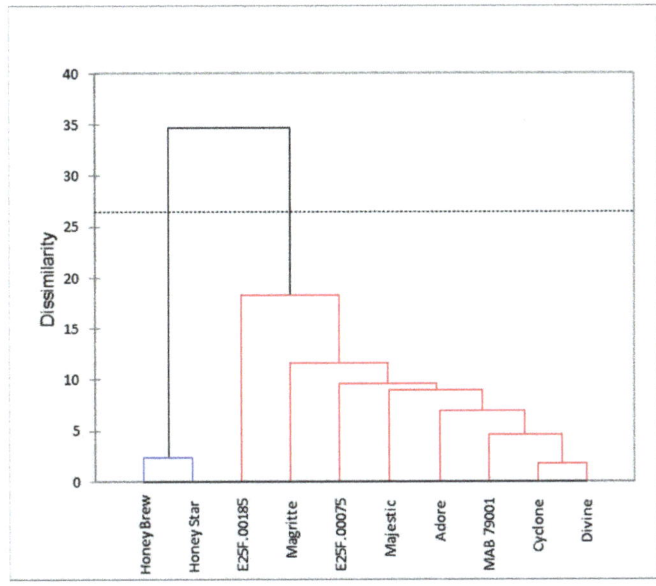

Figure 4. Dendrogram indicating the interrelatedness of the cultivars based on the measured fruit quality at harvest.

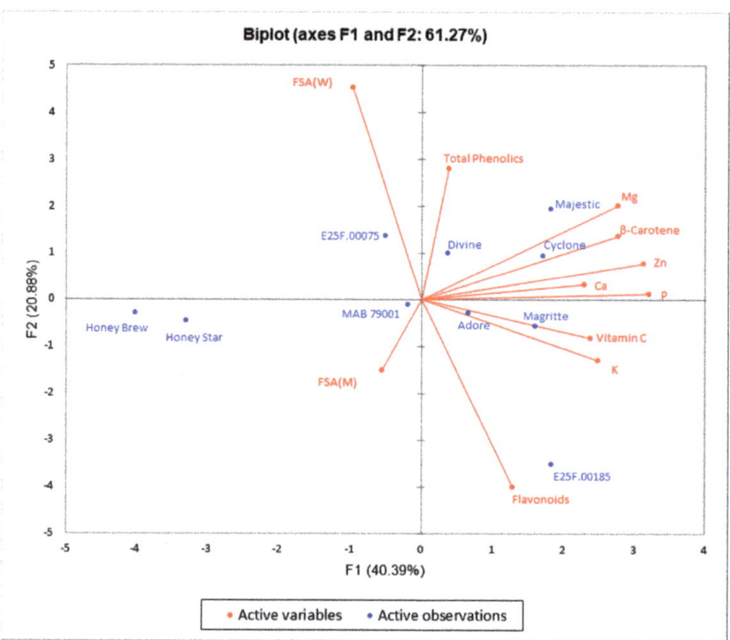

Figure 5. Principal component biplot of fruit nutritional and phytochemical qualities for the evaluated honeydew and cantaloupe melon cultivars at harvest. FSA (W) = free radical scavenging activity of water extract; FSA (M) = free radical scavenging activity of 50% methanol extract.

2.2. Effect of Cultivar and Postharvest Storage on Fruit Nutritional and Phytochemical Quality

Cultivar and postharvest storage duration as individual factors, and interactively in some cases, significantly affected fruit quality, plus nutritional and phytochemical parameters (Table 3). Fruit flesh color a^* and b^* and total soluble solid (TSS) content were only affected by cultivar (Table 4). Cultivar MAB79001 had the highest color a^* and TSS content. The minimum commercially acceptable TSS content for sweet melons is 9%; premium markets may require 11% or more [24]. Based on TSS content alone, all cultivars in this study were of satisfactory quality for local or international markets.

Postharvest storage duration also affected the TSS content (Table 3). The longer the storage duration, the higher the TSS (Figure 6). Thus, a 4% increase in TSS was recorded after 7 days of storage while it increased by 10% after 14 days of storage (Table S3). An increase in the TSS content has been reported in several fruit, which could be due to the alteration in cell wall structure, fruit climacteric nature and/or breakdown of complex carbohydrates into sucrose and other simple sugars including glucose and fructose [25–29]. An increase in postharvest storage duration, however, resulted in a decrease in total phenolic content (Figure 6, Table S3). Cultivar and postharvest storage duration did not affect electrical conductivity (average of 3.8 mS cm^{-1}). On the other hand, cultivar and postharvest storage duration had a significant interaction effect on fruit color L^*, fruit juice pH, vitamin C, β-carotene and flavonoid contents and free radical scavenging activity (Table 3). Fruit color L^* and fruit juice pH of almost all cultivars increased over the postharvest storage duration (Figure 7). The cantaloupe cultivars particularly had a relatively high increase ($\geq$14%) in fruit color L^* after 14 days of storage in comparison to the honeydew cultivars (Table S4). The increase in the lightness of the fruit over the 14 days of storage indicates the absence of browning reactions [15]. The pH of a fruit has an inverse relationship with organic acids in the fruit juice. An increased pH in fruit during postharvest storage is accompanied by a decrease in organic acids due to hydrolysis, which subsequently increases the fruit sweetness and decreases sourness [30].

Table 3. Analysis of variance for the effect of cultivar and storage duration on sweet melon fruit quality parameters, plus nutritional and phytochemical contents.

Source	df	Mean Squares										
		TPC	Flav	Vit. C	β-Car	FSA		Color			TSS	pH
						W	AM	L^*	a^*	b^*		
Rep	2	0.061	0.001	3.345	0.028	1.70	4.43	5.54	0.70	62.14	0.05	0.50
C	4	0.542 ns	0.466 **	6435.795 **	387.061 **	68.25 **	222.26 **	61.85 **	488.75 **	33.77 *	3.64 **	0.85 ns
S	2	2.927**	0.807 **	5836.557 **	17.369 **	617.22 **	3.48 ns	93.32 **	1.23 ns	23.72 ns	4.29 **	1.78 **
C × S	8	0.134 ns	0.085 **	417.577 **	5.009 **	53.74 **	22.18 **	6.96 *	1.47 ns	2.95 ns	0.44 ns	0.41 *
Error	20	0.171	0.000	1.168	0.025	1.68	3.01	2.57	2.24	24.03	0.22	0.11
Total	44											

ns, *, ** = not significant, or significant at $p < 0.05$ or $p < 0.001$, respectively. Rep = replicate; C = cultivar; S = storage; TPC = total phenolics content; Flav = flavonoid; Vit. C = vitamin C; β-Car = β-carotene; FSA = free radical scavenging activity; W = water extract; AM = aqueous methanol extract; TSS = total soluble solids; L^* = lightness; a^* = red/green; b^* = blue/yellow.

Table 4. Effect of cultivar on flesh color and total soluble solids of melon juice.

Cultivar	Color		Total Soluble Solids (% Brix)
	a^*	b^*	
Cyclone	7.3 ± 0.7 b	13.4 ± 2.8 a,b	10.1 ± 0.3 c
Honey Brew	−5.4 ± 0.8 c	11.1 ± 1.6 b	11.4 ± 0.3 b
Honey Star	−5.8 ± 0.7 c	12.1 ± 1.5 b	11.2 ± 0.3 b
MAB79001	9.3 ± 0.7 a	10.9 ± 1.8 b	11.8 ± 0.5 a
Majestic	6.7 ± 1.1 b	15.6 ± 3.8 a	11.2 ± 0.4 b
$LSD_{0.05}$	1.6	2.8	0.3

Values (±standard error of means) in each column followed by different letters are significantly different at $p \leq 0.05$ (Fisher's protected LSD test, $n = 3$); a^* = red/green; b^* = blue/yellow.

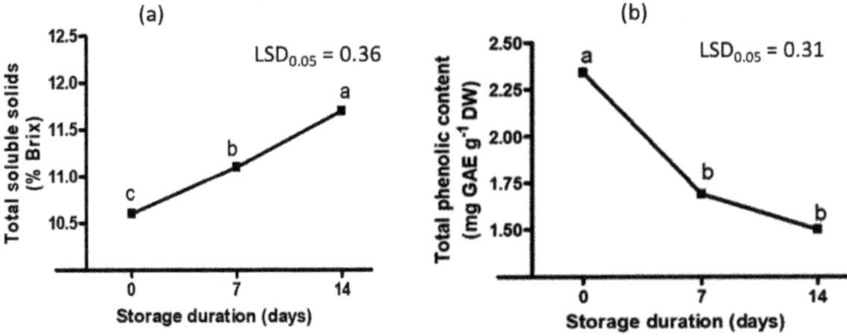

Figure 6. Effect of postharvest storage duration on sweet melon total soluble solids (**a**) and total phenolic content (**b**). Points on the line with different letters in each graph are significantly different at $p \leq 0.05$ (Fisher's protected LSD test, $n = 3$). GAE = gallic acid equivalent; DW = dry weight.

All cultivars exhibited a decrease in vitamin C concentration with an increase in storage duration (Figure 8). At the end of the 14 days of storage, cv. Honey Star had the greatest decrease (46%) in vitamin C (Table S4). Cultivar Cyclone consistently had the highest vitamin C content during storage while cvs. Honey Brew and Honey Star had the lowest concentrations (Figure 8). A decline in vitamin C content of non-netted green-fleshed and orange-fleshed sweet melons collected from the field, and the non-netted green-fleshed Honey Brew, had the lowest concentration on day 17 of storage [14]. In the current study, an increase in storage duration resulted in a decline in the vitamin C content of sweet melons regardless of the cultivar type (Table S4). During postharvest storage, β-carotene concentration may remain stable, or decrease, depending on storage temperature [31]. In the current study, fruit were stored at the same temperature. A decrease in β-carotene content occurred in cantaloupe melons with an increase in postharvest storage duration (Figure 8, Table S4). The β-carotene concentration of cvs. Honey Star and Honey Brew, although the lowest, remained fairly stable during postharvest storage. Cultivar and flesh color influence β-carotene concentration among the sweet melon fruit while storage period and cultivar determine its retention. Unlike vitamin C and β-carotene concentrations, flavonoid content of fruit increased during storage in almost all cultivars (Figure 8, Table S4). Cultivar MAB79001 had the highest increase in flavonoid concentration (Table S4), which was 7-fold after 14 days of storage compared to its flavonoid concentration at harvest. With the exception of MAB79001 methanolic extract, there was an increase in antioxidant activity during storage of water and methanolic extracts of all cultivars (Figure 9, Table S4).

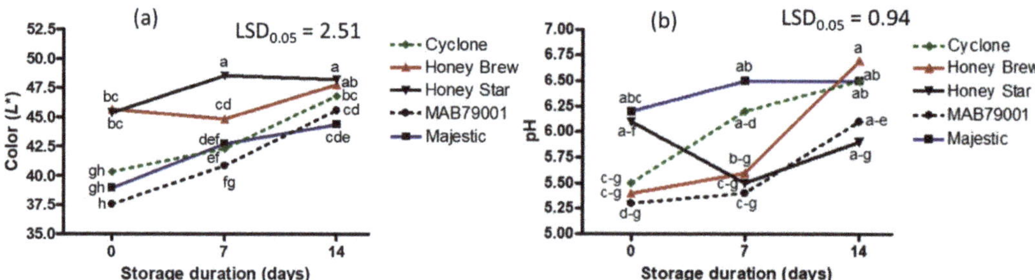

Figure 7. Interaction effect of sweet melon cultivars and storage duration on fruit color (**a**) and fruit juice pH (**b**). Points on lines with different letters in each graph are significantly different at $p \leq 0.05$ (Fisher's protected LSD test, $n = 3$). L^* = lightness.

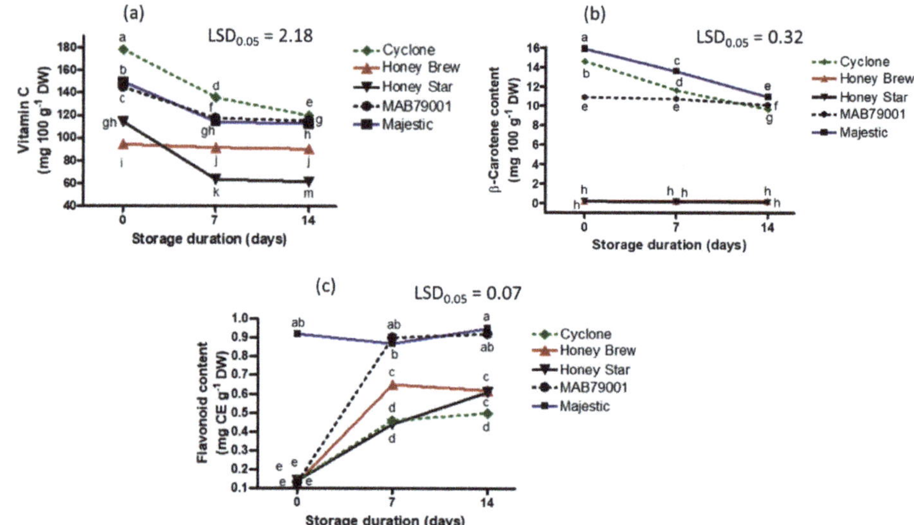

Figure 8. Interaction effect of sweet melon cultivars and storage duration on vitamin C (**a**), β-carotene (**b**) and flavonoid (**c**) contents. Points on lines with different letters in each graph are significantly different at $p \leq 0.05$ (Fisher's protected LSD test, $n = 3$). CE = catechin equivalent; DW = dry weight.

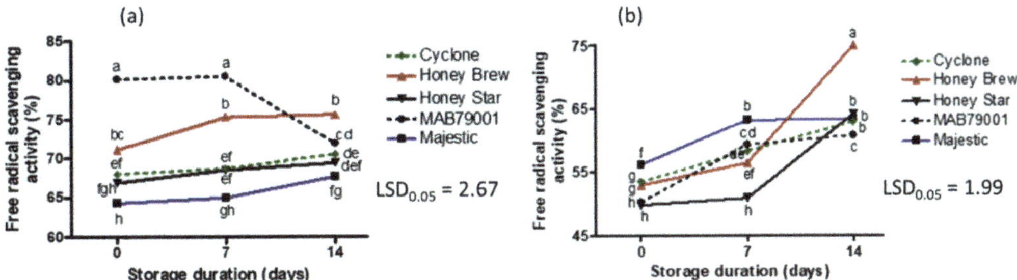

Figure 9. Interaction effect of sweet melon cultivars and storage duration on free radical scavenging activity of 50% methanol (**a**) and water (**b**) extracts. Points on lines with different letters in each graph are significantly different at $p \leq 0.05$ (Fisher's protected LSD test, $n = 3$).

3. Materials and Methods

3.1. Effect of Cultivars on Fruit Nutritional and Phytochemical Quality at Harvest

The procedures used to establish and maintain the plants from September 2016 until January 2017 were as described by Pulela et al. [10]. Ten fruit per cultivar per plot were randomly harvested at physiological maturity from sweet melon cultivars arranged in a randomized complete block design with four replicates (Figure S1). The cantaloupe cvs. Majestic, Magritte, Divine, Cyclone, MAB 79001, Adore, E25F.00075 and E25F.00185 and honeydew cvs. Honey Brew and Honey Star types were used [10]. The experiment was carried out in a non-temperature controlled, high plastic tunnel at the Agricultural Research Council—Vegetables, Industrial and Medicinal Plants, Roodeplaat, Pretoria, South Africa. The fruit were peeled with a sharp knife, sliced and homogenized. Fruit samples were kept at −80 °C, freeze-dried and ground into fine powders. The nutritional, phytochemical and antioxidant qualities of the fruit samples from each cultivar were analyzed as described below.

3.1.1. Nutritional Analysis

Freeze-dried sweet melon samples were analyzed for phosphorus (P), potassium (K), calcium (Ca), magnesium (Mg) and zinc (Zn). The sample analysis was carried out with an aliquot of digested solution using inductively coupled plasma–optical emission spectrometry (ICP-OES) as described by Mahlangu et al. [32]. Each determination was in quadruplicate.

β-Carotene content was determined according to Moyo et al. [33]. Ice-cold hexane:acetone (1:1 v/v) was used to extract freeze-dried sweet melon samples. The pooled organic extract was gravity filtered through a 0.45 μm syringe filter before injection into an HPLC (Prominence-*i*-HPLC-PDA with LC-2030C sample cooler (Shimadzu, Kyoto, Japan)). There were four replicates for each determination at harvest, while analysis performed for postharvest experiment was carried out in triplicate.

Extraction and high-performance liquid chromatography (HPLC) quantification of vitamin C content were as described by Moyo et al. [33]. The extracted solution was gravity filtered through a 0.45 μm syringe filter before injection into an HPLC (Prominence-*i* HPLC-PDA, equipped with sample cooler LC-2030C (Shimadzu, Kyoto, Japan)). A calibration curve was plotted using L-ascorbic acid. Each determination was carried out in quadruplicate for analysis of fruit at harvest, while the analysis conducted for postharvest experiment was performed in triplicate.

3.1.2. Phytochemical Analysis

Freeze-dried sweet melon samples (0.2 g) were extracted with 10 mL of 50% methanol in a sonication bath for 20 min. The mixtures were centrifuged and the supernatant was used for total phenolic and flavonoid analysis. The total phenolic content was determined using the Folin Ciocalteu colorimetric method [34]. Reaction mixes were incubated for

40 min at 25 °C and total phenolics content was determined by measuring absorbance at 725 nm, against a blank using a spectrophotometer (SPECORD® 210 PLUS, Analytik Jena, Jena, Germany). Gallic acid (0.1 mg/mL^{-1}) was used as a standard for preparing the calibration curve and the results were reported in mg gallic acid equivalent (GAE) per g dry weight. The assay was in quadruplicate for fruit analyzed at harvest, and triplicate for the postharvest experiment.

Flavonoid content quantification was carried out using the aluminum chloride method [35]. Absorbance was measured against a freshly prepared reagent blank and read at 510 nm using a spectrophotometer. Suitable aliquots of catechin (0.1 mg mL^{-1}) were used to plot a standard calibration curve. The quantification was carried out in quadruplicate for fruit analyzed at harvest, and triplicates for the postharvest experiment. The results were presented in mg catechin equivalent (CE) per g dry weight.

3.1.3. Free Radical Scavenging Activity

To determine free radical scavenging activity, freeze-dried, powdered samples were extracted with 50% methanol or distilled water at 20 mL g^{-1} in a sonication bath containing ice-cold water for 1 h. The 50% methanol extracts were concentrated *in vacuo* after vacuum filtration through Whatman No. 1 filter paper. Concentrated 50% methanol extracts were air-dried at 25 °C. Water extracts were vacuum filtered through Whatman No. 1 filter paper and lyophilized. Dried extracts (50% methanol and water extracts) were dissolved in methanol at a known sample concentration in glass vials to give a final assay concentration of 200 µg mL^{-1}. Antioxidant activity was measured using the 2,2-diphenyl-1-picryl hydrazyl (DPPH) scavenging assay [36] with slight variations. The reaction mix contained 300 µL of each sample, 450 µL of methanol and 750 µL of DPPH solution (0.1 mM). The reaction mix was allowed to stand in the dark for 30 min at 25 °C after which absorbance at 517 nm was recorded. Ascorbic acid was used as a standard antioxidant (positive control) while methanol served as the negative control.

3.2. *Effects of Cultivar and Postharvest Storage on Fruit Nutritional and Phytochemical Quality*

Effects of cultivar and postharvest storage duration on fruit quality parameters and nutritional and phytochemical profile were determined with fruit harvested from cvs. Majestic, Cyclone and MAB 79001 (cantaloupe) and cvs. Honey Brew and Honey Star (honeydew) (Figure S2). Ten fruit per cultivar were selected. Fruit were packed into crates, arranged in a randomized complete block design with three replicates inside a cold room at 10 °C and approximately 95% relative humidity. Fruit were stored for 0, 7 and 14 days. At the end of each storage period, fruit were evaluated for vitamin C, β-carotene, total phenolic and flavonoid contents and free radical scavenging activity as described above. The following fruit quality parameters were also determined: fruit color chroma, total soluble solids (TSSs), pH and electrical conductivity (EC). Fruit flesh color was determined using a chromameter (CR400, Minolta Sensing Inc., Konica, Japan). The chromameter was calibrated with a standard white tile before measurements. Color changes were quantified in the L^*, a^* and b^* color space. To determine TSS, pH and EC, fruit were sliced and blended to produce a puree. The puree was gravity filtered through a cheesecloth to produce sweet melon juice. The pH and EC of the juice were measured using a combo pH and EC meter (Hanna Instruments® Inc., Curepipe, Mauritius). The TSS content of the sweet melon juice was determined using a hand-held refractometer (ATAGO, Tokyo, Japan) and expressed as % Brix.

3.3. *Statistical Analysis*

Data of fruit analysis at harvest were subjected to a one-way analysis of variance (ANOVA) using GenStat® (ver. 11.1, VSN, Rothamsted, UK). Where significant differences were established, cultivar mean values were separated using Fisher's protected least significant difference (LSD) test. To determine possible associations between measured parameters at harvest, data were subjected to Pearson correlation analysis using XLSTAT (ver. 17.04.36025 Add-in-soft, New York, NY, USA). Principal component and agglomerative hierarchical clustering analyses were also carried out to establish quality traits and cultivar associations. To determine the effects of cultivar and postharvest storage duration on physiochemical qualities and bioactive compounds of melons, data from the experiment involving postharvest duration and cultivars were subjected to a two-way analysis of variance using GenStat®. Significant interactions were used to explain results. Where interactions were not significant, mean values of main effects were separated using Fisher's protected least significant difference test.

4. Conclusions

At harvest, the orange-fleshed sweet melon cultivars had higher mineral element (K, P, Mg, Ca and Zn), vitamin C and β-carotene contents than the green-fleshed cultivars. Thus, with some limitations, cultivar and fruit flesh color may be used as indicators of nutrient-rich sweet melons. Cultivar and postharvest storage duration had significant interaction effects on flavonoid, vitamin C and β-carotene concentrations, antioxidant activity, pH and color L^* of the fruit. Total soluble solids, flavonoid content and antioxidant activity generally increased during postharvest storage, whereas the reverse was the case with vitamin C and β-carotene contents. Although some nutritional traits decreased as postharvest storage duration lengthened, sweet melon remains a comparatively healthy food in the human diet. Cultivars Cyclone and Majestic relatively outperformed other cultivars in terms of some postharvest storage nutritional qualities. In selecting the right cultivars, variation in nutritional traits of sweet melon cultivars as affected by postharvest storage should be an important consideration.

Supplementary Materials: The following supporting information can be downloaded at: https://www.mdpi.com/article/10.3390/plants11162136/s1, Figure S1: Experimental layout of sweet melon cultivar trial in a non-temperature controlled plastic tunnel; Figure S2: Sweet melon cultivars used for postharvest storage; Table S1: Eigenvectors from principal component (PC) analysis based on the nutritional and phytochemical quality at harvest; Table S2: Contribution (%) of each variable to the principal components; Table S3: Changes (%) in total soluble solid and total phenolic contents as influenced by postharvest storage duration; Table S4: Changes (%) in fruit quality parameters, nutritional and phytochemical contents as influenced by postharvest storage duration.

Author Contributions: Conceptualization, M.M.M., P.S. and S.O.A.; methodology, B.L.P. and S.O.A.; investigation, B.L.P.; writing—original draft preparation, B.L.P. and S.O.A.; writing—review and editing, M.M.M., P.S. and S.O.A.; supervision, M.M.M., P.S. and S.O.A.; funding acquisition, M.M.M., P.S. and S.O.A. All authors have read and agreed to the published version of the manuscript.

Funding: The Agricultural Research Council, South Africa, funded this research and Tshwane University of Technology funded the APC.

Institutional Review Board Statement: Not applicable.

Informed Consent Statement: Not applicable.

Data Availability Statement: Not applicable.

Acknowledgments: The technical assistance provided by Hlabana A. Seepe and Liesl Morey during the laboratory work and statistical analysis, respectively, is gratefully acknowledged.

Conflicts of Interest: The authors declare no conflict of interest. The funders had no role in the design of the study; in the collection, analyses or interpretation of data; in the writing of the manuscript, or in the decision to publish the results.

References

1. Louvet, J. The relationship between substrates and plant diseases. *Acta Hortic.* **1982**, *126*, 147–152. [CrossRef]
2. Riviere, L.M.; Caron, J. Research on substrates: State of the art and need for the coming 10 years. *Acta Hortic.* **2001**, *548*, 29–41. [CrossRef]
3. Rodriguez, J.C.; Cantliffe, D.J.; Shaw, N.L. Soilless media and container for greenhouse production of 'Galia' type muskmelon. *HortScience* **2006**, *41*, 1200–1205. [CrossRef]

4. Maboko, M.M.; Du Plooy, C.P.; Bertling, I. Performance of tomato cultivars in temperature and non-temperature controlled plastic tunnels. *Acta Hortic.* **2012**, *927*, 405–411. [CrossRef]
5. Cantliffe, G.J.; Vansickle, J.J. Competitiveness of the Spanish and Dutch greenhouse industries with the Florida fresh vegetable industry. *Proc. Fla. State Hortic. Soc.* **2001**, *114*, 283–287. [CrossRef]
6. Rodriguez, J.C.; Shaw, N.L.; Cantliffe, D.J. Influence of plant density on yield and fruit quality of greenhouse-grown Galia muskmelons. *HortTechnology* **2007**, *17*, 580–585. [CrossRef]
7. Massantini, F.; Favilli, R.; Magnani, G.; Oggiano, N. Soilless culture-biotechnology for high quality vegetables. *Soilless Cult.* **1988**, *4*, 27–40.
8. Perdigones, A.; Pascual, V.; García, J.L.; Nolasco, J.; Pallares, D. Interactions of crop and cooling equipment on greenhouse climate. *Acta Hortic.* **2005**, *691*, 203–208. [CrossRef]
9. Jensen, M.H.; Malter, A.J. *Protected Agriculture: A Global Review*; World Bank Technical Paper, No. 253; The World Bank: Washington, DC, USA, 1995.
10. Pulela, B.L.; Maboko, M.M.; Soundy, P.; Amoo, S.O. Development, yield and quality of Cantaloupe and Honeydew melon in soilless culture in a non-temperature controlled high tunnel. *Int. J. Veg. Sci.* **2020**, *26*, 292–301. [CrossRef]
11. Lester, G. Melon (*Cucumis melo* L.) fruit nutritional quality and health functionality. *HortTechnology* **1997**, *7*, 222–227. [CrossRef]
12. Lester, G.E.; Crosby, K.M. Ascorbic acid, folic acid and potassium content in postharvest green-fleshed honeydew muskmelons: Influence of cultivar, fruit size, soil type, and year. *J. Am. Soc. Hortic. Sci.* **2002**, *127*, 843–847. [CrossRef]
13. Lester, G.E.; Hodges, D.M. Antioxidants associated with fruit senescence and human health: Novel orange fleshed non-netted honeydew melon genotype comparisons following different seasonal productions and cold storage durations. *Postharvest Biol. Technol.* **2008**, *48*, 347–354. [CrossRef]
14. Hodges, D.M.; Lester, G.E. Comparison between orange-and green-fleshed non-netted and orange-fleshed netted muskmelons: Antioxidant changes following different harvest and storage periods. *J. Am. Soc. Hortic. Sci.* **2006**, *131*, 110–117. [CrossRef]
15. Lester, G.E.; Saftner, R.A.; Hodges, D.M. Market quality attributes of orange-fleshed, non-netted honeydew melon genotypes following different growing seasons and storage temperature durations. *HortTechnology* **2007**, *17*, 346–352. [CrossRef]
16. Laur, L.M.; Tian, L. Provitamin A and vitamin C contents in selected California-grown cantaloupe and honeydew melons and imported melons. *J. Food Compos. Anal.* **2011**, *21*, 194–201. [CrossRef]
17. Ritchie, H.; Roser, M. Micronutrient Deficiency. OurWorldInData.org. 2020. Available online: https://ourworldindata.org/micronutrient-deficiency (accessed on 10 January 2021).
18. Portnoy, V.; Benyamini, Y.; Bar, E.; Harel-Beja, R.; Gepstein, S.; Giovannoni, J.J.; Schaffer, A.A.; Burger, J.; Tadmor, Y.; Lewinsohn, E.; et al. The molecular and biochemical basis for varietal variation in sesquiterpene content in melon (*Cucumis melo* L.) rinds. *Plant Mol. Biol.* **2008**, *66*, 647–661. [CrossRef]
19. Lester, G.E.; Eischen, F. Beta-carotene content of postharvest orange-fleshed muskmelon fruit: Effect of cultivar, growing location and fruit size. *Plant Foods Hum. Nutr.* **1996**, *49*, 191–197. [CrossRef]
20. Ionica, M.E.; Nour, V.; Trandafir, I. Evolution of some physical and chemical characteristics during growth and development of muskmelon (*Cucumis melo* L.). *Pak. J. Agric. Sci.* **2015**, *52*, 265–271.
21. Zeb, A. Phenolic profile and antioxidant activity of melon (*Cucumis melo* L.) seeds from Pakistan. *Foods* **2016**, *5*, 67. [CrossRef]
22. Singh, J.; Singh, V.; Shukla, S.; Rai, A.K. Phenolic content and antioxidant capacity of selected cucurbit fruits extracted with different solvents. *J. Nutr. Food Sci.* **2016**, *6*, 6. [CrossRef]
23. Menon, S.V.; Rao, R.T.V. Nutritional quality of muskmelon fruit as revealed by its biochemical properties during different rates of ripening. *Int. Food Res. J.* **2012**, *19*, 1621–1628.
24. Paris, H.S.; Amar, Z.; Lev, E. Medieval emergence of sweet melons, *Cucumis melo* (Cucurbitaceae). *Ann. Bot.* **2012**, *110*, 23–33. [CrossRef]
25. Augustin, M.A.; Osman, A.; Azudin, M.N.; Mohamed, S. Physico-chemical changes in muskmelon (*Cucumis melo* L.) during storage. *Pertanika* **1988**, *11*, 203–209.
26. Parveen, S.; Ali, M.A.; Asghar, M.; Khan, A.R.; Salam, A. Physico-chemical changes in muskmelon (*Cucumis melo* L.) as affected by harvest maturity stage. *J. Agric. Res.* **2012**, *50*, 249–260.
27. Youn, A.R.; Kwon, K.H.; Kim, B.S.; Kim, S.H.; Noh, B.; Cha, H.S. Changes in quality of muskmelon (*Cucumis melo* L.) during storage at different temperatures. *Korean J. Food Sci. Technol.* **2009**, *41*, 251–257.
28. Munira, A.Z.; Rosnah, S.; Zaulia, O.; Russly, A.R. Effect of postharvest storage of whole fruit on physico-chemical and microbial changes of fresh-cut cantaloupe (*Cucumis melo* L. *reticulatus* cv. Glamour). *Int. Food Res. J.* **2013**, *20*, 501–508.
29. Supapvanich, S.; Tucker, G.A. Cell wall hydrolysis in netted melon fruit (*Cucumis melo* var. *reticulatus* L. Naud) during storage. *Chiang Mai J. Sci.* **2013**, *40*, 447–458.
30. Singh, K.; Sharma, M. Review on biochemical changes associated with storage of fruit juice. *Int. J. Curr. Microbiol. Appl. Sci.* **2017**, *6*, 236–245. [CrossRef]
31. Pénicaud, C.; Achir, N.; Dhuique-Mayer, C.; Dornier, M.; Bohuon, P. Degradation of β-carotene during fruit and vegetable processing or storage: Reaction mechanisms and kinetic aspects: A review. *Fruits* **2011**, *66*, 417–440. [CrossRef]
32. Mahlangu, R.I.S.; Maboko, M.M.; Sivakumar, D.; Soundy, P.; Jifon, J. Lettuce (*Lactuca sativa* L.) growth, yield and quality response to nitrogen fertilization in a non-circulating hydroponic system. *J. Plant Nutr.* **2016**, *39*, 1766–1775. [CrossRef]

33. Moyo, M.; Amoo, S.O.; Aremu, A.O.; Gruz, J.; Subrtova, M.; Jarosova, M.; Tarkowski, P.; Doležal, K. Determination of mineral constituents, phytochemicals and antioxidant qualities of *Cleome gynandra*, compared to *Brassica oleracea* and *Beta vulgaris*. *Front. Chem.* **2018**, *5*, 128. [CrossRef] [PubMed]
34. Fawole, O.A.; Ndhlala, A.R.; Amoo, S.O.; Finnie, J.F.; Van Staden, J. Anti-inflammatory and phytochemical properties of twelve medicinal plants used for treating gastro-intestinal ailments in South Africa. *J. Ethnopharmacol.* **2009**, *123*, 237–243. [CrossRef] [PubMed]
35. Zhishen, J.; Mengcheng, T.; Jianming, W. The determination of flavonoid contents in mulberry and their scavenging effects on superoxide radicals. *Food Chem.* **1999**, *64*, 555–559. [CrossRef]
36. Amoo, S.O.; Aremu, A.O.; Moyo, M.; Van Staden, J. Antioxidant and acetylcholinesterase-inhibitory properties of long-term stored medicinal plants. *BMC Complement. Altern. Med.* **2012**, *12*, 87. [CrossRef]

 plants

Article

Maturation and Post-Harvest Resting of Fruits Affect the Macronutrients and Protein Content in Sweet Pepper Seeds

Lidiane Fernandes Colombari [1], Larissa Chamma [1], Gustavo Ferreira da Silva [1,*], Willian Aparecido Leoti Zanetti [2], Fernando Ferrari Putti [2] and Antonio Ismael Inácio Cardoso [1]

1 Department of Crop Science, School of Agriculture, São Paulo State University (UNESP), Botucatu 18610-034, Brazil
2 Department of Biosystems Engineering, School of Sciences and Engineering, São Paulo State University (UNESP), Tupã 17602-496, Brazil
* Correspondence: gustavo.ferreira@unesp.br

Abstract: There are few studies about the influence of fruit maturation and post-harvest resting on seed composition, which can be necessary for seedling development and future establishment. Thus, the objective of this study was to evaluate the effect of maturation and post-harvest resting of fruits on the macronutrient and protein content of sweet pepper seeds. The experimental design was a randomized block, with eight treatments, in a 4 × 2 factorial arrangement. The first factor was fruit maturation stages (35, 50, 65 and 80 days after anthesis), and the second, with and without post-harvest resting of the fruits for 7 days. The characteristics evaluated in seeds were the dry weight of one thousand seeds, macronutrient content, and content of albumin, globulin, prolamin and glutelin proteins. There were reductions in K, Ca and Mg content, and an increase in seed content of albumin, globulin and prolamins as a function of the fruit maturation stage. Post-harvest resting of the fruits provided higher Ca content and protein albumin in seeds. The decreasing order of macronutrients and protein content in seeds, independent of fruit maturation and resting stage of the fruits, was N > K > P > Mg > S > Ca, and albumin > globulin ≈ glutelin > prolamine, respectively.

Keywords: *Capsicum annuum* L.; maturation; nutritional quality; seed chemical composition

Citation: Colombari, L.F.; Chamma, L.; da Silva, G.F.; Zanetti, W.A.L.; Putti, F.F.; Cardoso, A.I.I. Maturation and Post-Harvest Resting of Fruits Affect the Macronutrients and Protein Content in Sweet Pepper Seeds. *Plants* **2022**, *11*, 2084. https://doi.org/10.3390/plants11162084

Academic Editors: Georgia Ouzounidou, Lord Abbey, Josephine Ampofo and Mason MacDonald

Received: 20 May 2022
Accepted: 26 July 2022
Published: 10 August 2022

1. Introduction

The main compounds that seeds store are carbohydrates, lipids, proteins, minerals, vitamins and plant hormones [1]. During plant development, nutrients are translocated to the fruits, and later, to seeds. The nutritional requirement of plants becomes intense in the reproductive phase, being more critical during seed formation, where the stored compounds will influence the formation of the embryo, and consequently, the metabolism, vigor and storage capacity of seeds [2,3]. Nitrogen is the most accumulated nutrient in seeds, followed by potassium and phosphorus, but it depends on the species [4,5]. As for protein reserves, two classes are mainly found in seeds: albumins and globulins, or prolamins [6].

Seed proteins are the main sources of nitrogen and sulfur, indispensable for the synthesis of new proteins, nucleic acids and secondary compounds [6–8]. Subsequently, the proteins that were synthesized and stored in the seeds will be broken down into amino acids for biosynthesis and energy generation, and, together with the other reserves, will be mobilized during germination for the development of the embryo, until the seedling manages to emerge above the ground and becomes photosynthetically active [6,9,10].

In the production of sweet pepper seeds, the fruits must be harvested at physiological maturity, when the maximum accumulation of dry matter occurs. If harvested before this point, the fruits must remain at rest after harvesting before extracting seeds for seven to ten days. This procedure allows early harvesting, reducing the time the fruits are exposed to unfavorable climatic conditions and the attack of pests and diseases [11,12]. During the post-harvest resting, the complete formation of the biochemical, morphological, and structural systems of seeds occurs [12,13].

During the maturation period and fruit post-harvest resting, the weight of seeds increases [11,12], probably because of translocation and accumulation of reserve compounds in seeds. However, there is no research showing the changes of this species in the chemical composition of seeds at each stage of fruit maturation and during fruit rest after harvest. Knowledge of these factors can help identify the best fertilization management approach, as it can indicate which nutrients are most important at each stage of plant development. Thus, the objective of this study was to evaluate the effect of maturation and post-harvest resting of fruits on the macronutrient and protein content of sweet pepper seeds.

2. Results

2.1. Water Content, Dry Weight of One Thousand and Macronutrient Content of Seeds

There was only a significant interaction between factors (maturation periods and post-harvest resting of the fruits) for nitrogen and sulfur content in the seeds. Therefore, for all other parameters, the factors were analyzed separately (Tables 1 and 2).

Table 1. Dry weight of one thousand seeds (DWTS), phosphorus (P), potassium (K), calcium (Ca) and magnesium (Mg) content in sweet pepper seeds as a function of post-harvest resting of fruits.

Post-Harvest Resting	DWTS	P	K	Ca	Mg
			g kg^{-1} of Dry Matter		
Without	4.3b	4.7a	12.4a	0.87b	2.4a
With	5.2a	4.4b	12.3a	1.02a	2.4a
CV (%)	2.5	1.41	4.07	8.79	7.10

Means followed by the same letter do not differ from each other by the F-test at 5% probability.

Table 2. Nitrogen (N) and sulfur (S) content of sweet pepper seeds without and with post-harvest resting of fruits at each maturation stage.

Post-Harvest Resting	N Content (g kg^{-1} of Dry Matter)				S Content (g kg^{-1} of Dry Matter)			
	Days after Anthesis							
	35	50	65	80	35	50	65	80
Without	30a	30a	29b	28b	1.9b	2.1a	2.0a	1.9a
With	28b	30a	31a	30a	2.1a	1.9b	1.8b	1.8a
CV (%)	3.65				6.43			

Means followed by the same letter in each column and for each nutrient do not differ by the F-test at 5% probability.

Seed water content was adjusted to the decreasing linear model as a function of the maturation stage. The highest values were obtained at 35 DAA, and they reduced over time to a minimum of 56% and 53% at 80 DAA, without and with post-harvest fruit resting, respectively (Figure 1A).

A linear increase in dry weight of one thousand seeds (DWTS) was observed in fruit maturation, reaching a maximum value of 6.6 and 7.7 g at 80 DAA in seeds without and with post-harvest resting, respectively (Figure 1B).

There was a reduction of N content in seeds without fruit rest as a function of the maturation stage, with a minimum content of 28.5 g kg^{-1} of dry matter (DM) at 80 DAA (Figure 1C). The N content had a quadratic response with fruit rest, with a maximum estimated at 30.5 g kg^{-1} of DM, at 64 DAA.

Seed P contents were adjusted to the quadratic model, being estimated at a maximum of 5.1 g kg^{-1} of DM at 62 DAA, without rest, and 4.2 g kg^{-1} of DM at 74 DAA, with post-harvest resting of the fruits (Figure 1D).

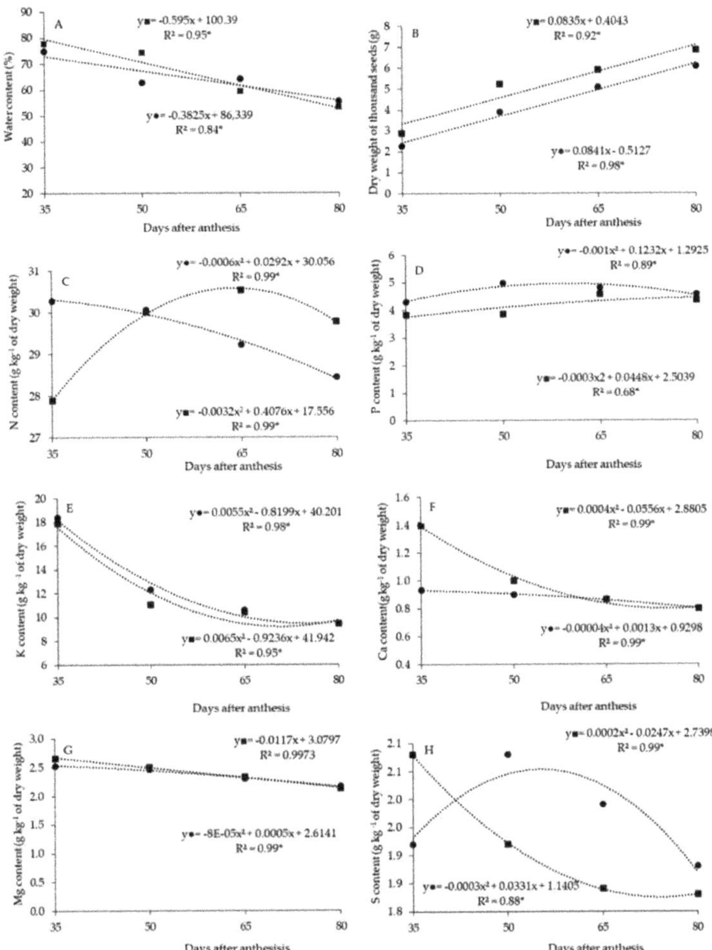

Figure 1. Seed water content (**A**), dry weight of one thousand seeds (**B**), nitrogen—N (**C**), phosphorus—P (**D**), potassium—K (**E**), calcium—Ca (**F**), magnesium—Mg (**G**) and sulfur—S (**H**) content in sweet pepper seeds, as a function of maturation stage, without (•) and with (■) post-harvest resting of fruits. * significant difference at 5% probability.

For K, Ca and Mg contents in the seeds, there were reductions during most periods of maturation, obtaining minimum contents of 9.6, 0.81 and 2.2 g kg^{-1} of DM at 74, 80 and 80 DAA, respectively, without post-harvest resting. For fruits with post-harvest resting, the minimum contents were 9.1, 0.95 and 2.1 g kg^{-1} of DM at 71, 70 and 80 DAA for K, Ca and Mg, respectively (Figure 1E–G).

S contents in seeds were also adjusted to the quadratic model; however, without fruit rest, there was an increase up to 50 DAA, with a reduction after this date. On the other hand, there was a reduction in contents up to 65 DAA with post-harvest resting, and a slight increase in values after this period (Figure 1H).

Comparing the factors post-harvest resting of the fruits, the presence of rest for seven days provided higher DWTS (Table 1). The post-harvest resting of the fruits also provided higher Ca content. Conversely, it was observed that the P content was higher without rest (Table 1).

Seed N content at 35 DAA was higher without fruit rest; however, at 65 and 80 DAA, higher contents were obtained after rest (Table 2). Without post-harvest resting, the N content in seeds reduced linearly with fruit maturation (Figure 1C); with rest, there was an increase until 64 DAA, and after this, a small decrease was observed.

The S content of seeds at 35 DAA was higher with fruit rest, but at 50 and 65 DAA, the contents were higher without fruit rest (Table 2).

Nutrient contents in the sweet pepper seeds followed the following decreasing order: N > K > P > Mg > S > Ca.

2.2. Seed Protein Content

For proteins, the content of albumin and prolamine had quadratic responses, with maximum estimated values without fruit rest of 64.7 and 7.9 mg g^{-1} of DM, at 66 and 53 DAA, respectively. Post-harvest resting, the maximum albumin content was estimated at 65.9 mg g^{-1} of DM at 64 DAA, and prolamine was estimated at 7.1 mg g^{-1} of DM at 55 DAA (Figure 2A,C).

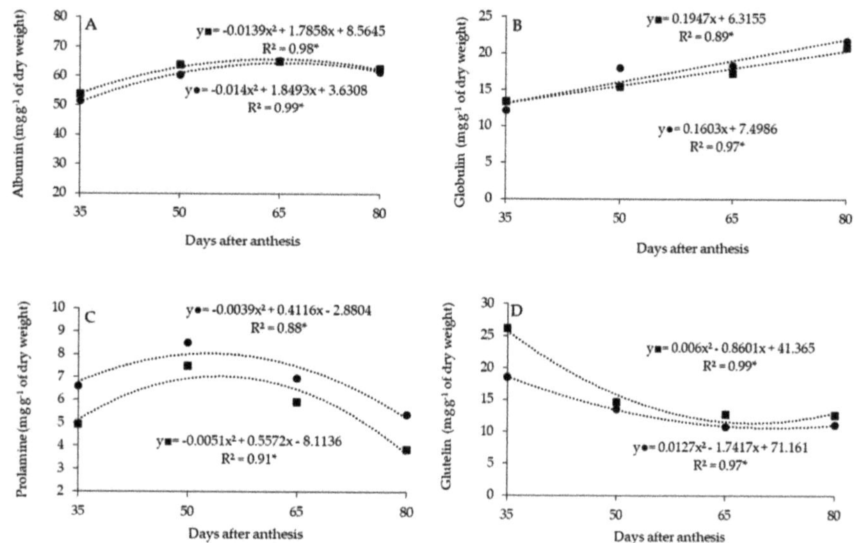

Figure 2. Albumin (**A**), globulin (**B**), prolamine (**C**) and glutelin (**D**) content in sweet pepper seeds, as a function of maturation stage, without (•) and with (■) post-harvest resting of fruits. * significant difference at 5% probability.

Globulin protein had a linear increase, with maximum contents of 21.9 and 20.3 mg g^{-1} of DM, at 80 DAA, without and with fruit rest, respectively (Figure 2B).

Glutelin content adjusted to the quadratic model according to the maturation stage; however, there were reductions over the maturation stage, with the minimum without fruit rest estimated at 10.5 mg g^{-1} of DM at 72 DAA, and with rest estimated at 11.4 mg g^{-1} of DM at 68 DAA (Figure 2D).

It was observed that the post-harvest resting of the fruits enabled higher contents of albumin and glutelin. However, the prolamine protein content was higher without fruit resting (Table 3). Protein content in the sweet pepper seeds followed the following decreasing order: albumin > globulin ≈ glutelin > prolamine.

Table 3. Albumin, globulin, prolamine and glutelin content in sweet pepper seeds as a function of post-harvest resting of the fruits.

Post-Harvest Resting	Albumin	Globulin	Prolamine	Glutelin
	mg g^{-1} of Dry Matter			
Without	59.6b	17.5a	6.9a	13.5b
With	61.3a	16.7a	5.8b	16.2a
CV (%)	2.2	8.5	8.8	7.7

Means followed by the same letter do not differ from each other by the F-test at 5% probability.

3. Discussion

3.1. Water Content, Dry Weight of One Thousand and Macronutrients Content of Seeds

During maturation inside the fruits, seeds maintain a high water content (35 to 40%) and undergo dry matter accumulation. Water is considered the vehicle for photoassimilate translocation from the plant to the seeds, so it is necessary to synthesize their reserves [6,14]. However, the water content necessary is lower than that needed to initiate germination [11,12].

Studies conducted by refs. [13,14] determined the minimum water content in 'Magda' sweet pepper seeds to be 54% at 70 DAA, and 47% at 75 DAA in sweet yellow pepper, without fruit rest. The reduction in seed water content as a function of the fruit maturation stage was also reported in other species of the *Capsicum* genus [15,16].

An increase in the dry weight of one thousand seeds (DWTS) occurred because seeds tend to increase the dry weight until physiological maturity during the maturation process [17].

The beginning of seed development was characterized by the relatively slow accumulation of dry mass. This phase predominates the cell division and expansion, which are responsible for the constitution of the adequate structure to receive the substances transferred from the mother plant. Soon after, the replacement of water content with dry matter begins after the initial seed growth [17,18].

The stage of fruit harvest for better physiological seed quality can change according to species, cultivar and environmental conditions [13,19–22]. For the DWTS, higher values were obtained after post-harvest resting. Similar results were reported in other pepper seeds [23–25].

The post-harvest temporary storage of fruits before extraction allows the seeds to complete their physiological maturation [12,13,26,27]. Thus, the reserves continue to be metabolized and translocated to the seeds, allowing increases in weight and improving the physiological and nutritional quality of the seed.

Post-harvest resting of fruits is especially important in species with an indeterminate growth habit that produce fleshy fruits, such as pepper, cucumber, tomato and other species. The post-harvest resting in species with indeterminate growth habits is helpful to improve the uniformity generated by continuous flowering, reducing the number of harvests and the exposure of fruits and seeds to unfavorable field conditions [24].

An increase in the DWTS during the maturation stage is associated with the amount of reserve accumulated during seed maturation. Studies of the sweet pepper cultivar Amarela Comprida demonstrated increases in DWTS up to 75 DAA, with a maximum weight of 6.6 g [28]. However, in the pepper cultivar 'Malagueta', the maximum DWTS obtained was 3.2 g at 80 DAA, and in the 'Biquinho' pepper, it was 2.5 g at 70 DAA [25], thus showing that values can vary according to genotype.

The decreasing order of macronutrient content in the seeds was N > K > P > Mg > S > Ca. N was the most accumulated in sweet pepper seeds. Several studies reported that the N content in the seeds was always higher than the other nutrients [4,5,29,30]. Nitrogen occupies a prominent place in the plant metabolism system because all vital plant processes are associated with proteins, in which N is an essential constituent [31,32].

N is one of the most easily translocated nutrients from leaves to fruits [33]. However, there are no studies of this translocation from fruits at rest to seeds, as may have occurred in this research, mainly in older fruits (more than 65 DAA). In younger fruits, with rest, there may have been a "dilution effect"; that is, the increase in dry seed weight with rest may have been greater than the translocation of N to seeds. In comparison, in older fruits, it is possible that the N translocation rate from fruit to seeds was more intense, favoring the increase of N content (Figure 1C and Table 2). "Nutrient dilution" is characterized when the DM growth rate is higher than the nutrient absorption rate. Similarly, it can occur in the seeds. With the advance of the maturation stage, together with the rest of the fruits, it allows the continuity of the seed maturation process and DM accumulation.

P was the third most accumulated nutrient in seeds. Phosphorous compounds are important in several reactions observed in seeds [34]. Furthermore, phosphorus is a constituent of the nucleic acid molecule, related to protein synthesis. Additionally, P is present in phospholipids and phosphate sugars, nucleotides and phytin, which is a salt with calcium and magnesium in seeds [30,35]. The requirement of this macronutrient for seeds may be associated with the fact that this nutrient provides faster initial root growth, improving initial seedling establishment. Seeds usually contain enough P to ensure maximum seedling growth for several weeks after germination [36].

The content of P in seeds is probably regulated by plants so that there is no deficiency of this nutrient during seed germination and the beginning of plant development. When there is a lack of P in the soil, it is translocated from the leaves to the fruits and seeds [37].

Regarding K, Ca and Mg content, the behavior was different from the dry matter accumulation in seeds. The content of these macronutrients had a reduction during most periods of the maturation stage (Figure 1E–G). According to [38], as the plant grows, nutrients are diluted, reducing the concentration in the tissues. The same probably happens in seeds as the maturation stages advance.

K was the second most accumulated nutrient in seeds, confirming the importance of this element in seed formation. The physiological role of K during fruit formation and maturation is mainly expressed in carbohydrate metabolism [39], which also makes it one of the most accumulated nutrients of sweet pepper seeds.

Ca was the nutrient with the lowest content in sweet pepper seeds, probably due to its low mobility in the phloem. According to ref. [29], Ca accumulation in seeds occurs only by absorption and transport during their maturation process, with no redistribution. Therefore, although fruit resting increases seed DM, it does not provide continuity of translocation of all macronutrients. This translocation occurs at a lower intensity than DM accumulation and is called the "dilution effect".

Mg in seeds is associated with proteins. However, its role is not yet fully understood, whether related to protein formation or whether this is a consequence of increased amino acid translocation from leaves to drains [40].

Without post-harvest resting, S content during maturation stages increased until 50 DAA, and decreased in older fruits (Figure 1H and Table 2), similarly to N, probably because of the "dilution effect", when the DM growth rate is higher than the nutrient absorption rate. With post-harvest resting, there is DM accumulation due to the maturation stage and to resting; as such, the S content decreases during all maturation stages, similarly to most nutrients. S regulates seed metabolism in terms of carbohydrates and storage proteins. S content in sweet pepper seeds was higher than Ca; S is responsible for carbohydrate regulation and storage of protein seed metabolism [41]. It is essential to mention that the S content in the seeds is low in most species, and that the mineral composition of the seeds can vary according to the species [42], emphasizing the importance of studying the mineral composition of seeds in different species. However, in cauliflower (*Brassica oleracea*), S was the second most accumulated nutrient in the seeds [30].

3.2. Seed Protein Content

Although proteins are part of seed reserves, not all groups are found in the seeds of a determined species. Albumin and globulin are standard in dicotyledonous seeds [8]. They are probably translocated in greater proportions to sweet pepper seeds, favoring their concentration, with albumin, the main storage protein, being present in greater quantity (Figure 2). The present study observed an increase in albumin content during maturation stages (Figure 2A), and, according to ref. [43], there are relationships between protein content, especially albumin, and seed physiological quality.

Glutelins are common proteins in cereals, and prolamins are common in grasses [8]; this may justify the lower prolamine content in sweet pepper seeds. Thus, plants accumulate reserves such as carbohydrates, oils and proteins during the maturation process, which are essential for germination and establishment [44]. As such, at the beginning of the germination process, the seeds are soaked in water, which begins the mobilization of a food reserve. The storage organs (cotyledons and endosperm) provide essential energy to nourish the seedling until stabilization [45]. Thus, the present result obtained in the study corroborates the literature. Since the reduction of storage proteins is degraded by the offending action and exopeptidases, proteolytic enzymes convert the storage proteins into soluble peptides that are further hydrolyzed into free amino acids, which are then mobilized to the embryonic axis to support growth [46,47].

4. Materials and Methods

4.1. Site Description

The experiment was conducted in a protected environment at the experimental area of São Paulo State University (UNESP), in São Manuel—SP (22°46′ S, 48°34′ W and altitude of 740 m). During the experiment, the average maximum daily temperature was 28.7 °C, the minimum was 22.6 °C, and the maximum and minimum relative humidity were 73 and 54%, respectively. The experiment was conducted in pots (13 L); the soil used in the pots was fertilized and corrected with limestone as is recommended by Bulletin 100 [44], and the top dressing by fertigation according to ref. [48].

The chemical characteristics of the soil used were: pH ($CaCl_2$): 4.4; organic matter: 5 g dm^{-3}; P(resin): 2 mg dm^{-3}; H+Al: 26 mmol$_c$ dm^{-3}; K: 1.1 mmol$_c$ dm^{-3}; Ca: 33 mmol$_c$ dm^{-3}; Mg: 4 mmol$_c$ dm^{-3}; sum of bases: 39 mmol$_c$ dm^{-3}; capacity of exchange cation: 64 mmol$_c$ dm^{-3}; base saturation: 60%.

4.2. Experiment Conduction and Experimental Design

Sowing was performed on 24th July 2017, and seedling transplantation at 47 days after sowing. An inbred line (SK 1730) from Sakata Seeds was used in this study. The management approach involved the withdrawal of sprouts until the appearance of the first flower, drip irrigation twice a day, and chemical pest and disease control when needed.

The experiment design was randomized in blocks, in a 4 $\times$ 2 factorial arrangement, and in four replications. The first factor comprised four maturation periods (35, 50, 65 and 80 days after anthesis (DAA)), and the second, the fruit post-harvest management (with and without the fruit rest for seven days after harvesting at laboratory conditions (25 $\pm$ 2 °C)). Ten plants were evaluated per plot, and all fruits fixed on the plants were harvested without thinning.

To determine the maturation period, all flowers were marked on the day of their anthesis. The harvests were performed when the fruits had the maturity stage corresponding to 35, 50, 65 and 80 DAA. Half of the fruits had their seeds extracted on the day of harvesting (without rest), and then half remained post-harvest, resting before seed extraction. At harvest time, the visual appearance of the fruits was: fully green fruits at 35 DAA; fruits with transient coloration from green to yellow at 50 DAA; fruits with 75% bright yellow color at 65 DAA; and at 80 DAA, the fruits were 100% yellow, but opaque and with less pulp firmness (Figure 3). Harvests were carried out manually, using scissors to separate the fruit from the mother plant.

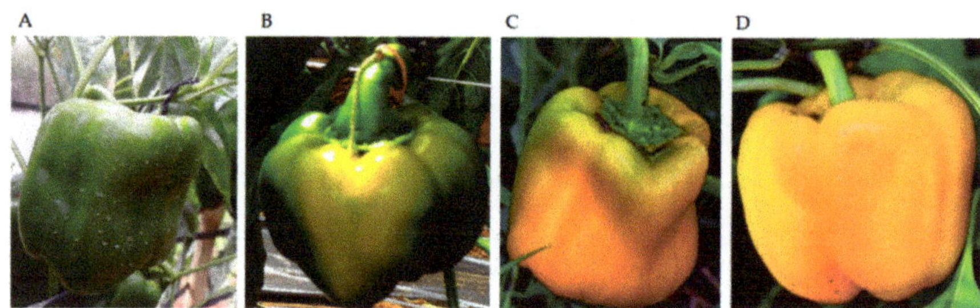

Figure 3. Visual aspects of sweet pepper fruit at 35 (**A**), 50 (**B**), 65 (**C**) and 80 (**D**) days after anthesis.

4.3. Seed Analysis

Seed water content was determined immediately after fruit extraction by the oven method at 105 ± 3 °C for 24 h, using 10 g of seeds [46]. After extraction, the seeds were put in a dry chamber (40% relative humidity and 20 °C) to reduce the seed water content to approximately 8% for storage.

The seed characteristics evaluated were the dry weight of one thousand seeds (DWTS), macronutrients (N, P, K, Ca, Mg and S) and protein (albumin, globulin, prolamine and glutelin) content.

To determine the dry weight of one thousand seeds (DWTS), seeds were dried in a forced-air oven at 65 °C until they reached a constant weight and weighed on a scale of 0.0001 g [49].

Sulfuric digestion was used to obtain the extract in order to determine the content of N, while P, K, Ca, Mg and S content were extracted by nitroperchloric acid digestion and determined by atomic absorption spectrophotometry, as described by AOAC [50]. To determine the contents of albumin, globulin, glutelin and prolamine proteins, the methodology proposed by ref. [51] was used. For all determinations, four replications were used.

4.4. Data Analysis

The data were submitted for analysis of variance (F test) and, when significant, to compare the post-harvest resting periods of the fruits, the means were considered different by the F-test ($p < 0.05$). The effects of maturation periods were analyzed by regression analysis ($p < 0.005$).

5. Conclusions

The determination of sweet pepper seed quality is affected by the stage of maturation and the presence of rest for 7 days. This is mainly due to the reduction of K, Ca and Mg, essential macronutrients for good germination. However, the reserve protein content (globulin and prolamine) increased due to the presence of maturation.

In this way, obtaining a good seed production with post-harvest rest helps with the accumulation of calcium, albumin and glutelin. Therefore, harvesting the sweet pepper fruit and resting is a management strategy that can improve the quality of seeds.

Author Contributions: Conceptualization, L.F.C. and A.I.I.C.; methodology, L.F.C., L.C., G.F.d.S. and W.A.L.Z.; formal analysis, L.F.C. and F.F.P.; investigation, L.F.C. and A.I.I.C.; writing—original draft preparation, L.F.C., G.F.d.S. and L.C.; writing—review and editing, G.F.d.S., L.C., F.F.P. and A.I.I.C.; supervision, A.I.I.C.; funding acquisition, L.F.C. and A.I.I.C. All authors have read and agreed to the published version of the manuscript.

Funding: This research was funded by the National Council for Scientific and Technological Development (CNPq), grant number 142044/2016-4.

Institutional Review Board Statement: Not applicable.

Informed Consent Statement: Not applicable.

Data Availability Statement: Not applicable.

Acknowledgments: The authors would like to acknowledge the School of Agriculture (UNESP/Botucatu) for supporting the development of this research.

Conflicts of Interest: The authors declare no conflict of interest.

References

1. Aguirre, M.; Kiegle, E.; Leo, G.; Ezquer, I. Carbohydrate reserves and seed development: An overview. *Plant Reprod.* **2018**, *31*, 263–290. [CrossRef] [PubMed]
2. Bang, T.C.; Husted, S.; Laursen, K.H.; Persson, D.P.; Schjoerring, J.K. The molecular–physiological functions of mineral macronutrients and their consequences for deficiency symptoms in plants. *New Phytol.* **2021**, *229*, 2446–2469. [CrossRef] [PubMed]
3. Ikram, A.; Saeed, F.; Afzaal, M.; Imran, A.; Niaz, B.; Tufail, T.; Hussain, M.; Anjum, F.M. Nutritional and end-use perspectives of sprouted grains: A comprehensive review. *Food Sci. Nutr.* **2021**, *9*, 4617–4628. [CrossRef] [PubMed]
4. Kołton, A.; Wojciechowska, R.; Leja, M. The effect of various light conditions and different nitrogen forms on nitrogen metabolism in pepper fruits. *Folia Hortic.* **2012**, *24*, 153–160. [CrossRef]
5. Zhang, J.; Lv, J.; Dawuda, M.M.; Xie, J.; Yu, J.; Li, J.; Zhang, X.; Tang, C.; Wang, C.; Gan, Y. Appropriate ammonium-nitrate ratio improves nutrient accumulation and fruit quality in pepper (*Capsicum annuum* L.). *Agronomy* **2019**, *9*, 683. [CrossRef]
6. Mouzo, D.; Bernal, J.; López-Pedrouso, M.; Franco, D.; Zapata, C. Advances in the biology of seed and vegetative storage proteins based on two-dimensional electrophoresis coupled to mass spectrometry. *Molecules* **2018**, *23*, 2462. [CrossRef]
7. Shewry, P.R.; Napier, J.A.; Tatham, A.S. Seed storage proteins: Structures and biosynthesis. *Plant Cell* **1995**, *7*, 945–956. [CrossRef]
8. Tan-Wilson, A.L.; Wilson, K.A. Mobilization of seed protein reserves. *Physiol. Plant.* **2012**, *145*, 140–153. [CrossRef]
9. Koornneef, M.; Bentsink, L.; Hilhorst, H. Seed dormancy and germination. *Curr. Opin. Plant Biol.* **2002**, *5*, 33–36. [CrossRef]
10. Han, C.; Yang, P. Studies on the molecular mechanisms of seed germination. *Proteomics* **2015**, *15*, 1671–1679. [CrossRef]
11. Nakada-Freitas, P.G.; Lanna, N.B.L.; Silva, P.N.L.; Bardiviesso, E.M.; Tavares, A.E.B.; Claudio, M.T.R.; Cardoso, A.I.I.; Magro, F.O.; Araujo, H.S. The physiological quality of "chilli pepper" seeds, extracted from fruits harvested at different stages of maturation, with and without post-harvest rest. *Aust. J. Crop Sci.* **2020**, *14*, 739–743. [CrossRef]
12. Colombari, L.F.; da Silva, G.F.; Chamma, L.; Chaves, P.P.N.; Martins, B.N.M.; Jorge, L.G.; de Lima Silva, P.N.; Putti, F.F.; Cardoso, A.I.I. Maturation and resting of sweet pepper fruits on physiological quality and biochemical response of seeds. *Brazilian Arch. Biol. Technol.* **2021**, *64*, e21200733. [CrossRef]
13. Medeiros, A.D.; Zavala-León, M.J.; da Silva, L.J.; Oliveira, A.M.S.; dos Santos Dias, D.C.F. Relationship between internal morphology and physiological quality of pepper seeds during fruit maturation and storage. *Agron. J.* **2020**, *112*, 25–35. [CrossRef]
14. Ohanenye, I.C.; Tsopmo, A.; Ejike, C.E.C.C.; Udenigwe, C.C. Germination as a bioprocess for enhancing the quality and nutritional prospects of legume proteins. *Trends Food Sci. Technol.* **2020**, *101*, 213–222. [CrossRef]
15. Cisternas-Jamet, J.; Salvatierra-Martínez, R.; Vega-Gálvez, A.; Stoll, A.; Uribe, E.; Goñi, M.G. Biochemical composition as a function of fruit maturity stage of bell pepper (*Capsicum annum*) inoculated with Bacillus amyloliquefaciens. *Sci. Hortic.* **2020**, *263*, 109107. [CrossRef]
16. Yildirim, K.C.; Canik Orel, D.; Okyay, H.; Gursan, M.M.; Demir, I. Quality of immature and mature pepper (*Capsicum annuum* L.) seeds in relation to bio-priming with endophytic pseudomonas and *Bacillus* spp. *Horticulturae* **2021**, *7*, 75. [CrossRef]
17. Bewley, J.D.; Bradford, K.J.; Hilhorst, H.W.M.; Nonogaki, H. *Seeds*; Springer: New York, NY, USA, 2013.
18. Wolny, E.; Betekhtin, A.; Rojek, M.; Braszewska-Zalewska, A.; Lusinska, J.; Hasterok, R. Germination and the early stages of seedling development in *Brachypodium distachyon*. *Int. J. Mol. Sci.* **2018**, *19*, 2916. [CrossRef]
19. Linkies, A.; Leubner-Metzger, G. Beyond gibberellins and abscisic acid: How ethylene and jasmonates control seed germination. *Plant Cell Rep.* **2012**, *31*, 253–270. [CrossRef]
20. Yan, D.; Duermeyer, L.; Leoveanu, C.; Nambara, E. The functions of the endosperm during seed germination. *Plant Cell Physiol.* **2014**, *55*, 1521–1533. [CrossRef]
21. Bortey, H.M.; Dzomeku, B.M. Fruit and seed quality of okra [*Abelmoschus esculentus* (L.) Moench] as influenced by harvesting stage and drying method. *Indian J. Agric. Res.* **2016**, *50*, 330–334. [CrossRef]
22. Singkaew, J.; Miyagawa, S.; Wongs-Aree, C.; Vichitsoonthonkul, T.; Sokaokha, S.; Photchanachai, S. Season, fruit maturity, and storage affect on the physiological quality of F1 hybrid 'VTM580' tomato seeds and seedlings. *Hortic. J.* **2017**, *86*, 121–131. [CrossRef]
23. Lima, J.M.E.; Smiderle, O.J. Physiologic quality of pepper seeds obtained from to fruit maturation and storage and storage. *Semin. Ciências Agrárias* **2014**, *35*, 251. [CrossRef]
24. Vidigalde Souza Vidigal, D.; Dias, D.C.F.S.; de Rezende Von Pinho, E.V.; dos Santos Dias, L.A. Physiological and enzymatic changes during pepper seeds (*Capsicum annuum* L.) maturation. *Rev. Bras. Sementes* **2009**, *31*, 129–136. [CrossRef]
25. Caixeta, F.; Von Pinho, É.V.R.; Catão, R.M.G.; Pereira, P.H.A.R.; Catão, H.C.R.M. Physiological and biochemical alterations during germination and storage of habanero pepper seeds. *African J. Agric. Res.* **2014**, *9*, 627–635. [CrossRef]

26. Nogueira, J.L.; da Silva, B.A.; Mógor, Á.F.; de Souza Grzybowski, C.R.; Panobianco, M. Quality of organically produced bell pepper seeds. *J. Seed Sci.* **2017**, *39*, 100–105. [CrossRef]
27. Pinheiro, D.T.; de Oliveira, R.M.; de Souza Silveira, A.; León, M.J.Z.; Brum, L.B.T.L.; dos Santos Dias, D.C.F. Antioxidant enzyme activity and physiological potential of *Capsicum baccatum* var. baccatum seeds as a function of post-harvest storage of fruit. *J. Seed Sci.* **2020**, *42*, e202042028. [CrossRef]
28. de Souza Vidigal, D.; dos Santos Dias, D.C.F.; dos Santos Dias, L.A.; Finger, F.L. Changes in seed quality during fruit maturation of sweet pepper. *Sci. Agric.* **2011**, *68*, 535–539. [CrossRef]
29. Kano, C.; Cardoso, A.I.I.; Villas Bôas, R.L. Macronutrient content in lettuce affected by potassium side dressing. *Hortic. Bras.* **2010**, *28*, 287–291. [CrossRef]
30. Cardoso, A.I.; Claudio, M.T.; Nakada-Freitas, P.G.; Magro, F.O.; Tavares, A.E. Phosphate fertilization over the accumulation of macronutrients in cauliflower seed production. *Hortic. Bras.* **2016**, *34*, 196–201. [CrossRef]
31. Masclaux-Daubresse, C.; Daniel-Vedele, F.; Dechorgnat, J.; Chardon, F.; Gaufichon, L.; Suzuki, A. Nitrogen uptake, assimilation and remobilization in plants: Challenges for sustainable and productive agriculture. *Ann. Bot.* **2010**, *105*, 1141–1157. [CrossRef]
32. Sandhu, N.; Sethi, M.; Kumar, A.; Dang, D.; Singh, J.; Chhuneja, P. Biochemical and genetic approaches improving nitrogen use efficiency in cereal crops: A review. *Front. Plant Sci.* **2021**, *12*, 657629. [CrossRef] [PubMed]
33. Doyle, J.W.; Nambeesan, S.U.; Malladi, A. Physiology of nitrogen and calcium nutrition in blueberry (*Vaccinium* sp.). *Agronomy* **2021**, *11*, 765. [CrossRef]
34. Marschner, P. *Mineral Nutrition of Higher Plants*; Academic Press: London, UK, 2012.
35. Schachtman, D.P.; Reid, R.J.; Ayling, S.M. Phosphorus uptake by plants: From soil to cell. *Plant Physiol.* **1998**, *116*, 447–453. [CrossRef] [PubMed]
36. White, P.J.; Veneklaas, E.J. Nature and nurture: The importance of seed phosphorus content. *Plant Soil* **2012**, *357*, 1–8. [CrossRef]
37. Görlach, B.M.; Sagervanshi, A.; Henningsen, J.N.; Pitann, B.; Mühling, K.H. Uptake, subcellular distribution, and translocation of foliar-applied phosphorus: Short-term effects on ion relations in deficient young maize plants. *Plant Physiol. Biochem.* **2021**, *166*, 677–688. [CrossRef]
38. Lemaire, G.; Sinclair, T.; Sadras, V.; Bélanger, G. Allometric approach to crop nutrition and implications for crop diagnosis and phenotyping. A review. *Agron. Sustain. Dev.* **2019**, *39*, 27. [CrossRef]
39. Sawan, Z.M.; Fahmy, A.H.; Yousef, S.E. Effect of potassium, zinc and phosphorus on seed yield, seed viability and seedling vigor of cotton (*Gossypium barbadense* L.). *Arch. Agron. Soil Sci.* **2011**, *57*, 75–90. [CrossRef]
40. Ceylan, Y.; Kutman, U.B.; Mengutay, M.; Cakmak, I. Magnesium applications to growth medium and foliage affect the starch distribution, increase the grain size and improve the seed germination in wheat. *Plant Soil* **2016**, *406*, 145–156. [CrossRef]
41. Chandra, N.; Pandey, N. Role of sulfur nutrition in plant and seed metabolism of *Glycine max* L. *J. Plant Nutr.* **2016**, *39*, 1103–1111. [CrossRef]
42. Mondal, S.; Pramanik, K.; Panda, D.; Dutta, D.; Karmakar, S.; Bose, B. Sulfur in Seeds: An Overview. *Plants* **2022**, *11*, 450. [CrossRef]
43. Dos Reis, A.R.; Boleta, E.H.M.; Alves, C.Z.; Cotrim, M.F.; Barbosa, J.Z.; Silva, V.M.; Porto, R.L.; Lanza, M.G.D.B.; Lavres, J.; Gomes, M.H.F.; et al. Selenium toxicity in upland field-grown rice: Seed physiology responses and nutrient distribution using the μ-XRF technique. *Ecotoxicol. Environ. Saf.* **2020**, *190*, 110147. [CrossRef]
44. Ramakrishna, V. Mobilization of albumins and globulins during germination of Indian bean (*Dolichos lablab* L. var. lignosus) seeds. *Acta Bot. Croat.* **2007**, *66*, 135–142.
45. Zhao, M.; Zhang, H.; Yan, H.; Qiu, L.; Baskin, C.C. Mobilization and role of starch, protein, and fat reserves during seed germination of six wild grassland species. *Front. Plant Sci.* **2018**, *9*, 234. [CrossRef]
46. Schlereth, A.; Standhardt, D.; Mock, H.-P.; Müntz, K. Stored cysteine proteinases start globulin mobilization in protein bodies of embryonic axes and cotyledons during vetch (*Vicia sativa* L.) seed germination. *Planta* **2001**, *212*, 718–727. [CrossRef]
47. Müntz, K.; Belozersky, M.A.; Dunaevsky, Y.E.; Schlereth, A.; Tiedemann, J. Stored proteinases and the initiation of storage protein mobilization in seeds during germination and seedling growth. *J. Exp. Bot.* **2001**, *52*, 1741–1752. [CrossRef]
48. Trani, P.E.; Tivelli, S.W.; Carrijo, O.A. *Fertirrigação em Hortaliças*, 2nd ed.; Instituto Agronômico: Campinas, Brazil, 2011.
49. Brasil. *Regras Para Análise de Sementes*; MAPA/ACS: Brasília, Brazil, 2009; 399p.
50. AOAC. *Official Methods of Analysis of AOAC International*; AOAC International: Rockville, MD, USA, 2016.
51. Bradford, M. A Rapid and sensitive method for the quantitation of microgram quantities of protein utilizing the principle of protein-dye binding. *Anal. Biochem.* **1976**, *72*, 248–254. [CrossRef]

MDPI

Article

Effect of Pyroligneous Acid on the Productivity and Nutritional Quality of Greenhouse Tomato

Raphael Ofoe [1], Dengge Qin [1], Lokanadha R. Gunupuru [1], Raymond H. Thomas [2] and Lord Abbey [1,*]

[1] Department of Plant, Food, and Environmental Sciences, Faculty of Agriculture, Dalhousie University, Halifax, NS B2N 5E3, Canada; Raphael.ofoe@dal.ca (R.O.); dn376773@dal.ca (D.Q.); lk811170@dal.ca (L.R.G.)
[2] School of Science and the Environment, Grenfell Campus, Memorial University of Newfoundland, Corner Brook, NL A2H 5G4, Canada; rthomas@grenfell.mun.ca
* Correspondence: loab07@gmail.com

Abstract: Pyroligneous acid (PA) is a reddish-brown liquid obtained through the condensation of smoke formed during biochar production. PA contains bioactive compounds that can be utilized in agriculture to improve plant productivity and quality of edible parts. In this study, we investigated the biostimulatory effect of varying concentrations of PA (i.e., 0%, 0.25%, 0.5%, 1%, and 2% PA/ddH_2O (v/v)) application on tomato (*Solanum lycopersicum* 'Scotia') plant growth and fruit quality under greenhouse conditions. Plants treated with 0.25% PA exhibited a significantly ($p < 0.001$) higher sub-stomatal CO_2 concentration and a comparable leaf transpiration rate and stomatal conductance. The total number of fruits was significantly ($p < 0.005$) increased by approximately 65.6% and 34.4% following the application of 0.5% and 0.25% PA, respectively, compared to the control. The 0.5% PA enhanced the total weight of fruits by approximately 25.5%, while the 0.25% PA increased the elemental composition of the fruits. However, the highest PA concentration of 2% significantly ($p > 0.05$) reduced plant growth and yield, but significantly ($p < 0.001$) enhanced tomato fruit juice Brix, electrical conductivity, total dissolved solids, and titratable acidity. Additionally, total phenolic and flavonoid contents were significantly ($p < 0.001$) increased by the 2% PA. However, the highest carotenoid content was obtained with the 0.5% and 1% PA treatments. Additionally, PA treatment of the tomato plants resulted in a significantly ($p < 0.001$) high total ascorbate content, but reduced fruit peroxidase activity compared to the control. These indicate that PA can potentially be used as a biostimulant for a higher yield and nutritional quality of tomato.

Keywords: *Solanum lycopersicum*; biostimulant; pyroligneous acid; vegetable production; post-harvest

Citation: Ofoe, R.; Qin, D.; Gunupuru, L.R.; Thomas, R.H.; Abbey, L. Effect of Pyroligneous Acid on the Productivity and Nutritional Quality of Greenhouse Tomato. *Plants* **2022**, *11*, 1650. https://doi.org/10.3390/plants11131650

Academic Editors: Tika Adhikari and Fermin Morales

Received: 24 May 2022
Accepted: 20 June 2022
Published: 22 June 2022

1. Introduction

Tomato (*Solanum lycopersicum*) is among the most cultivated greenhouse vegetable crops worldwide [1], and is known to be a rich source of health-promoting phytochemicals including carotenoids, phenolics, flavonoids, and ascorbic acid [2]. These phytochemicals exhibit antioxidant properties, which protect cells against oxidative stress by scavenging reactive oxygen species. Its antioxidant properties are known to induce anticancer, anti-inflammatory, and chemo-preventive effects. Thus, contributing largely to the prevention of chronic diseases such as cardiovascular, cancer, atherosclerosis, and neurodegenerative disorders [2,3]. The flavor and dietary qualities of food, which strongly influence consumers preference, are usually associated with physical characteristics (e.g., chewability and texture) and chemical composition (pH, °Brix, elements, carotenoids, phenolics, and flavonoid) [4]. These properties can be influenced by growing conditions, environmental factors, and the genetic characteristics of the plant. As a result, current greenhouse producers seek alternative inputs which rely mostly on organic amendments to improve the yield and quality of tomato fruits. One such input is the use of pyroligneous acid (PA), which is a natural and environmentally friendly by-product of pyrolysis of plant biomass [5].

During pyrolysis, organic biomass is burnt at a high temperature under the presence of limited oxygen and the gaseous and smoke phase is condensed to produce a liquid smoke [6]. The condensed liquid smoke is stabilized by allowing it to stand for six months, which results in the formation of wood tar at the bottom, light oil at the top and condensed aqueous translucent PA. This aqueous translucent PA is also known as wood vinegar, bio-oil or liquid smoke [6]. PA has a smoky odor and the color may vary from light yellow to reddish-brown depending on the feedstock [7]. It is a complex mixture containing 80–90% water as a major component and over 200 water-soluble chemical compounds including nitrogen, phenolics, organic acids, sugar derivates, alcohols, and esters [6,8,9]. The chemical composition of PA mainly depends on the temperature, heating rate, feedstock, and residence time, and has been widely used in diverse areas including agriculture, food and medicine [6,10]. Evidence revealed that PA also contains a butanolide, a biologically active compound, that belongs to a new family of phytohormones known as karrikinolide or karrikins [11,12]. Interestingly, the signaling mechanism and mode of action of karrikins are analogous to that of known phytohormones [11,13,14], suggesting that PA at an appropriate concentration can positively influence plant growth and productivity. Furthermore, karrikins are thermal resistant, hydrophilic, and long lasting and can therefore remain highly potent at a wide range of concentrations. Several studies revealed that karrikins stimulate seed germination and regulate seedling photomorphogenesis by enhancing seedling sensitivity to light [11,12,15–18].

PA is commonly used as a biostimulant to improve plant growth and productivity [6]. Depending on the concentration, PA can be used as an antimicrobial agent [19,20], a herbicide [21], a soil enhancer [22], and an insect repellent [23] or promote root development [24,25] and microbial activities [26] when diluted. Recent studies reported that PA enhances seed germination rate, vegetative and reproductive growth of several plants species [6,24,25,27–29]. However, the concentration of PA applied to promote plant growth varied between studies. For instance, it was reported that the application of 1:500 (v/v) increased tomato yield but did not affect fruit nutritional quality, whereas according to Mungkunkamchao et al. [27], 1:800 PA enhanced the growth and yield of tomato. Similarly, soil drench with 20% PA increased the growth and yield of rockmelon (*Cucumis melo* var. *cantalupensis*) [30]. These suggest that the effectiveness of PA is dependent on its concentration, type of crop, and mode of application. Generally, the high acidity of PA necessitates its use at low concentrations for plant growth and productivity [6]. As such, an appropriate concentration can contain the right proportions of several bioactive compounds which induce beneficial effects on crop growth and quality [17]. Furthermore, phenolic compounds in PA induce high reactive oxygen species scavenging, reducing power activities and anti-lipid peroxidation capacity [8,31]. However, the chemical composition and individual chemical activities can be influenced by the pyrolytic temperature, as a high pyrolytic temperature between 311 and 550 °C was demonstrated to exhibit the strongest antioxidant activity [8]. It was amply demonstrated that a high PA concentration increases the availability of phenolics and organic acids that could adversely affect plant growth performance [32]. All these studies demonstrated the use of PA as a natural biostimulant with high efficacy for crop production but this was not extensively explored.

Accordingly, most studies on PA efficacy and use in crop production have focused on seed priming and foliar application. There is limited information on the efficacy of drench application on crop yield and especially on crop quality [6]. Additionally, agricultural use of PA in Canada and globally is in its infant stage due to limited studies on its efficacy for growth promotion and because recommended applications rate have not been clearly established. An understanding of how PA can regulate plant growth, yield and quality of tomato under greenhouse conditions is crucial not only to growers but also to consumers and researchers. In this study, we investigated the biostimulatory effect of varying concentrations of PA for production and increase in nutritional quality of tomato 'Scotia' under greenhouse conditions.

2. Results

2.1. PA Chemical Composition

The chemical composition of PA is presented in Supplementary Table S1. The most significant elements were nitrate, nitrite, calcium and potassium. Significant amounts of organic acids (i.e., salicylic acid, oxalic acid, propionic acid, and malic acid) and small amounts of shikimic acid and acylcarnithines were also present.

2.2. Morpho-Physiological Response

PA application had no significant ($p > 0.05$) effect on plant height, stem diameter, and the number of branches and flowers (Table 1). Plant height non-significantly increased slightly with low PA concentrations, i.e., 0.25% and 0.5% PA, by *ca.* 5% compared to the control. The highest stem diameter was recorded with 0.25% PA followed by with 2% PA but was not statistically different from that of other treatments. Additionally, plants treated with 0.5% PA increased numbers of branches and flowers by *ca.* 13% and 8%, respectively, compared to that of the control although they were not statistically different ($p > 0.05$). Similarly, PA treatments had no significant ($p > 0.05$) effect on F_v/F_m, F_v/F_o, and chlorophyll content (Table 2). The effect of PA on F_v/F_m and F_v/F_o was comparable to the control. Likewise, PA had no significant ($p > 0.05$) effect on leaf intracellular CO_2 and photosynthetic rate (Table 2). However, leaf transpiration rate, sub-stomatal CO_2, and stomatal conductance were significantly ($p < 0.001$) reduced by PA compared to the control. Plants treated with 0.25% and 0.5% PA showed significant ($p < 0.001$) reductions in these physiological characteristics except for sub-stomatal CO_2, which was increased by *ca.* 3% with 0.25% PA compared to the control. On the other hand, plants treated with 1% and 2% PA exhibited significant ($p < 0.001$) reductions in leaf transpiration rate, sub-stomatal CO_2, and stomatal conductance compared to the other PA treatments.

Table 1. Morphological response of tomato 'Scotia' plants treated with pyroligneous acid (PA).

Treatment	Plant Height (cm)	Stem Diameter (mm)	Branch Number	Flower Number
Control	57.82 ± 2.86 a	9.60 ± 0.80 a	6.52 ± 0.58 a	33.50 ± 7.23 a
0.25% PA	60.50 ± 5.79 a	10.02 ± 0.61 a	5.81 ± 0.96 a	27.25 ± 7.80 a
0.5% PA	60.62 ± 5.23 a	9.32 ± 0.77 a	7.04 ± 0.82 a	38.00 ± 8.41 a
1% PA	57.71 ± 6.40 a	9.51 ± 0.92 a	5.70 ± 1.73 a	31.00 ± 11.86 a
2% PA	56.07 ± 2.97 a	9.81 ± 0.49 a	6.38 ± 1.50 a	32.25 ± 11.41 a
p-value	0.565	0.689	0.480	0.622

Values are the means ± SD of four replicates and different letters indicate significant ($p < 0.05$) difference according to Fisher's least significant difference (LSD) post hoc test.

Table 2. Physiological response of tomato 'Scotia' plants treated with pyroligneous acid (PA).

Treatment	F_v/F_o	F_v/F_m	SPAD	Intra Cellular CO_2 (μmol mol^{-1})	A (μmol m^{-2} s^{-1})	E (mol m^{-2} s^{-1})	Ci (μmol mol^{-1})	g_s (mol m^{-2} s^{-1})
Control	4.16 ± 0.41 a	0.80 ± 0.01 a	34.14 ± 5.80 a	410.56 ± 6.13 a	2.15 ± 0.60 a	2.53 ± 0.52 a	360.70 ± 30.46 ab	0.11 ± 0.02 a
0.25% PA	4.06 ± 0.27 a	0.81 ± 0.01 a	36.59 ± 3.74 a	417.74 ± 8.72 a	1.80 ± 0.84 a	2.16 ± 0.60 ab	370.27 ± 19.04 a	0.09 ± 0.01 ab
0.5% PA	3.96 ± 0.33 a	0.80 ± 0.01 a	34.07 ± 2.96 a	410.85 ± 6.61 a	2.19 ± 0.80 a	1.95 ± 0.71 b	343.01 ± 35.68 b c	0.08 ± 0.03 b
1% PA	4.07 ± 0.34 a	0.80 ± 0.01 a	35.57 ± 5.14 a	413.55 ± 13.84 a	1.80 ± 0.79 a	1.28 ± 1.03 c	325.23 ± 42.80 c	0.05 ± 0.04 c
2% PA	3.91 ± 0.24 a	0.80 ± 0.01 a	37.13 ± 6.32 a	415.68 ± 14.65 a	1.84 ± 1.38 a	1.35 ± 0.59 c	332.50 ± 41.80 c	0.05 ± 0.04 c
p-value	0.196	0.188	0.262	0.226	0.534	<0.001	<0.001	<0.001

A: photosynthetic rate; E: transpiration rate; g_s: stomatal conductance; Ci: sub-stomatal CO_2. Values are the means ± SD of four replicates and different letters indicate significant ($p < 0.05$) difference according to Fisher's least significant difference (LSD) post hoc test.

The application of 0.25% PA increased above-ground fresh weight but similar to the control (Figure 1A). However, tomato plants treated with 0.5% and 1% PA reduced the above-ground fresh weight by *ca.* 13% compared to the control. The above-ground dry weight of the tomato plant treated with 0.25% PA was significantly ($p < 0.005$) increased by *ca.* 11% compared to the control (Figure 1B). In contrast, the 0.5% and 1% PA reduced the above-ground plant dry weight but was not significantly ($p > 0.05$) different from those of the control and the 2% PA treatment.

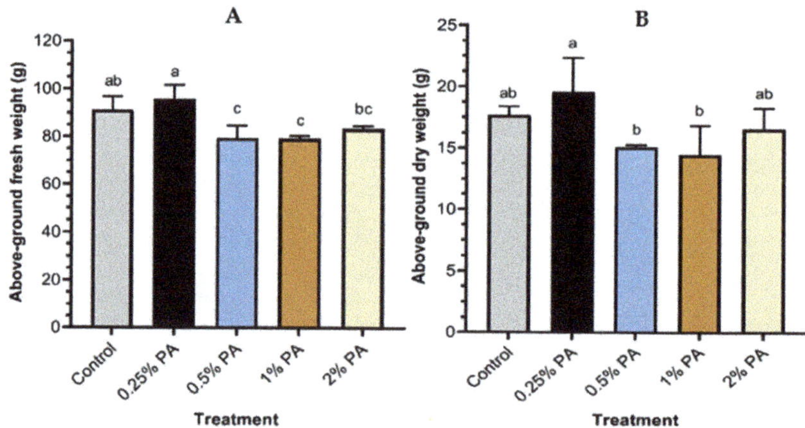

Figure 1. Pyroligneous acid effect on tomato plant above-ground biomass: (**A**) fresh weight and (**B**) dry weight. Values are the means of four replicates and different lowercase alphabetical letters indicate significant ($p < 0.05$) difference according to Fisher's least significant difference (LSD) post hoc test. Error bars show the standard deviations.

2.3. *Fruit Yield and Quality*

The 0.5% PA treatment increased the total fruit weight by *ca.* 26% although not significantly different from that of 0.25% PA and the control (Figure 2A). However, 2% PA had a significant reduction in total fruit weight, which is not different from that of the 1% PA-treated plants. Similarly, the number of fruits was significantly ($p < 0.005$) increased by *ca.* 66% and *ca.* 34% by 0.5% and 0.25% PA, respectively, compared to the control (Figure 2B). Nevertheless, the application of 2% PA and e control reduced the number of fruits compared to the other PA treatments. Fruit morphological characteristics including polar (Figure 2C) and equatorial diameters (Figure 2D) were not significantly ($p > 0.05$) affected by PA treatment. Tomato fruit juice pH, °Brix, salinity, electric conductivity (EC), total dissolved solids (TSS) and titratable acidity (TA) were significantly ($p < 0.001$) affected by PA treatment (Table 3). Juice pH was significantly ($p < 0.001$) increased by *ca.* 3.3% and 1.3% following the application of 0.25% and 0.5% PA to the plants, respectively, compared to the control. An increase in PA concentration from 1% to 2% did not alter fruit juice pH. The °Brix content of the fruits was increased by *ca.* 13% following the application of 2% PA compared to the control (Table 3). However, °Brix content was significantly ($p < 0.001$) reduced by *ca.* 45% in fruits following the application of 0.25% PA compared to the control.

A significantly ($p < 0.001$) high fruit juice salinity was noticed with the 2% PA treatment compared to the control, while the 0.25% PA recorded the least salinity (Table 3). A considerable increase in fruit electrical conductivity was recorded with the 2% PA, while the least PA of 0.25% reduced fruit juice electrical conductivity. Likewise, the 2% PA recorded the highest fruit juice total dissolved solids (Table 3). Moreover, fruit titratable acidity was significantly ($p < 0.001$) increased by *ca.* 39% upon the application of 2% PA compared to the control (Table 3). Nevertheless, the 0.25% PA had a significant ($p < 0.001$) reduction on fruit TA, which was not different from those of 0.5% PA and 1% PA treatments.

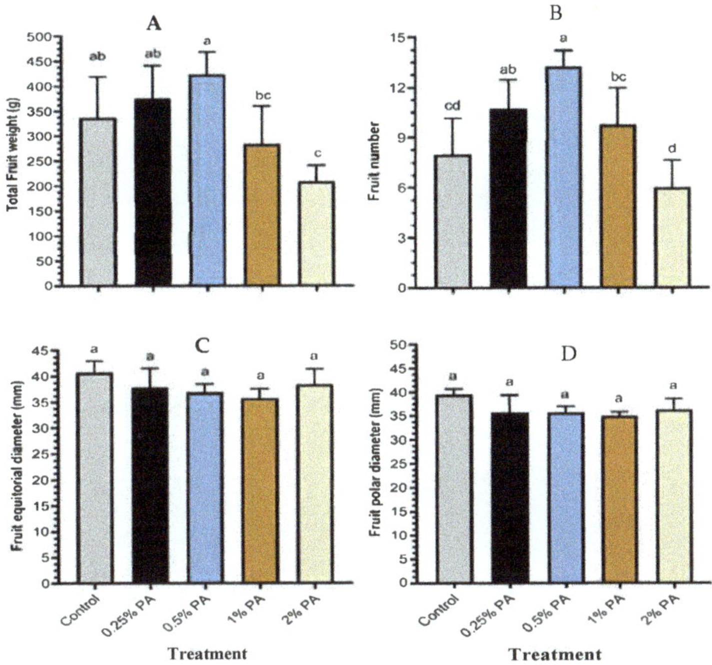

Figure 2. Fruit yield of tomato 'Scotia' in response to pyroligneous acid treatment: (**A**) total fruit weight, (**B**) fruit number, (**C**) fruit polar diameter, and (**D**) fruit equatorial diameter. Values are the means of four replicates and different lowercase alphabetical letters indicate significant ($p < 0.05$) difference according to Fisher's least significant difference (LSD) post hoc test. Error bars show the standard deviations.

Table 3. Chemical quality of tomato 'Scotia' fruits from plants treated with pyroligneous acid (PA).

Treatment	Juice pH	°Brix	Salinity (g L^{-1})	EC (mS)	TDS (g L^{-1})	TA (% Citric Acid)
Control	3.60 ± 0.04 c	5.67 ± 0.05 b	2.95 ± 0.02 b	5.42 ± 0.03 b	3.80 ± 0.01 b	0.26 ± 0.01 b
0.25% PA	3.72 ± 0.01 a	3.12 ± 0.05 d	1.66 ± 0.02 e	3.11 ± 0.08 e	2.20 ± 0.02 e	0.23 ± 0.03 c
0.5% PA	3.67 ± 0.01 b	5.62 ± 0.13 b	2.68 ± 0.03 c	5.01 ± 0.07 c	3.44 ± 0.07 c	0.24 ± 0.01 b c
1% PA	3.62 ± 0.03 c	5.20 ± 0.14 c	2.50 ± 0.03 d	4.65 ± 0.04 d	3.22 ± 0.02 d	0.23 ± 0.01 b c
2% PA	3.62 ± 0.03 c	6.42 ± 0.10 a	3.01 ± 0.03 a	5.71 ± 0.04 a	3.94 ± 0.04 a	0.36 ± 0.01 a
p-value	<0.001	<0.001	<0.001	<0.001	<0.001	<0.001

EC: electrical conductivity; TDS: total dissolved solids; TA: titratable acidity. Values are the means ± SD of four replicates and different letters indicate significant ($p < 0.05$) difference according to Fisher's least significant difference (LSD) post hoc test.

2.4. Fruit Biochemicals and Peroxidase Activities

Carotenoid was significantly ($p < 0.05$) increased by the 0.5% PA and 1% PA by *ca.* 20% and 22%, respectively, compared to that of the control (Figure 3A). The carotenoid contents of the 0.5% and 1% PA fruits were not statistically ($p > 0.05$) different from that of the 2% PA, while the carotenoid content of the 0.25% PA fruits was low and comparable to the control. Tomato fruit total phenolics (Figure 3B) and flavonoid were significantly ($p < 0.001$) influenced with PA treatment (Figure 3C). The application of 2% PA exhibited a considerably higher fruit total phenolic compounds (*ca.* 23%) and flavonoid content (*ca.* 39%) compared to the control. The 0.5% PA reduced fruit TPC and flavonoid contents. Total ascorbate was increased by *ca.* 377%, *ca.* 177%, *ca.* 165% and *ca.* 129% following the application of 2%, 0.25%, 1% and 0.5% PA, respectively, compared to the control (Figure 3D).

Although 0.5% PA had the highest impact on total fruit protein, it was not statistically ($p > 0.05$) different from those of the 0.25% PA and the control treatments (Figure 3E). However, the 2% PA significantly ($p < 0.001$) reduced total fruit protein content compared to the control. Furthermore, PA caused a significant ($p < 0.001$) reduction in total fruit sugar content (Figure 3F). The 1% PA-treated plants exhibited the least total fruit sugar content, while the 2% PA slightly increased total fruit sugar but was *ca.* 5% lower than that of the control. Furthermore, PA application exhibited a significant ($p < 0.001$) reduction in fruit peroxidase activity (Figure 4). The reduction in peroxidase activity was more obvious in the 0.25% PA fruits followed by the 1% PA and the 2% PA fruits.

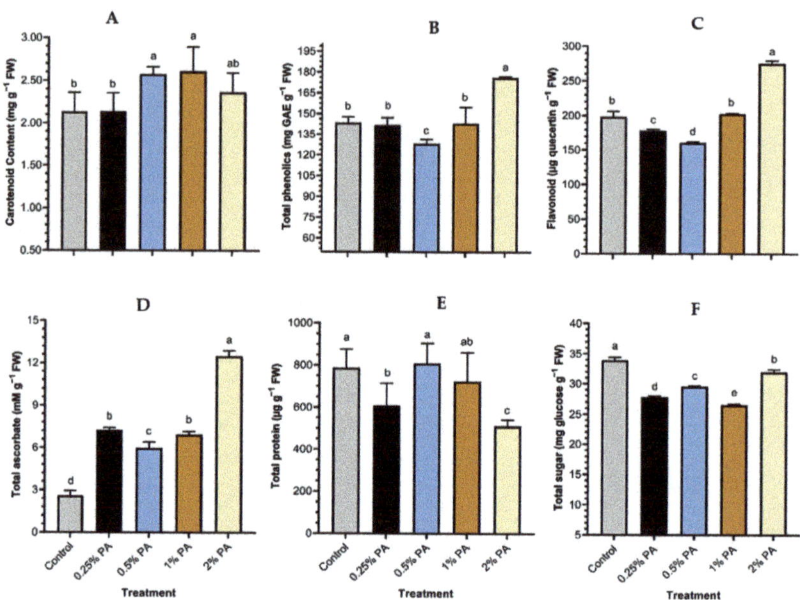

Figure 3. Tomato 'Scotia' fruit biochemical content in response to pyroligneous acid treatment: (**A**) carotenoid content, (**B**) total phenolic content, (**C**) flavonoid content, (**D**) total ascorbate content, (**E**) total protein content, and (**F**) total sugar content. Values are the means of four replicates and different lowercase alphabetical letters indicate significant ($p < 0.05$) difference according to Fisher's least significant difference (LSD) post hoc test. Error bars show the standard deviations.

2.5. Fruit Elemental Composition

Tomato 'Scotia' fruit N content was increased by *ca.* 10% upon plant application with 0.25% PA compared to the control but was reduced by the 0.5% PA (Table 4). Fruit Ca was markedly increased by *ca.* 29% upon plant treatment with the 1% PA, but was reduced by the 0.5% PA. Generally, PA had no effect on fruit K compared to the control. However, fruit Mg was increased by *ca.* 13% with the 0.25% PA but was reduced by *ca.* 12% with the 0.5% PA compared to the control. Fruit P content was increased slightly by the 2% PA, which was similar to the effect of the 0.25% PA but was reduced by the 0.5% PA treatment. Fruit Na content increased by *ca.* 59% following the application of 1% PA compared to the control, but was reduced by the 0.5% PA. Variation in PA concentration did not change fruit B content. Overall, Fe, Zn, Mn and Cu, contents were increased with the application of 0.25% PA by *ca.* 8%, *ca.* 8%, *ca.* 9% and *ca.* 15%, respectively, compared to the control. However, the 0.5% PA markedly reduced these four elements in the fruits.

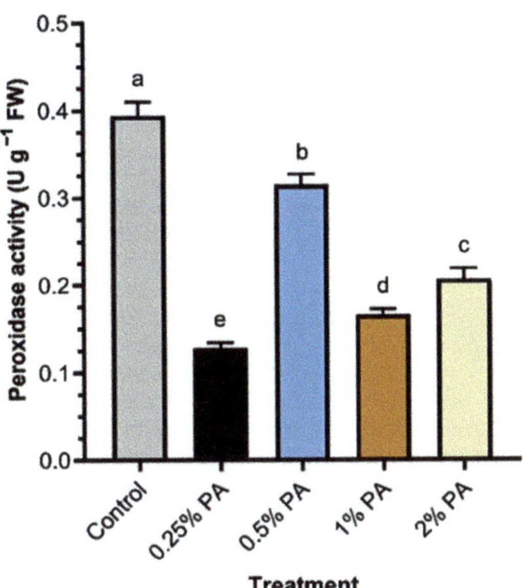

Figure 4. Peroxidase activity of tomato 'Scotia' fruit in response to pyroligneous acid treatment. Values are the means of four replicates and different lowercase alphabetical letters indicate significant ($p < 0.05$) difference according to Fisher's least significant difference (LSD) post hoc test. Error bars show the standard deviations.

Table 4. Tomato 'Scotia' fruit elemental composition in response to pyroligneous acid (PA) treatments.

Element	Treatment						
	Control	0.25% PA	0.5% PA	1% PA	2% PA	Mean	CV (%)
Nitrogen (N %)	1.68	1.84	1.44	1.61	1.56	1.63	9.12
Calcium (Ca %)	0.24	0.29	0.23	0.31	0.28	0.27	12.39
Potassium (K %)	2.68	2.27	2.67	2.32	2.59	2.51	7.91
Magnesium (Mg %)	0.17	0.19	0.15	0.17	0.17	0.17	8.21
Phosphorus (P %)	0.44	0.46	0.41	0.42	0.47	0.44	5.60
Sodium (Na %)	0.02	0.03	0.02	0.04	0.02	0.02	26.28
Boron (B mg L^{-1})	12.61	13.61	12.59	13.91	13.62	13.27	4.69
Copper (Cu mg L^{-1})	7.51	8.86	5.98	6.53	7.02	7.18	15.29
Iron (Fe mg L^{-1})	42.46	49.87	40.08	43.37	45.00	44.16	8.28
Manganese (Mn mg L^{-1})	26.14	28.37	22.58	27.49	25.06	25.93	8.71
Zinc (Zn mg L^{-1})	14.81	17.71	14.80	14.68	16.25	15.65	8.44

CV = coefficient of variation.

2.6. Association between Morpho-Physiological Properties of Tomato Plants and Productivity and Quality in Response to PA Application

Pearson's correlation coefficient (r) was used to further assess the association amongst the morpho-physiological, yield and quality of tomato plants in response to PA application (Table S2). The PCA biplot showed a projection of response variables in the factor spaces and explained *ca.* 69% of the total variations in the data set. The results revealed that the number of suckers had a significantly ($p < 0.05$) stronger positive correlation with the number of flowers (r = 0.903) and fruit K content (r = 0.914), while SPAD had a significantly ($p < 0.05$) stronger positive association with leaf intracellular CO_2 content (r = 0.927) and a negative correlation with photosynthetic rate (r = −0.891). Similarly, leaf transpiration had a significantly strong positively correlated with sub-stomatal CO_2

content (r = 0.888) and stomatal conductance (r = 0.996) and moderate association with photosynthetic rate (r = 0.608) and total fruit weight (r = 0.651) although this was not statistically significant. Total fruit weight exhibited a significant ($p < 0.05$) and strong positive correlation with plant height (r = 0.943) and fruit number (r = 0.887). However, it had a significantly ($p < 0.05$) strong negative interaction with total phenolics (r = −0.915) and flavonoid content (r = −0.953). Additionally, fruit number has a similar association with plant height (r = 0.915), total phenolics (r = −0.897) and flavonoid content (r = −0.906). Fruit salinity content showed a significantly strong positive correlation with EC (r = 0.998), TDS (r = 0.999) and Brix (r = 0.979), and a negative association with pH (r = −0.864).

3. Discussion

Current crop production practices make use of natural products that can boost plant growth and the desirable dietary and nutritional quality without compromising the environment and agroecological systems. Therefore, the functional properties of various natural materials such as PA have recently attracted the interest of farmers and researchers. In this study, although the drench application of PA had no statistically significant effect on tomato 'Scotia' plant morphological parameters, they were slightly increased by 0.25% and 0.5% PA concentrations. These results agree with other studies where the foliar application of PA influenced the morphological growth of several plant species including tomato [27], soybean [33], rockmelon [30], and rapeseed [21]. The discovery of karrikins in PA has revolutionized its use in crop production because its signaling and biophysiological activities in plants mimic that of known phytohormones [11,12,15,16]. Moreover, karrikins have been demonstrated to stimulate seed germination and plant growth [12,18]. Hence, the increase in plant growth, although not significant, can be ascribed to the presence of karrikins. Compared to the other elements, N required for vegetative plant growth was considerably high in the PA used for this study. Therefore, the increase in plant growth with PA treatment was reflected in the above-ground fresh and dry weights, which can be attributed to increased nutrient uptake and promotion of cell division and elongation [27].

Stomatal conductance and transpiration rate play a pivotal role in thermoregulation and photosynthesis [34,35]. It was demonstrated that PA and other biostimulants affect stomatal conductance in plants under both stress and non-stress conditions [21,36]. We observed that lower concentrations of PA, i.e., 0.25% or 0.5% PA, exhibited a comparable stomatal conductance and leaf transpiration effect while higher PA concentrations, i.e., >1%, reduced these parameters drastically. A reduction in stomatal conductance is an adaptive strategy used by plants to minimize water loss during water-deficit and other related climatic stress conditions. This scenario adversely affects CO_2 diffusion and net photosynthesis [37]. Although the photosynthesis rate in the present study was not affected by PA treatment, we surmised that the reduction in stomatal conductance with PA treatment could be due to adaptive thermoregulation of the photosynthesis system and stress mitigation mechanism [35], which will require further investigation.

Plant productivity (i.e., the total number of fruits and yield) increased with PA application as widely reported by many authors [18,21,27,30]. The composition of PA is complex and consists of numerous bioactive compounds including organic acids, phenolics, alcohol, alkane, and ester [18,21]. This suggests that plants with varying genotypic characteristics will respond differently to PA application. In the present study, an increase in the number of tomato fruits and fruit yield were observed with the application of 0.5% PA. The application of 0.5% and 0.25% PA may be considered less toxic to root systems and may promote root growth, thereby enhancing plant nutrient uptake and utilization [25]. Although data on trusses number were not considered, the increase in fruit number in plants treated with lower PA concentrations could suggest that fruit setting was higher in low-PA-treated plants compared to those treated with higher PA concentrations. This was reflected in the correlation analyses where total fruit weight had a strong association with fruit number. From the farmer's perspective, a slight increase in total fruit yield is considered significant improvement to the overall cashflow. Furthermore, the chemical components of PA might have interacted with and stimulated the activities of various phytohormones including gibberellin, cytokinin, auxin, and various enzymes to enhance plant growth and development as previously reported [21].

Interestingly, determinants of fruit quality such as °Brix, titratable acidity, flavonoid, phenolics, and ascorbate were increased by the 2% PA. This suggests that PA could be used to enhance crop quality for human health and nutritional purposes. These results are inconsistent with the report by Kulkarni et al. [38]. The discrepancies may be due to differences in the tested concentration, time of application, and tomato variety. Generally, tomato fruits are considered an excellent source of phytochemicals including phenolics, flavonoids, and ascorbates, which exhibit high antioxidant properties by scavenging reactive oxygen species (ROS) radicals [2]. Studies demonstrated that higher PA concentration increases the availability of phenolics and organic acids that could affect plant growth [32]. Thus, the increased tomato fruits antioxidants in the present study was highly expected since previous studies have demonstrated that phenolic compounds in PA exhibited high ROS-scavenging activities, reducing power, and anti-lipid peroxidation capacity [8,31].

Accordingly, the present finding may be attributed to the increased phenolics and organic acids as reported in *Citrus limon* [39] and *Olea europaea* [40]. The ROS-scavenging abilities of these phytochemicals protect cells against oxidative stress, which are crucial for preventing chronic diseases including cancers, atherosclerosis, and inflammation disorders [2,3,41]. Moreover, fruit carotenoids are lipophilic pigments essential for human health [42]. Carotenoid content was higher in fruits harvested from plants that were treated with 0.5% and 1% PA compared to the control. This beneficial effect of PA can be attributed to the activation of pathways involved in N metabolism [43]. Furthermore, most plants adapt to stress conditions by accumulating these compounds, which ultimately enhances fruit dietary and nutritional quality. For instance, salinity stress increase TDS, sugar, and antioxidant compounds in tomato fruits [44,45]. Hence, it is plausible that although the 2% PA did not alter the growth of the tomato plants, it stimulated the plants to accumulate these phytochemicals in the fruits.

Mineral elements represent a minute fraction of the fruit dry matter content but constitute a vital component of the quality and nutritional profile of vegetables [46]. The present study demonstrated that the application of 0.25% PA enhanced tomato fruit N, Mg, P, and all the analyzed micronutrients except B. Additionally, the 1% PA increased Ca and Na in the tomato fruits. Some possible explanations could be (1) PA increased the uptake and translocation of mineral elements due to enhanced root growth and root functional activities [24]; (2) PA activated and promoted the expression of transporter genes in root cells for efficient nutrient element transport (not determined); and (3) some bioactive compounds in PA intensified the sink effect resulting in continuous flow and accumulation of these elements [21,47]. Therefore, it can be suggested that the optimal application rate of PA for enhancing tomato fruit elemental composition may range between 0.25% and 1% PA. Similar observations were made following the application of other biostimulants that enhanced the elemental composition of numerous crops including tomato [46,48] and eggplant [49]. Therefore, increased yield and dietary and nutrition quality of tomato can be obtained when the appropriate concentration of PA is applied in a greenhouse production system.

4. Materials and Methods

4.1. Plant Material and Growing Condition

This research was carried out in the greenhouse located in the Department of Plant, Food, and Environmental Sciences, Faculty of Agriculture, Dalhousie University between November 2020 and February 2021 and repeated in March (spring) and July (summer) 2021. Tomato (*Solanum lycopersicum*) cultivar 'Scotia' seeds were purchased from Halifax Seeds (Halifax, Canada). Seeds were sterilized with 10% sodium hypochlorite (NaClO) for 10 min, and thoroughly washed three times with sterile distilled water (ddH$_2$O) followed by 70% ethanol sterilization for 5 min, and subsequently washed 5 times with sterile distilled water.

The sterilized seeds were germinated in a 32-cell pack containing Pro-Mix® BX (Premier Tech Horticulture, Québec, Canada) and grown for 30 days in a growth chamber with a day/night temperature regime of 25 °C, 16/8 h d^{-1} illumination, 300 μmol $m^{-2}\cdot s^{-1}$ light intensity and 70% relative humidity. The seedlings were transplanted at the third to fourth true-leaf stage into 11.35 L-plastic pots containing approximately 1.5 kg of Pro-Mix® BX peat-based soilless medium. The plants were climate hardened for a week before the first treatment application under greenhouse conditions at 28 °C/20 °C (day/night cycle) temperature and 70% relative humidity with a 16 h photoperiod. Supplemental lighting was provided by a 600 W HS2000 high-pressure sodium lamp with NAH600.579 ballast (P.L. Light Systems, Beamsville, Canada) throughout the planting duration.

4.2. Experimental Treatment and Design

The five experimental treatments were arranged in a completely randomized design with four replications. The experimental treatments consisted of 0.25%, 0.5%, 1%, and 2% PA, and distilled water was used as a negative control. The PA derived from white pine biomass was obtained from Proton Power Inc. (Lenoir City, TN, USA). The company (Proton Power Inc.) produces and sells graphene and biochar and not PA. The PA is a by-product to them. So, our study, which was funded by the federal agency, was to test this by-product for potential commercialization in the future by which time it will be available to purchase. At present, PA samples may be obtained from Proton Power for only research purposes before it can be available later for purchase. The chemical composition of the PA used in this study is listed in Table S1. All the treatments were applied biweekly as a soil drench to field capacity, and water-soluble compound fertilizer nitrogen-phosphorus-potassium (20:20:20) was applied at 20-day intervals. Pots were rearranged weekly on the bench to offset unpredictable occurrences due to variations in the environment. The entire study was repeated twice.

4.3. Plant Growth and Yield Components

Plant growth parameters were measured at 50 days after transplanting (DAT). Plant height was measured from the stem collar to the highest leaf tip with a ruler and the stem girth (i.e., diameter of the main stem) was measured at 10 cm from the collar with Vernier calipers (Mastercraft®, Ontario, Canada). Total numbers of flowers and suckers (i.e., branching) were recorded for each treatment. Intracellular carbon dioxide concentration, net photosynthetic rate, and stomatal conductance were determined from the same four fully expanded leaves per plant using LC*i* portable photosynthesis system (ADC BioScientific Ltd., Hoddesdon, UK). Chlorophyll fluorescence indices including maximum quantum efficiency (F_v/F_m) and potential photosynthetic capacity (F_v/F_o) were measured on the same leaves using a Chlorophyll fluorometer (Optical Science, Hudson, NH, USA) [50]. Chlorophyll content was measured on the same leaves using a chlorophyll meter (SPAD 502-plus, Spectrum Technologies, Inc., Aurora, IL, USA). The total fresh weight of the above-ground tissues (i.e., leaves and shoot) was measured with a portable balance (Ohaus navigator®, ITM Instruments Inc., Sainte-Anne-de-Bellevue, QC, Canada) and subsequently oven-dried at 65 °C for 72 h for dry weight determination. Tomato fruit yield, determined as the total fresh weight of ripe fruits per plant, was recorded using the XT portable balance. The equatorial and polar diameters of the harvested fruits were measured with the digital Vernier caliper.

4.4. Fruit Quality and Phytochemical Analysis

At harvest (75DAT), seven representative ripe fruits based on size and color were randomly selected and surface-sterilized with 70% ethanol. The pericarp (containing the epidermis) was excised from the longitudinal part of each fruit using a sterile scalpel blade. The pericarp was immediately frozen in liquid nitrogen and stored in a −80 °C freezer while the remaining fruits were frozen at −20 °C until further analysis. All frozen fruits were thawed at room temperature and fruit total soluble solids (TSS) were determined using a handheld refractometer (Atago, Japan). Briefly, ripe fruits were cut, placed in a clear Ziploc bag and hand squashed. The juice was poured into a 50 mL beaker and 500 μL was used for TSS determination expressed as degree Brix (°Brix). Fruit juice qualities including pH, salinity, total dissolved solids (TDS), and electrical conductivity (EC) were determined with a multi-purpose pH meter (EC 500 ExStik II S/N 252957, EXTECH Instrument, Nashua, New Hampshire, USA). For titratable acidity, 10 mL of juice from each treatment was diluted in 50 mL distilled water, and titratable acidity was determined at an endpoint of pH 8.1 with 0.1 N sodium hydroxide (NaOH). The mean titratable acidity was expressed in citric acid percentage [1]. The elemental composition of the fruits was determined at the Nova Scotia Department of Agriculture Laboratory Services, Truro, using inductively coupled plasma mass spectrometry (PerkinElmer 2100DV, Wellesley, Massachusetts, USA) [51].

4.4.1. Fruit Carotenoid Content

Fruit carotenoid content was determined as described by Lichtenthaler [52]. Briefly, 0.2 g of ground fruit pericarp was homogenized in 2 mL of 80% acetone. The homogenate was centrifuged at 15,000× g for 15 min and the absorbance of the supernatant was measured at 646.8, 663.2, and 470 nm using a UV–Vis spectrophotometer with 80% acetone alone as the blank. Total carotenoid content was expressed as μg g^{-1} fresh weight (FW) of the sample.

4.4.2. Total Ascorbate Content

Total ascorbate was measured following the method described by Ma et al. [53] with little modification. Approximately 0.2 g of ground fruit pericarp was homogenized in 1.5 mL ice-cold freshly prepared 5% trichloroacetic acid (TCA). The mixture was vortexed for 2 min and centrifuged at 12,000× g for 10 min at 4 °C. A volume of 100 μL of the supernatant was transferred into a new tube and 400 μL phosphate buffer (150 mM potassium dihydrogen phosphate (KH_2PO_4) (pH 7.4), 5 mM Ethylenediaminetetraacetic acid (EDTA)) was added. A volume of 100 μL of 10 mM Dithiothreitol (DTT) was added and vortexed for 30 s. A reaction mixture containing 400 μL of 10% (w/v) trichloroacetic acid (TCA), 400 μL of 44% orthophosphoric acid, 400 μL of 4% (w/v) α,α-dipyridyl in 70% ethanol and 200 μL of 30 g/L ferric chloride ($FeCl_3$) was added to obtain color. The mixture was incubated at 40 °C for 60 min in a shaking incubator and the absorbance was measured at 525 nm. The total ascorbate content was determined using a standard L-ascorbic acid curve and expressed as μmol g^{-1} FW.

4.4.3. Soluble Sugar Content

The total sugar content of the tomato fruits was estimated following the phenol-sulfuric acid method described by Dubois et al. [54]. An amount of 0.2 g of ground fruit pericarp was homogenized in 10 mL of 90% ethanol and the mixture was incubated in a water bath at 60 °C for 60 min. The final volume of the mixture was adjusted to 5 mL with 90% ethanol and centrifuged at 12,000 rpm for 3 min. An aliquot of 1 mL was transferred into a thick-walled glass test tube containing 1 mL of 5% phenol and mixed thoroughly. A volume of 5 mL of concentrated sulfuric acid was added to the reaction mixture, vortexed for 20 s, and incubated in the dark for 15 min. The mixture was cooled to room temperature and the absorbance was measured at 490 nm against a blank. Total sugar was calculated using a standard sugar curve and expressed as μg of glucose g^{-1} FW.

4.4.4. Total Phenolics Content

Total phenolics content (TPC) was determined by the Folin–Ciocalteu assay described by Ainsworth and Gillespie [55] with little modification. An amount of 0.2 g of ground fruit pericarp was homogenized in 1.5 mL of ice-cold 95% methanol and incubated in the dark at room temperature for 48 h. The mixture was centrifuged at 13,000× g for 5 min before mixing 100 μL of the supernatant to 200 μL of 10% (v/v) Folin–Ciocalteau reagent. The mixture was vortexed for 5 min, mixed with 800 μL 700 mM Na_2CO_3, and incubated in the dark at room temperature for 2 h. The absorbance of the supernatant was measured at 765 nm against a blank. TPC was calculated using a gallic acid standard curve and expressed as mg gallic acid equivalents per g FW (mg GAE g^{-1} FW).

4.4.5. Total Flavonoid Content

Total flavonoid was estimated following the colorimetric method described by Chang et al. [56]. An amount of 0.2 g of ground fruit pericarp was homogenized in 1.5 mL of ice-cold 95% methanol followed by centrifugation at 15,000× *g* for 10 min. A volume of 500 µL of supernatant was added to a reaction mixture containing 1.5 mL of 95% methanol, 0.1 mL of 10% aluminum chloride ($AlCl_3$), 0.1 mL of 1 M potassium acetate, and 2.8 mL distilled water. The mixture was incubated at room temperature for 30 min and the absorbance was measured at 415 nm against a blank lacking $AlCl_3$. Total flavonoid content was estimated using quercetin equivalents and expressed as percentage flavonoid using the formula:

$$\text{Total flavonoid} = \frac{([\text{flavonoids}](\mu g/mL) \times \text{ total volume of methanolic extract } (mL))}{\text{mass of extract } (g)}$$

4.4.6. Protein Content and Peroxidase Activity

For protein content and antioxidant enzyme activity, approximately 0.2 g of ground sample was homogenized in 3 mL ice-cold extraction buffer (50 mM potassium phosphate buffer (pH 7.0), 1% polyvinylpyrrolidone, and 0.1 mM EDTA). The homogenate was centrifuged at 15,000× *g* for 20 min at 4 °C. The supernatant (crude enzyme extract) was transferred to a new microfuge tube on ice and the protein content was measured at 595 nm after 5 min of mixing with Bradford's reagent [57]. The protein content was estimated from a standard curve of bovine serum albumin (200–900 µg mL^{-1}). Peroxidase (POD, EC 1.11.1.7) activity was determined using Pyrogallol as substrate according to Chance and Maehly [58] with little modification. The reaction mixture consisted of 100 mM potassium-phosphate buffer (pH 6.0), 5% pyrogallol, 0.5 % H_2O_2 and 100 µL of crude enzyme extract. Following reaction mixture incubation at 25 °C for 5 min, 1 mL of 2.5 N H_2SO_4 was added to stop the reaction and the absorbance was read at 420 nm against a blank (ddH_2O). One unit of POD forms 1 mg of purpurogallin from pyrogallol in 20 s at pH 6.0 at 20 °C.

4.5. Statistical Analysis

All data obtained were subjected to one-way analysis of variance (ANOVA) with the averages of the two experiments using Minitab statistical software version 20 (Minitab Inc., State College, PA, USA). Treatment means were compared using Fisher's least significant difference (LSD) post hoc test at $p \leq 0.05$. Pearson's correlation analysis was performed using XLSTAT version 19.1 (Addinsoft, New York, NY, USA).

5. Conclusions

In conclusion, the drench application of low PA concentrations of 0.25% and 0.5% increases the morpho-physiological response of tomato plants. Overall, the application of 0.5% PA enhances the number of fruits and yield of tomato but reduces the quality of the fruits. Alternatively, the application of 0.25% PA will increase the elemental composition of tomato fruits. Additionally, the drench application of 2% PA can be considered stressful to tomato plants, but significantly enhanced fruit phytochemical contents including total phenolics and flavonoids and can be adopted to improve the nutritional and health benefits of tomato fruits. Hence, PA represents a novel natural product for improvement of plant growth, productivity, and nutritional content of tomato and other plants. However, further investigation is required to elucidate the molecular basis of the effect of PA on different plant species.

Supplementary Materials: The following supporting information can be downloaded at: https://www.mdpi.com/article/10.3390/plants11131650/s1, Table S1: Chemical composition of PA obtained from White pine; Table S2: Pearson's correlation between the morpho-physiological, yield and quality of tomato plants in response to PA application.

Author Contributions: Conceptualization, L.A. and R.O.; formal analysis, R.O. and L.R.G.; funding acquisition, L.A. and R.H.T.; investigation, R.O., L.R.G. and D.Q.; methodology, R.O., L.R.G. and L.A.; project administration, L.A. resources, L.A.; supervision, R.H.T. and L.A.; validation, R.O., L.R.G. and L.A.; writing—original draft, R.O.; writing—review and editing, R.O., L.R.G. and L.A. All authors have read and agreed to the published version of the manuscript.

Funding: This work was financially supported by the Natural Sciences and Engineering Research Council of Canada (NSERC), Grant #CRDPJ532183-18.

Institutional Review Board Statement: Not applicable.

Informed Consent Statement: Not applicable.

Data Availability Statement: Not applicable.

Acknowledgments: The lead author wishes to thank all her laboratory team for their unflinching support and suggestions during this study.

Conflicts of Interest: The authors declare no conflict of interest.

References

1. Gyimah, L.A.; Amoatey, H.M.; Boatin, R.; Appiah, V.; Odai, B.T. The impact of gamma irradiation and storage on the physico-chemical properties of tomato fruits in Ghana. *Food Qual. Saf.* **2020**, *4*, 151–157. [CrossRef]
2. Chaudhary, P.; Sharma, A.; Singh, B.; Nagpal, A.K. Bioactivities of phytochemicals present in tomato. *J. Food Sci. Technol.* **2018**, *55*, 2833–2849. [CrossRef] [PubMed]
3. Nowak, D.; Gośliński, M.; Wojtowicz, E.; Przygoński, K. Antioxidant properties and phenolic compounds of vitamin C-rich juices. *J. Food Sci.* **2018**, *83*, 2237–2246. [CrossRef] [PubMed]
4. Rodrigues, M.; Baptistella, J.L.C.; Horz, D.C.; Bortolato, L.M.; Mazzafera, P. Organic plant biostimulants and fruit quality—A review. *Agronomy* **2020**, *10*, 988. [CrossRef]
5. Pan, G.; Li, L.; Liu, X.; Cheng, K.; Bian, R.; Ji, C.; Zheng, J.; Zhang, X.; Zheng, J. Industrialization of biochar from biomass pyrolysis: A new option for straw burning ban and green agriculture of China. *Sci. Technol. Rev.* **2015**, *33*, 92–101. [CrossRef]
6. Grewal, A.; Abbey, L.; Gunupuru, L.R. Production, prospects and potential application of pyroligneous acid in agriculture. *J. Anal. Appl. Pyrolysis* **2018**, *135*, 152–159. [CrossRef]
7. Theapparat, Y.; Chandumpai, A.; Leelasuphakul, W.; Laemsak, N.; Ponglimanont, C. Physicochemical characteristics of wood vinegars from carbonization of Leucaena leucocephala, Azadirachta indica, Eucalyptus camaldulensis, Hevea brasiliensis and Dendrocalamus asper. *Agric. Nat. Resour.* **2014**, *48*, 916–928.
8. Wei, Q.; Ma, X.; Zhao, Z.; Zhang, S.; Liu, S. Antioxidant activities and chemical profiles of pyroligneous acids from walnut shell. *J. Anal. Appl. Pyrolysis* **2010**, *88*, 149–154. [CrossRef]
9. Zheng, H.; Sun, C.; Hou, X.; Wu, M.; Yao, Y.; Li, F. Pyrolysis of *Arundo donax* L. to produce pyrolytic vinegar and its effect on the growth of dinoflagellate Karenia brevis. *Bioresour. Technol.* **2018**, *247*, 273–281. [CrossRef]
10. Cai, K.; Jiang, S.; Ren, C.; He, Y. Significant damage-rescuing effects of wood vinegar extract in living *Caenorhabditis elegans* under oxidative stress. *J. Sci. Food Agric.* **2012**, *92*, 29–36. [CrossRef]
11. Dixon, K.W.; Merritt, D.J.; Flematti, G.R.; Ghisalberti, E.L. Karrikinolide—A phytoreactive compound derived from smoke with applications in horticulture, ecological restoration and agriculture. *Acta Hortic.* **2009**, *813*, 155–170. [CrossRef]
12. Chiwocha, S.D.S.; Dixon, K.W.; Flematti, G.R.; Ghisalberti, E.L.; Merritt, D.J.; Nelson, D.C.; Riseborough, J.-A.M.; Smith, S.M.; Stevens, J.C. Karrikins: A new family of plant growth regulators in smoke. *Plant Sci.* **2009**, *177*, 252–256. [CrossRef]
13. Van Staden, J.; Sparg, S.G.; Kulkarni, M.G.; Light, M.E. Post-germination effects of the smoke-derived compound 3-methyl-2H-furo [2, 3-c] pyran-2-one, and its potential as a preconditioning agent. *Field Crops Res.* **2006**, *98*, 98–105. [CrossRef]
14. Gomez-Roldan, V.; Fermas, S.; Brewer, P.B.; Puech-Pagès, V.; Dun, E.A.; Pillot, J.-P.; Letisse, F.; Matusova, R.; Danoun, S.; Portais, J.-C. Strigolactone inhibition of shoot branching. *Nature* **2008**, *455*, 189–194. [CrossRef]
15. Van Staden, J.; Jager, A.K.; Light, M.E.; Burger, B.V. Isolation of the major germination cue from plant-derived smoke. *S. Afr. J. Bot.* **2004**, *70*, 654–659. [CrossRef]
16. Flematti, G.R.; Ghisalberti, E.L.; Dixon, K.W.; Trengove, R.D. Identification of alkyl substituted 2 H-furo [2, 3-c] pyran-2-ones as germination stimulants present in smoke. *J. Agric. Food Chem.* **2009**, *57*, 9475–9480. [CrossRef]
17. Nelson, D.C.; Flematti, G.R.; Ghisalberti, E.L.; Dixon, K.W.; Smith, S.M. Regulation of seed germination and seedling growth by chemical signals from burning vegetation. *Annu. Rev. Plant Biol.* **2012**, *63*, 107–130. [CrossRef]

18. Kulkarni, M.G.; Ascough, G.D.; Verschaeve, L.; Baeten, K.; Arruda, M.P.; Van Staden, J. Effect of smoke-water and a smoke-isolated butenolide on the growth and genotoxicity of commercial onion. *Sci. Hortic.* **2010**, *124*, 434–439. [CrossRef]
19. Mourant, D.; Yang, D.-Q.; Lu, X.; Roy, C. Anti-fungal properties of the pyroligneous liquors from the pyrolysis of softwood bark. *Wood Fiber Sci.* **2007**, *37*, 542–548.
20. Jung, K.-H. Growth inhibition effect of pyroligneous acid on pathogenic fungus, *Alternaria mali*, the agent of alternaria blotch of apple. *Biotechnol. Bioprocess Eng.* **2007**, *12*, 318–322. [CrossRef]
21. Zhu, K.; Gu, S.; Liu, J.; Luo, T.; Khan, Z.; Zhang, K.; Hu, L. Wood Vinegar as a Complex Growth Regulator Promotes the Growth, Yield, and Quality of Rapeseed. *Agronomy* **2021**, *11*, 510. [CrossRef]
22. Lashari, M.S.; Liu, Y.; Li, L.; Pan, W.; Fu, J.; Pan, G.; Zheng, J.; Zheng, J.; Zhang, X.; Yu, X. Effects of amendment of biochar-manure compost in conjunction with pyroligneous solution on soil quality and wheat yield of a salt-stressed cropland from Central China Great Plain. *Field Crops Res.* **2013**, *144*, 113–118. [CrossRef]
23. Mmojieje, J.; Hornung, A. The potential application of pyroligneous acid in the UK agricultural industry. *J. Crop Improv.* **2015**, *29*, 228–246. [CrossRef]
24. Wang, Y.; Qiu, L.; Song, Q.; Wang, S.; Wang, Y.; Ge, Y. Root proteomics reveals the effects of wood vinegar on wheat growth and subsequent tolerance to drought stress. *Int. J. Mol. Sci.* **2019**, *20*, 943. [CrossRef] [PubMed]
25. Ofoe, R.; Gunupuru, L.R.; Wang-Pruski, G.; Fofana, B.; Thomas, R.H.; Abbey, L. Seed priming with pyroligneous acid mitigates aluminum stress, and promotes tomato seed germination and seedling growth. *Plant Stress* **2022**, *4*, 100083. [CrossRef]
26. Steiner, C.; Das, K.C.; Garcia, M.; Förster, B.; Zech, W. Charcoal and smoke extract stimulate the soil microbial community in a highly weathered xanthic Ferralsol. *Pedobiologia* **2008**, *51*, 359–366. [CrossRef]
27. Mungkunkamchao, T.; Kesmala, T.; Pimratch, S.; Toomsan, B.; Jothityangkoon, D. Wood vinegar and fermented bioextracts: Natural products to enhance growth and yield of tomato (*Solanum lycopersicum* L.). *Sci. Hortic.* **2013**, *154*, 66–72. [CrossRef]
28. Lei, M.; Liu, B.; Wang, X. Effect of adding wood vinegar on cucumber (*Cucumis sativus* L) seed germination. In Proceedings of the Institute of Physics (IOP) Conference Series: Earth and Environmental Science, Beijing, China, 28–31 December 2017; p. 012186.
29. Shan, X.; Liu, X.; Zhang, Q. Impacts of adding different components of wood vinegar on rape (*Brassica napus* L.) seed germiantion. In Proceedings of the IOP Conference Series: Earth and Environmental Science, Beijing, China, 28–31 December 2017; p. 012183.
30. Zulkarami, B.; Ashrafuzzaman, M.; Husni, M.O.; Ismail, M.R. Effect of pyroligneous acid on growth, yield and quality improvement of rockmelon in soilless culture. *Aust. J. Crop Sci.* **2011**, *5*, 1508–1514.
31. Loo, A.; Jain, K.; Darah, I. Antioxidant and radical scavenging activities of the pyroligneous acid from a mangrove plant, *Rhizophora apiculata. Food Chem.* **2007**, *104*, 300–307. [CrossRef]
32. Loo, A.Y.; Jain, K.; Darah, I. Antioxidant activity of compounds isolated from the pyroligneous acid, *Rhizophora apiculata. Food Chem.* **2008**, *107*, 1151–1160. [CrossRef]
33. Travero, J.T.; Mihara, M. Effects of pyroligneous acid to growth and yield of soybeans (*Glycine max*). *Int. J. Environ. Rural Dev.* **2016**, *7*, 50–54.
34. Tuzet, A.; Perrier, A.; Leuning, R. A coupled model of stomatal conductance, photosynthesis and transpiration. *Plant Cell Environ.* **2003**, *26*, 1097–1116. [CrossRef]
35. Urban, J.; Ingwers, M.; McGuire, M.A.; Teskey, R.O. Stomatal conductance increases with rising temperature. *Plant Signal. Behav.* **2017**, *12*, e1356534. [CrossRef]
36. Francesca, S.; Raimondi, G.; Cirillo, V.; Maggio, A.; Barone, A.; Rigano, M.M. A Novel Plant-Based Biostimulant Improves Plant Performances under Drought Stress in Tomato. *Biol. Life Sci. Forum* **2021**, *4*, 52. [CrossRef]
37. Hasanuzzaman, M.; Parvin, K.; Bardhan, K.; Nahar, K.; Anee, T.I.; Masud, A.A.C.; Fotopoulos, V. Biostimulants for the Regulation of Reactive Oxygen Species Metabolism in Plants under Abiotic Stress. *Cell* **2021**, *10*, 2537. [CrossRef]
38. Kulkarni, M.G.; Ascough, G.D.; Van Staden, J. Smoke-water and a smoke-isolated butenolide improve growth and yield of tomatoes under greenhouse conditions. *Horttechnology* **2008**, *18*, 449–454. [CrossRef]
39. Rouina, B.; Sensoy, S.; Boukhriss, M. Saline water irrigation effects on fruit development, quality, and phenolic composition of virgin olive oils, cv. Chemlali. *J. Agric. Food Chem.* **2009**, *57*, 2803–2811. [CrossRef]
40. Pedrero, F.; Allende, A.; Gil, M.I.; Alarcón, J.J. Soil chemical properties, leaf mineral status and crop production in a lemon tree orchard irrigated with two types of wastewater. *Agric. Water Manag.* **2012**, *109*, 54–60. [CrossRef]
41. Das, K.; Roychoudhury, A. Reactive oxygen species (ROS) and response of antioxidants as ROS-scavengers during environmental stress in plants. *Front. Environ. Sci.* **2014**, *2*, 53. [CrossRef]
42. Young, A.J.; Lowe, G.L. Carotenoids—Antioxidant properties. *Antioxidants* **2018**, *7*, 28. [CrossRef]
43. Ertani, A.; Pizzeghello, D.; Francioso, O.; Sambo, P.; Sanchez-Cortes, S.; Nardi, S. *Capsicum chinensis* L. growth and nutraceutical properties are enhanced by biostimulants in a long-term period: Chemical and metabolomic approaches. *Front. Plant Sci.* **2014**, *5*, 375. [CrossRef]
44. Li, J.; Chen, J.; Qu, Z.; Wang, S.; He, P.; Zhang, N. Effects of alternating irrigation with fresh and saline water on the soil salt, soil nutrients, and yield of tomatoes. *Water* **2019**, *11*, 1693. [CrossRef]
45. Yin, Y.-G.; Kobayashi, Y.; Sanuki, A.; Kondo, S.; Fukuda, N.; Ezura, H.; Sugaya, S.; Matsukura, C. Salinity induces carbohydrate accumulation and sugar-regulated starch biosynthetic genes in tomato (*Solanum lycopersicum* L. cv. 'Micro-Tom') fruits in an ABA- and osmotic stress-independent manner. *J. Exp. Bot.* **2010**, *61*, 563–574. [CrossRef]

46. Abou Chehade, L.; Al Chami, Z.; De Pascali, S.A.; Cavoski, I.; Fanizzi, F.P. Biostimulants from food processing by-products: Agronomic, quality and metabolic impacts on organic tomato (*Solanum lycopersicum* L.). *J. Sci. Food Agric.* **2018**, *98*, 1426–1436. [CrossRef]
47. Calvo, P.; Nelson, L.; Kloepper, J.W. Agricultural uses of plant biostimulants. *Plant Soil.* **2014**, *383*, 3–41. [CrossRef]
48. Caruso, G.; De Pascale, S.; Cozzolino, E.; Cuciniello, A.; Cenvinzo, V.; Bonini, P.; Colla, G.; Rouphael, Y. Yield and nutritional quality of Vesuvian Piennolo tomato PDO as affected by farming system and biostimulant application. *Agronomy* **2019**, *9*, 505. [CrossRef]
49. Alicja, P.; Grabowska, A.; Kalisz, A.; Sekara, A. The eggplant yield and fruit composition as affected by genetic factor and biostimulant application. *Not. Bot. Horti Agrobot. Cluj-Napoca* **2019**, *47*, 929–938. [CrossRef]
50. Maxwell, K.; Johnson, G.N. Chlorophyll fluorescence—A practical guide. *J. Exp. Bot.* **2000**, *51*, 659–668. [CrossRef]
51. Donohue, S.J.; Aho, D.W.; Plank, C.O. Determination of P, K, Ca, Mg, Mn, Fe, Al, B, Cu and Zn in plant tissue by inductively coupled plasma (ICP) emission spectroscopy. *Plant Anal. Ref. Proced. South. Reg. USA* **1992**, *368*, 34–37.
52. Lichtenthaler, H.K. Chlorophylls and carotenoids: Pigments of photosynthetic biomembranes. *Meth. Enzymol.* **1987**, *148*, 350–382. [CrossRef]
53. Ma, Y.-H.; Ma, F.-W.; Zhang, J.-K.; Li, M.-J.; Wang, Y.-H.; Liang, D. Effects of high temperature on activities and gene expression of enzymes involved in ascorbate–glutathione cycle in apple leaves. *Plant Sci.* **2008**, *175*, 761–766. [CrossRef]
54. Dubois, M.; Gilles, K.A.; Hamilton, J.K.; Rebers, P.A.t.; Smith, F. Colorimetric method for determination of sugars and related substances. *Anal. Chem.* **1956**, *28*, 350–356. [CrossRef]
55. Ainsworth, E.A.; Gillespie, K.M. Estimation of total phenolic content and other oxidation substrates in plant tissues using Folin–Ciocalteu reagent. *Nat. Protoc.* **2007**, *2*, 875–877. [CrossRef] [PubMed]
56. Chang, C.-C.; Yang, M.-H.; Wen, H.-M.; Chern, J.-C. Estimation of total flavonoid content in propolis by two complementary colorimetric methods. *J. Food Drug Anal.* **2002**, *10*, 178–182. [CrossRef]
57. Bradford, M.M. A rapid and sensitive method for the quantitation of microgram quantities of protein utilizing the principle of protein-dye binding. *Anal. Biochem.* **1976**, *72*, 248–254. [CrossRef]
58. Chance, B.; Maehly, A.C. Assay of catalases and peroxidases. *Meth. Enzymol.* **1955**, *2*, 764–775. [CrossRef]

Article

Plant Growth-Promoting Rhizobacteria Improve Growth and Fruit Quality of Cucumber under Greenhouse Conditions

Gerardo Zapata-Sifuentes [1,2], Luis G. Hernandez-Montiel [3,*], Jorge Saenz-Mata [4], Manuel Fortis-Hernandez [1], Eduardo Blanco-Contreras [2], Roberto G. Chiquito-Contreras [5] and Pablo Preciado-Rangel [1,*]

1 Tecnológico Nacional de México, Instituto Tecnológico de Torreón, Carretera Torreón-San Pedro km 7.5, Torreón 27170, Mexico; gerardo.zapata@uaaan.edu.mx (G.Z.-S.); manuel.fh@torreon.tecnm.mx (M.F.-H.)
2 Departamento de Agroecología, Universidad Autónoma Agraria Antonio Narro, Unidad Laguna, Carretera Periférico s/n, Col. Valle Verde, Torreón 27054, Mexico; eduardo.blanco@uaaan.edu.mx
3 Nanotechnology & Microbial Biocontrol Group, Centro de Investigaciones Biológicas del Noroeste, Av. Politécnico Nacional 195, Col. Playa Palo Santa Rita, La Paz 23090, Mexico
4 Facultad de Ciencias Biológicas, Universidad Juárez del Estado de Durango, Av. Universidad s/n, Col. Filadelfia, Gómez Palacio 35010, México; jsaenz_mata@ujed.mx
5 Facultad de Ciencias Agrícolas, Universidad Veracruzana, Circuito Universitario Gonzalo Aguirre Beltrán s/n, Zona Universitaria, Xalapa 91090, Mexico; rchiquito@uv.mx
* Correspondence: lhernandez@cibnor.mx (L.G.H.-M.); pablo.pr@torreon.tecnm.mx (P.P.-R.)

Abstract: Cucumber fruit is rich in fiber, carbohydrates, protein, magnesium, iron, vitamin B, vitamin C, flavonoids, phenolic compounds, and antioxidants. Agrochemical-based production of cucumber has tripled yields; however, excessive synthetic fertilization has caused problems in the accumulation of salts in the soil and has increased production costs. The objective of this study was to evaluate the effect of three strains of plant growth-promoting rhizobacteria (PGPR) on cucumber fruit growth and quality under greenhouse conditions. The rhizobacteria *Pseudomonas paralactis* (KBendo6p7), *Sinorhizobium meliloti* (KBecto9p6), and *Acinetobacter radioresistens* (KBendo3p1) was adjusted to 1×10^8 CFU mL^{-1}. The results indicated that the inoculation with PGPR improved plant height, stem diameter, root length, secondary roots, biomass, fruit size, fruit diameter, and yield, as well as nutraceutical quality and antioxidant capacity, significantly increasing the response of plants inoculated with *A. radioresistens* and *S. meliloti* in comparison to the control. In sum, our findings showed the potential functions of the use of beneficial bacteria such as PGPR for crop production to reduce costs, decrease pollution, and achieve world food safety and security.

Keywords: rhizobacteria; PGPR; nutraceutical quality

Citation: Zapata-Sifuentes, G.; Hernandez-Montiel, L.G.; Saenz-Mata, J.; Fortis-Hernandez, M.; Blanco-Contreras, E.; Chiquito-Contreras, R.G.; Preciado-Rangel, P. Plant Growth-Promoting Rhizobacteria Improve Growth and Fruit Quality of Cucumber under Greenhouse Conditions. *Plants* **2022**, *11*, 1612. https://doi.org/10.3390/plants11121612

Academic Editor: Lord Abbey

Received: 25 May 2022
Accepted: 14 June 2022
Published: 20 June 2022

1. Introduction

Cucumber (*Cucumis sativus* L.) is a plant of the family Cucurbitaceae which is produced worldwide in open fields and protected agriculture [1]. Cucumber fruit is widely consumed for its taste and freshness around the world, in addition, its nutritional contribution and nutraceutical properties have positive impacts on health, especially in people with diabetes, hypertension, cardiovascular, and Alzheimer's disease [2–4]. Moreover, it is also a high source of fiber, carbohydrates, proteins, magnesium, iron, vitamin B and C, flavonoids, phenolic compounds, and antioxidants [5,6]. In the last few years, agrochemical-based production schemes have helped to increase the yield of the cucumber crops by three, however, synthetic fertilization has caused problems with salt accumulation in soil and increased production costs [7–10].

Currently, beneficial microorganisms have been applied in the plants as a sustainable alternative for food production [11], likewise, the combination with synthetic fertilizers has led to increased plant growth and productivity [7,12]. Plant growth-promoting rhizobacteria (PGPR) are a microorganism group able to increase the shoot and root length, improve water and nutrient absorption, and improve fruit quality and productivity in plants [13]. Undoubtedly, among the promoting mechanisms the production of phytohormone (auxin and cytokinins), volatile compounds, siderophores, atmospheric nitrogen fixation processes, and solubilized phosphate are the most important [12,14].

Moreover, the common genera of PGPR reported are *Azospirillum*, *Pseudomonas*, *Bacillus*, *Azotobacter*, *Caulobacter*, *Flavobacterium*, *Enterobacter*, among others [14–16]. One of the most studied genera within PGPR is the *Pseudomonas* [17]. Furthermore, the *P. paralactis* has stood out in vitro for its capacity to solubilize phosphate, indoleacetic acid (IAA), and siderophores production, however, up to now, no previous study has investigated its effect in vivo [18]. *Acinetobacter radioresistens* is another PGPR that produces IAA, it can solubilize phosphate and produce siderophores, in turn, it has promoted the growth of plants such as *Ilex paraguariensis* and *Aloe vera* [19,20]. Another one is *Sinorhizobiun meliloti* characterized by nitrogen fixation [21], IAA production [22], phosphate solubilization [23], and the growth-promoting of the *Medicago sativa* L. and *Lactuca sativa* L. [21]. Based on the above, the role of PGPR on plants is important and necessary to incorporate into the agricultural systems production. In addition, the evaluation of their potential for fruit production and quality is just as important, therefore, the aim of the present study was to evaluate the effect of PGPR on the growth and fruit quality of cucumber fruit under greenhouse conditions.

2. Results

2.1. Morphological Parameters of Cucumber Plants Inoculated with PGPR

PGPR inoculation of cucumber significantly increased plant growth (Table 1). The height of plants inoculated with rhizobacteria; *S. meliloti*, *P. paralactis*, and *A. radioresistens* showed significant differences with respect to plants without microorganisms. Plants inoculated with *S. meliloti* and *A. radioresistens* showed an increase in stem diameter of 36 and 30%, respectively, and an increase in dry biomass of 59 and 83%, respectively. Regarding root length, *A. radioresistens* promoted an increase of 135%. Finally, *P. paralactis* increased secondary roots by 97%.

Table 1. Effect of PGPR on growth of C. *sativus* cultivation under greenhouse conditions.

Treatment	Plant Height (cm)	Stem Diameter (mm)	Root Length (cm)	Secondary Roots	Dry Biomass (g)	Fruit Length (cm)	Fruit Diameter (mm)	Yield (kg/plant)
S. meliloti	156.27 a	4.97 a	17.21 ab	13 ab	20.07 a	21.33 a	54.44 a	5.96 b
P. paralactis	164.60 a	4.70 ab	16.52 ab	22 a	14.13 b	21.33 a	52.02 a	6.35 b
A. radioresistens	163.33 a	4.75 a	20.61 a	15 ab	23.10 a	22.67 a	56.52 a	6.94 a
Control	146.38 b	3.65 b	8.75 b	11 b	12.57 b	17.01 b	46.03 b	4.56 c

Different letters indicate a significant difference ($p < 0.05$) according to Tukey's test.

2.2. Fruit Length, Diameter, and Yield

Fruit length and diameter increased significantly when plants were inoculated with the three rhizobacteria and *A. radioresistens*, increasing yield by 51.9% (Table 1).

2.3. Fruit Quality, Total Phenolic Contents, Total Flavonoids, Antioxidant Capacity, and Vitamin C Content

PGPR inoculation had a significant effect on the phenolic content in cucumber fruits, with the rhizobacterium *S. meliloti* promoting the greatest increase in phenolic content by 73% (Figure 1a), flavonoids by 126% (Figure 1b), and antioxidant capacity by 47% (Figure 1c). *A. radioresistens* increased the content of vitamin C by 112% (Figure 1d).

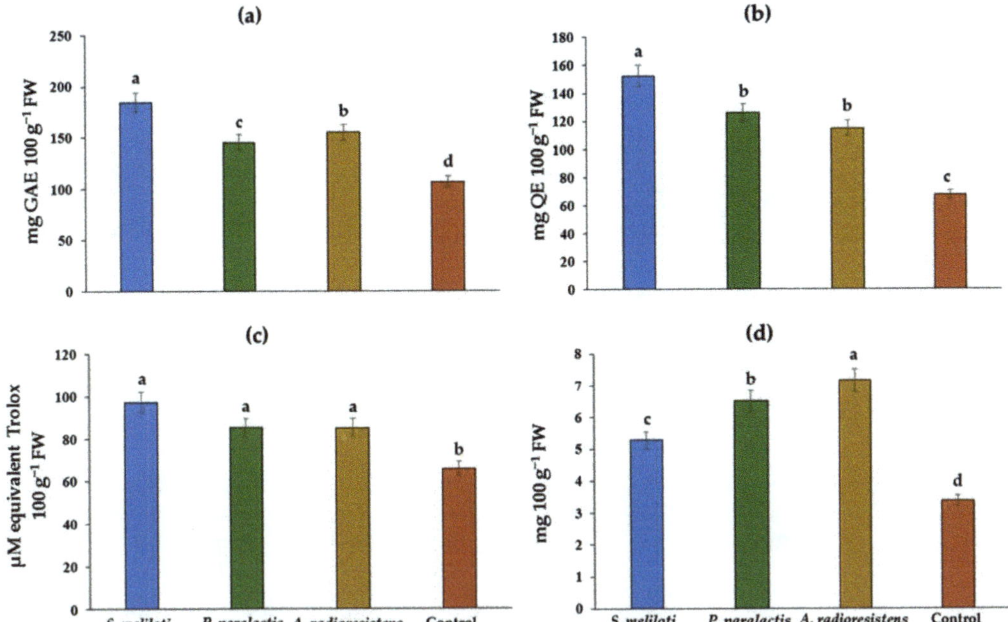

Figure 1. Effect of PGPR on the content of phenolic compounds (**a**), flavonoids (**b**), antioxidant capacity (**c**), and vitamin C (**d**) in *C. sativus* fruit under greenhouse conditions. Different letters indicate a significant difference ($p < 0.05$) according to Tukey's test.

2.4. Total Protein

Total protein values showed no significant difference, however, inoculation with *S. meliloti* increased by 64% with respect to the control (Figure 2).

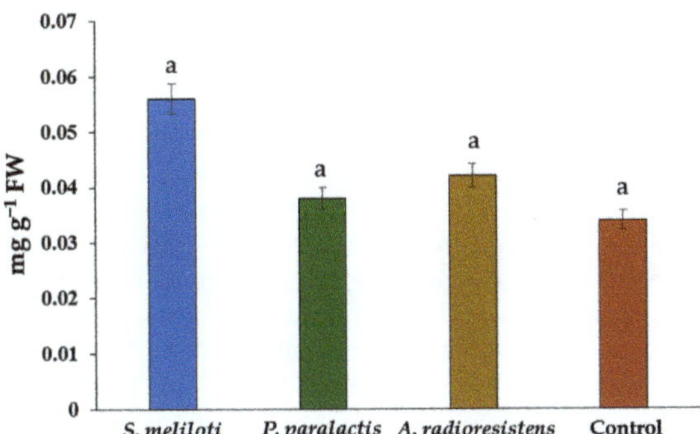

Figure 2. Effect of PGPR on total protein content in *C. sativus* fruit under greenhouse conditions. Data are shown as mean ± SD. Different letters indicate a significant difference ($p < 0.05$) according to Tukey's test.

2.5. Rhizobacterial Population

CFU counts indicated the presence of PGPR in the root of cucumber plants. There was a statistical difference between treatments, with *P. paralactis* being 409% higher than the control (Figure 3).

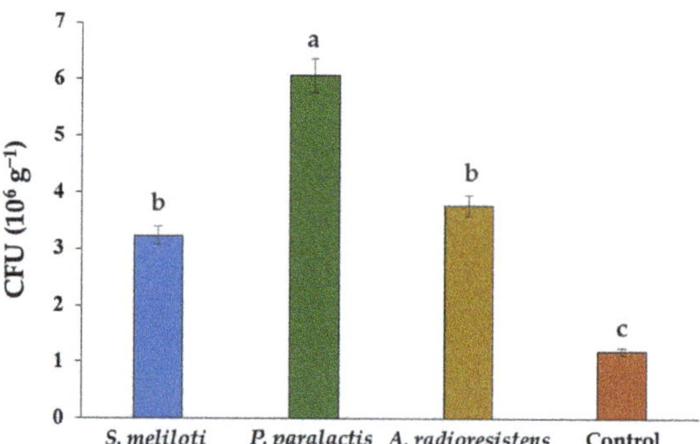

Figure 3. Rhizobacterial population on *C. sativus* L. Different letters indicate a significant difference ($p < 0.05$) according to Tukey's test.

3. Discussion

Pseudomonas paralactis, *Sinorhizobiun meliloti*, and *Acinetobacter radioresistens*, are considered PGPR due to their capability to promote growth, development, productivity, and fruit quality [21,24,25]. The results obtained on plant height and stem diameter when using *S. meliloti* and *A. radioresistens* as inoculant in *C. sativus* plants can be attributed to the increase in root length and the number of secondary roots, implying an effect on biomass accumulation and yield. In similar studies [26], using *Bacillus velezensis* as inoculant of *C. sativus*, it was observed that the host plant modifies the root structure, which improves water and nutrient uptake, promoting a higher photosynthetic rate. It has been suggested that the modification of the root structure of a cucumber plant is related to the phytohormone indoleacetic acid (IAA) produced by *S. meliloti*, *A. radioresistens*, and *P. paralactis* [11,14,27]. In addition, it has been reported that *S. meliloti* and *A. radioresistens* possess the ability to fix nitrogen and solubilize phosphate, resulting in an increase on the number of late-stage roots and root hairs, since these elements are essential for the plant to increase photosynthetic pigments and proteins, which improves photosynthetic activity and rate, having a positive effect on plant development and a greater accumulation of biomass [28,29]. Furthermore, it has been pointed out that IAA is associated with cell division and differentiation, which improves the structure of the root system [30–32].

Regarding nutraceutical quality, *S. meliloti* increased the contents of phenolic compounds, flavonoids and antioxidant capacity, which can be attributed to the ability of the microorganism's and/or its effect to induce phenolic compounds, osmolytes and organic acids in the plant, improving the availability of these compounds to be used by the plant [33]. It has been reported that in the cultivation of *Capsicum annuum*, the number of flavonoids increases when the plant is inoculated with *P. putida* because it uses quercetin and kaempferol catabolizing them as a carbon source [34], moreover, it has been observed that *S. meliloti* responds to the synthesis of proteins, mainly the Nod protein, excreting flavonoids [35]. Furthermore, if the antioxidant capacity is in dynamic equilibrium under normal growth conditions, it implies the decrease of enzymatic activity which results in the elimination of reactive oxygen, however, membrane lipid peroxidation is induced [29], and this could explain the increase of antioxidant capacity in the fruit of *C. sativus* in our results.

PGPR increases vitamin C in fruit as a non-enzymatic response to the retardation of the oxidation, also, the increase in total proteins may be due to oxidation; since the PGPR mitigate the effects of oxidation by allocating bioactive compounds to reduce the exposure of reactive oxygen species (ROS) with DNA and protein packaging, it can be inferred that the increased nutraceutical qualities of cucumber crop are due to a non-enzymatic response to natural oxidation of the plant and fruit [36].

The change in culturable bacterial subpopulation density can be attributed to the nature of each microorganism, host plant, plant genotype and environmental conditions [37], which may explain how PGPR decreases at a rate of 10^{-2} [38]. Although *P. paralactis* has been reported [39] as a PGPR due to IAA production and phosphate solubilization in vitro, this is the first study of its effect on growth promotion in cucumber plants.

4. Materials and Methods

4.1. Inoculation in Seed, Production of Seedlings, and Crop Management

The study was carried out through an agricultural cycle from Spring to Summer 2020 at Comarca Lagunera (101°40′ and 104°45′ W; 25°05′ and 26°54′ N). It was established at the Universidad Autónoma Agraria Antonio Narro, Unidad Laguna (UAAAN-UL), maintaining an average temperature of 25 °C and a relative humidity of 70%.

The genera of PGPR studied were *Pseudomonas paralactis* (KBendo6p7), *Sinorhizobium meliloti* (KBecto9p6), and *Acinetobacter radioresistens* (KBendo3p1) donated by the Microbial Ecology Laboratory of Biology Faculty from Universidad Juárez del Estado de Durango, México. After the reactivation of each bacterial strain, each bacterial strain was cultured in PBS al 0.5% [40] at 30 °C for 24 h with a constant shaking at 120 revolutions per minute (rpm). Subsequently, the concentration was adjusted to 1×10^8 colony-forming units (CFU) mL^{-1} with a spectrophotometer (VELAB VE-5100UV) at a wavelength of 540 nm (Absorbance = 1.0 unit).

Cucumber seeds (Poinset 76 variety) were prepared, seeds were then disinfected with 5% sodium hypochlorite for 1 h, the polystyrene germination plates were sterilized with the same solution and washed using distilled water. In addition, the peat moss was sterilized in autoclave at 15 psi for 1 h. Prior to sowing, around 200 seeds were inoculated with immersion in 50 mL of each bacteria for 24 h [41]. One seed per cavity was deposited in germination plates which were placed in a greenhouse. After 20 days, each seedling was transplanted into 10 L black polyethylene pots containing a substrate of sand and perlite in 80:20 ratio, respectively, the substrate was previously sanitized with a 5% sodium hypochlorite solution for 72 h. Eight plants per treatment were used in a completely randomized block design.

Irrigation was based on daily evapotranspiration according to evapotranspiration requirements [42] and was applied from transplanting to 150 days after emergence from a 100% Steiner nutrient solution [43] with an electrical conductivity (EC) of 2 dSm^{-1} and a pH of 5.5.

At 55 days after transplantation (DAT), plants were re-inoculated with 15 mL of each rhizobacterium at a concentration of 1×10^8 CFU mL^{-1} [40,41].

4.2. Parameter Evaluated

4.2.1. Morphology

At 45 DAT, the plant height (cm) and stem diameter (mm) were measured and at 150 DAT, root length (cm) and secondary roots were measured. For biomass (g), three plants per treatment were taken, placed in paper bags and placed in a drying oven at 70 °C for 72 h, determining the dry biomass (g) at the end.

4.2.2. Fruit Size, Fruit Diameter, and Yield

Three plants per treatment were harvested at 80 and 120 DAT to measure the fruit diameter using a digital vernier (H-7352, Uline) expressing its units in millimeters (mm), the length was measured (cm) taking the first point from the base to the appendix of the fruit and for the determination of yield the total weight of the fruit obtained by each plant (kg/plant) was recorded using a digital balance (VE-CB2000, VELAB).

4.3. Nutraceutical Quality, Phenolic Content, Flavonoids, Antioxidant Capacity, Vitamin C, and Total Protein

4.3.1. Extracts Preparation

The extracts were obtained as follows: from three plants per treatment, two fruits were taken at random, and then crushed to obtain a compound mixture. Afterwards, 2 g of fresh pulp were placed in 10 mL of 80% ethanol in a glass tube with screw cap, leaving them in a rotary shaker (Remi RS—24BL, Jayanti Scientific, Delhi, India) for 24 h at 70 rpm at room temperature. Finally, the tubes were centrifuged at 3000 rpm for 5 min, once the supernatant was extracted; the procedure was performed in triplicate for each treatment.

4.3.2. Total Phenolic Content

The total phenolic content was determined according to the Folin-Ciacalteau method [44] expressing the result in milligrams of gallic acid (GAE) per 100 g of fresh weight (FW) (mg GAE 100 g^{-1} FW).

4.3.3. Flavonoids

The total flavonoids were determined by the Lamaison and library technique [45], the results were expressed in milligrams of quercetin (QE) per 100 g of FW(mg QE 100 g^{-1} FW).

4.3.4. Antioxidant Capacity

The antioxidant capacity was determined with the DPPH++ method [46] and is expressed in μM equivalent Trolox 100 g^{-1} FW.

4.3.5. Vitamin C Content

Vitamin C content was determined according to Hernández-Hernández [47], the units of measurement correspond to milligrams per 100 g of FW (mg 100 g^{-1} FW).

4.3.6. Total Protein

Total soluble protein was determined by Bradford's method [48] expressed in milligrams per gram of fresh weight (mg g^{-1} FW).

4.4. Colonies Count of Rhizobacteria

The culturable bacterial subpopulation density was carried out on a sample composed of roots per treatment (10 g), macerating the roots with a ceramic mortar to homogenize them and then gauging to 100 mL with NaCl saline solution at 0.85%. 85% of the mixture was shaken for 5 min in a Benchmark vortex and allowed to stand for 20 min, then the samples were serially diluted to 10^{-4} and an aliquot of 100 μL was inoculated in triplicate, for each treatment, in Petri dishes with 20 mL of trypticase soy agar (TSA) culture medium and 1g of copper sulfate. They were incubated at 37 °C for 48 h and colony forming units (CFU) were counted [49].

4.5. Statistical Analysis

The study variables were analyzed by means of an analysis of variance with the SAS 9.4 statistical package; if a significant difference was found, a comparison of means was performed using the Tukey method ($p < 0.05$).

5. Conclusions

Inoculation of the plants with PGPR improved height, stem diameter, root length, secondary roots, biomass, size and fruit diameter, and yield. In addition, nutraceutical quality parameters in the plants inoculated with *P. paralactis*, *S. meliloti* and *A. radioresistens* were increase compared to the control. In the future, it is necessary studies on the effects of the inoculation of bacteria consortiums on growth, productivity, and fruit quality of cucumber.

Author Contributions: Conceptualization, P.P.-R.; methodology, G.Z.-S., E.B.-C. and M.F.-H.; software, J.S.-M. and R.G.C.-C.; validation, G.Z.-S., R.G.C.-C. and P.P.-R.; investigation, G.Z.-S., P.P.-R. and L.G.H.-M.; resources, E.B.-C.; writing—original draft preparation, G.Z.-S., P.P.-R. and L.G.H.-M.; writing—review and editing, P.P.-R. and L.G.H.-M.; visualization, E.B.-C., R.G.C.-C. and M.F.-H.; supervision, P.P.-R. and J.S.-M.; project administration, P.P.-R.; funding acquisition, P.P.-R. All authors have read and agreed to the published version of the manuscript.

Funding: This study was funded by Tecnológico Nacional de México, Consejo Nacional de Ciencia y Tecnología and Universidad Autónoma Agraria Antonio Narro (Project #11215.21P).

Institutional Review Board Statement: Not applicable.

Informed Consent Statement: Not applicable.

Data Availability Statement: Not applicable.

Acknowledgments: Consejo Nacional de Ciencia y Tecnología (CONACyT), Instituto Tecnológico de Torreón (ITT), Universidad Autónoma Agraria Antonio Narro Unidad Laguna (UAAAN UL) and technical support of María Sofía Ramos Galván and José Manuel Melero Astorga (members of Nanotechnology & Microbial Biocontrol Group) of the Centro de Investigaciones Biológicas del Noroeste.

Conflicts of Interest: The authors declare no conflict of interest.

References

1. Behera, T.K.; Boopalakrishnan, G.; Jat, G.S.; Das Munshi, A.; Choudhary, H.; Ravindran, A.; Kumari, S.; Kumari, R. Deriving stable tropical gynoecious inbred lines of slicing cucumber from American pickling cucumber using MABB. *Hortic. Environ. Biotechnol.* **2022**, *63*, 263–274. [CrossRef]
2. Achikanu, C.E.; Ani, O.N.; Akpata, E.I. Proximate, vitamin and phytochemical composition of Cucumis metuliferus seed. *Int. J. Food Sci. Nutr.* **2020**, *2*, 20–24.
3. Helal, M.; Sami, R.; Algarni, E.; Alshehry, G.; Aljumayi, H.; Al-Mushhin, A.A.; Benajiba, N.; Chavali, M.; Kumar, N.; Iqbal, A.; et al. Active Bionanocomposite Coating Quality Assessments of Some Cucumber Properties with Some Diverse Applications during Storage Condition by Chitosan, Nano Titanium Oxide Crystals, and Sodium Tripolyphosphate. *Crystals* **2022**, *12*, 131. [CrossRef]
4. Uthpala, T.G.G.; Marapana, R.A.U.J.; Lakmini, K.; Wettimuny, D.C. Nutritional bioactive compounds and health benefits of fresh and processed cucumber (*Cucumis sativus* L.). *SJB* **2020**, *3*, 75–82.
5. Preciado-Rangel, P.; Reyes-Pérez, J.J.; Ramírez-Rodríguez, S.C.; Salas-Pérez, L.; Fortis-Hernández, M.; Murillo-Amador, B.; Troyo-Diéguez, E. Foliar aspersion of salicylic acid improves phenolic and flavonoid compounds, and also the fruit yield in cucumber (*Cucumis sativus* L.). *Plants* **2019**, *8*, 44. [CrossRef]
6. Yuan, W.; Zhao, P.; Chen, H.; Wang, L.; Huang, G.; Cao, L.; Huang, Q. Natural green-peel orange essential oil enhanced the deposition, absorption and permeation of prochloraz in cucumber. *RSC Adv.* **2019**, *9*, 20395–20401. [CrossRef] [PubMed]
7. Geng, Y.; Bashir, M.A.; Zhao, Y.; Luo, J.; Liu, X.; Li, F.; Wang, H.; Raza, Q.; Rehim, A.; Zhang, X.; et al. Long-Term Fertilizer Reduction in Greenhouse Tomato-Cucumber Rotation System to Assess N Utilization, Leaching, and Cost Efficiency. *Sustainability* **2022**, *14*, 4647. [CrossRef]
8. Guan, X.; Liu, C.; Li, Y.; Wang, X.; Liu, Y.; Zou, C.; Chen, X.; Zhang, W. Reducing the environmental risks related to phosphorus surplus resulting from greenhouse cucumber production in China. *J. Clean. Prod.* **2022**, *332*, 130076. [CrossRef]
9. Tashiro, Y.; Sato, T.; Satitmunnaithum, J.; Kinouchi, H.; Li, J.; Tanabata, S. Case Study on the Use of the Leaf-Count Method for Drip Fertigation in Outdoor Cucumber Cultivation in Reconstructed Fields Devastated by a Tsunami. *Agriculture* **2021**, *11*, 656. [CrossRef]
10. Trejo Valencia, R.; Sánchez Acosta, L.; Fortis Hernández, M.; Preciado Rangel, P.; Gallegos Robles, M.Á.; Antonio Cruz, R.d.C.; Vázquez Vázquez, C. Effect of seaweed aqueous extracts and compost on vegetative growth, yield, and nutraceutical quality of cucumber (*Cucumis sativus* L.) fruit. *Agronomy* **2018**, *8*, 264. [CrossRef]
11. El-Mageed, A.; Taia, A.; El-Mageed, A.; Shimaa, A.; El-Saadony, M.T.; Abdelaziz, S.; Abdou, N.M. Plant Growth-Promoting Rhizobacteria Improve Growth, Morph-Physiological Responses, Water Productivity, and Yield of Rice Plants Under Full and Deficit Drip Irrigation. *Rice* **2022**, *15*, 16. [CrossRef] [PubMed]
12. Salim, H.A.; Kadhum, A.A.; Ali, A.F.; Saleh, U.N.; Jassim, N.H.; Hamad, A.R.; Attia, J.A.; Darwish, J.J.; Hassan, A.F. Response of cucumber plants to PGPR bacteria (Azospirillum brasilense, Pseudomonas fluorescens and Bacillus megaterium) and bread yeast (Saccharomyces cerevisiae). *Syst. Rev. Pharm.* **2021**, *12*, 969–975.
13. Kaloterakis, N.; Van Delden, S.H.; Hartley, S.; De Deyn, G.B. Silicon application and plant growth promoting rhizobacteria consisting of six pure Bacillus species alleviate salinity stress in cucumber (*Cucumis sativus* L). *Sci. Hortic.* **2021**, *288*, 110383. [CrossRef]
14. Raza, A.; Ejaz, S.; Saleem, M.S.; Hejnak, V.; Ahmad, F.; Ahmed, M.A.A.; Alotaibi, S.S.; El-Shehawi, M.A.; Alsubeie, M.S.; Zuan, A.T.K. Plant growth promoting rhizobacteria improve growth and yield related attributes of chili under low nitrogen availability. *PLoS ONE* **2021**, *16*, e0261468. [CrossRef] [PubMed]
15. George, P.; Gupta, A.; Gopal, M.; Thomas, L.; Thomas, G.V. Indigenous rhizobacteria possessing abiotic stress tolerant traits promote vigorous growth of coconut seedlings via increased nutrient uptake and positive plant–microbe feedback. *Proc. Indian Natl. Sci. Acad.* **2022**, *88*, 64–79. [CrossRef]

16. Gupta, K.; Dubey, N.K.; Singh, S.P.; Kheni, J.K.; Gupta, S.; Varshney, A. Plant Growth-Promoting Rhizobacteria (PGPR): Current and Future Prospects for Crop Improvement. In *Current Trends in Microbial Biotechnology for Sustainable Agriculture*; Yadav, A.N., Singh, J., Singh, C., Yadav, N., Eds.; Environmental and Microbial Biotechnology book series; Springer: Berlin/Heidelberg, Germany, 2021. [CrossRef]
17. Mihalache, G.; Zamfirache, M.M.; Hamburda, S.; Stoleru, V.; Munteanu, N.; Stefan, M. Synergistic effect of Pseudomonas lini and Bacillus pumilus on runner bean growth enhancement. *EEMJ* **2016**, *15*, 1823–1831.
18. Srivastava, P.; Jaggi, V.; Dasila, H.; Sahgal, M. Identification and characterisation of siderophore positive pseudomonas from north indian rosewood (dalbergia sissoo) roxb. Forest ecosystem. *IJASR* **2020**, *10*, 239–256.
19. Bergottini, V.M.; Otegui, M.B.; Sosa, D.A.; Zapata, P.D.; Mulot, M.; Rebord, M.; Wiss, F.; Benrey, B.; Junier, P. Bio-inoculation of yerba mate seedlings (Ilex paraguariensis St. Hill.) with native plant growth-promoting rhizobacteria: A sustainable alternative to improve crop yield. *Biol. Fertil. Soils* **2015**, *51*, 749–755. [CrossRef]
20. Meena, N.; Saharan, B.S. Plant growth promoting rhizobacteria improves growth in aloe vera. *J. Plant Dev. Sci.* **2017**, *9*, 811–815.
21. Galleguillos, C.; Aguirre, C.; Barea, J.M.; Azcón, R. Growth promoting effect of two Sinorhizobium meliloti strains (a wild type and its genetically modified derivative) on a non-legume plant species in specific interaction with two arbuscular mycorrhizal fungi. *Plant Sci.* **2000**, *159*, 57–63. [CrossRef]
22. Rosier, A.; Beauregard, P.B.; Bais, H.P. Quorum Quenching Activity of the PGPR Bacillus subtilis UD1022 Alters Nodulation Efficiency of Sinorhizobium meliloti on Medicago truncatula. *Front. Microbiol.* **2021**, *11*, 596299. [CrossRef] [PubMed]
23. Guiñazú, L.B.; Andrés, J.A.; Del Papa, M.F.; Pistorio, M.; Rosas, S.B. Response of alfalfa (*Medicago sativa* L.) to single and mixed inoculation with phosphate-solubilizing bacteria and Sinorhizobium meliloti. *Biol. Fertil. Soils* **2010**, *46*, 185–190. [CrossRef]
24. Lafi, F.F.; Alam, I.; Bisseling, T.; Geurts, R.; Bajic, V.B.; Hirt, H.; Saad, M.M. Draft genome sequence of the plant growth–promoting rhizobacterium Acinetobacter radioresistens strain SA188 isolated from the desert plant Indigofera argentea. *Genome Announc.* **2017**, *5*, e01708–e01716. [CrossRef] [PubMed]
25. Saidi, S.; Cherif-Silini, H.; Chenari Bouket, A.; Silini, A.; Eshelli, M.; Luptakova, L.; Alenezi, F.N.; Belbahri, L. Improvement of Medicago sativa crops productivity by the co-inoculation of Sinorhizobium meliloti–actinobacteria under salt stress. *Curr. Microbiol.* **2021**, *78*, 1344–1357. [CrossRef]
26. Wang, J.; Qu, F.; Liang, J.; Yang, M.; Hu, X. Bacillus velezensis SX13 promoted cucumber growth and production by accelerating the absorption of nutrients and increasing plant photosynthetic metabolism. *Sci. Hortic.* **2022**, *301*, 111151. [CrossRef]
27. He, A.; Niu, S.; Yang, D.; Ren, W.; Zhao, L.; Sun, Y.; Meng, L.; Zhao, Q.; Paré, P.W.; Zhang, J. Two PGPR strains from the rhizosphere of Haloxylon ammodendron promoted growth and enhanced drought tolerance of ryegrass. *Plant Physiol. Biochem.* **2021**, *161*, 74–85. [CrossRef]
28. Kumari, K.; Samantaray, S.; Sahoo, D.; Tripathy, B.C. Nitrogen, phosphorus and high CO2 modulate photosynthesis, biomass and lipid production in the green alga Chlorella vulgaris. *Photosynth. Res.* **2021**, *148*, 17–32. [CrossRef]
29. Su, B.; Wang, L.; Shangguan, Z. Morphological and physiological responses and plasticity in Robinia pseudoacacia to the coupling of water, nitrogen and phosphorus. *J. Plant Nutr. Soil. Sci.* **2021**, *184*, 271–281. [CrossRef]
30. Gallart, M.; Paungfoo-Lonhienne, C.; Gonzalez, A.; Trueman, S.J. Nitrogen source influences the effect of plant growth-promoting rhizobacteria (Pgpr) on macadamia integrifolia. *Agronomy* **2021**, *11*, 1064. [CrossRef]
31. Raffi, M.M.; Charyulu, P.B.B.N. Azospirillum-biofertilizer for sustainable cereal crop production: Current status. In *Recent Developments in Applied Microbiology and Biochemistry*; Academic Press: Cambridge, MA, USA, 2020; Volume 2, pp. 193–209. [CrossRef]
32. Sapre, S.; Gontia-Mishra, I.; Tiwari, S. Plant growth-promoting rhizobacteria ameliorates salinity stress in pea (*Pisum sativum*). *J. Plant Growth Regul.* **2022**, *41*, 647–656. [CrossRef]
33. Tith, S.; Duangkaew, P.; Laosuthipong, C.; Monkhung, S. Vermicompost efficacy in improvement of cucumber (*Cucumis sativus* L.) productivity, soil nutrients, and bacterial population under greenhouse condition. *APST* **2021**, *27*, APST-27.
34. Ureche, M.A.L.; Pérez-Rodriguez, M.M.; Ortiz, R.; Monasterio, R.P.; Cohen, A.C. Rhizobacteria improve the germination and modify the phenolic compound profile of pepper (*Capsicum annum* L.). *Rhizosphere* **2021**, *18*, 100334. [CrossRef]
35. Peck, M.C.; Fisher, R.F.; Long, S.R. Diverse flavonoids stimulate NodD1 binding to nod gene promoters in Sinorhizobium meliloti. *J. Bacteriol. Res.* **2006**, *188*, 5417–5427. [CrossRef] [PubMed]
36. Santoyo, G.; Urtis-Flores, C.A.; Loeza-Lara, P.D.; Orozco-Mosqueda, M.; Glick, B.R. Rhizosphere colonization determinants by plant growth-promoting rhizobacteria (PGPR). *Biology* **2021**, *10*, 475. [CrossRef] [PubMed]
37. Rosenblueth, M.; E, M.-R. Bacterial endophytes and their interactions with hosts. *MPMI* **2006**, *19*, 827–837. [CrossRef]
38. Akköprü, A.; Akat, Ş.; Özaktan, H.; Gül, A.; Akbaba, M. The long-term colonization dynamics of endophytic bacteria in cucumber plants, and their effects on yield, fruit quality and Angular Leaf Spot Disease. *Sci. Hortic.* **2021**, *282*. [CrossRef]
39. Navarro-Muñoz, C.E.; Balderas-Hernández, F.; Palacio-Rodríguez, R.; Sáenz-Mata, J. Caracterización de la capacidad de promoción de crecimiento vegetal de rizobacterias aisladas de plantas del Desierto Chihuahuense. *Rev. Mex. Biodivers.* **2021**, *96*.
40. Carrillo, A.; Puente, M.; Castellanos, T.; Bashan, Y. Aplicaciones biotecnológicas de ecología microbiana. In *Manual de Laboratorio. Biotechnological Applications of Microbial Ecology. Laboratory Manual*; 1998.
41. Akköprü, A.; Özaktan, H. Identification of rhizobacteria that increase yield and plant tolerance to angular leaf spot disease in cucumber. *Plant Protect. Sci.* **2018**, *54*, 67–73. [CrossRef]
42. Hargreaves, G.H.; Samani, Z.A. Estimating potential evapo-transpiration. *J. Irrig. Drain. Div.* **1982**, *108*, 225–230. [CrossRef]

43. Steiner, A.A. A universal method for preparing nutrient solutions of a certain desired composition. *Plant Soil* **1961**, *15*, 134–154. [CrossRef]
44. Singleton, V.L.; Orthofer, R.; Lamuela-Raventós, R.M. Analysis of total phenols and other oxidation substrates and antioxidants using a folic-ciocalteu reagent. *Methods Enzymol.* **1999**, *299*, 152–178. [CrossRef]
45. Lamaison, J.L.C.; Carnet, A. Contents in main flavonoid compounds of Crataegus Monogyna Jacq. and *Crataegus laevigata* (Poiret) DC flowers at different development stages. *Pharm. Acta Helv.* **1990**, *65*, 315–320.
46. Brand-Williams, W.; Cuvelier, M.E.; Berset, C.L.W.T. Use of a free radical method to evaluate antioxidant activity. *LWT-Food Sci. Technol.* **1995**, *28*, 25–30. [CrossRef]
47. Hernández-Hernández, H.; Quiterio-Gutiérrez, T.; Cadenas-Pliego, G.; Ortega-Ortiz, H.; Hernández-Fuentes, A.D.; Cabrera de la Fuente, M.; Valdés-Reyna, J.; Juárez-Maldonado, A. Impact of selenium and copper nanoparticles on yield, antioxidant system, and fruit quality of tomato plants. *Plants* **2019**, *8*, 355. [CrossRef] [PubMed]
48. Bradford, M.M. A rapid and sensitive method for the quantitation of microgram quantities of protein utilizing the principle of protein-dye binding. *Anal. Biochem.* **1976**, *72*, 248–254. [CrossRef]
49. Amaral, M.B.; Ribeiro, T.G.; Alves, G.C.; Coelho, M.R.R.; Matta, F.D.P.; Baldani, J.I.; Baldani, V.L.D. The occurrence of rhizobacteria from Paspalum genotypes and their effects on plant growth. *Sci. Agric.* **2021**, *79*, 1–10. [CrossRef]

 *plants*

Article

Ascorbic Acid Preconditioning Effect on Broccoli Seedling Growth and Photosynthesis under Drought Stress

Mason T. MacDonald *, Rajeswari Kannan and Renuga Jayaseelan

Faculty of Agriculture, Dalhousie University, Bible Hill, NS B2N 5E3, Canada; rj938452@dal.ca (R.K.); rn598954@dal.ca (R.J.)
* Correspondence: mason.macdonald@dal.ca; Tel.: +1-902-893-6626

Abstract: Drought is an abiotic stress that decreases crop photosynthesis, growth, and yield. Ascorbic acid has been used as a seed preconditioning agent to help mitigate drought in some species, but not yet in broccoli (*Brassica oleracea* var. *italica*). The objective was to investigate the effect of ascorbic acid on growth, photosynthesis, and related parameters in watered and drought-stressed broccoli seedlings. A 2 × 4 factorial experiment was designed where stress (watered or drought) was the first factor and ascorbic acid preconditioning (untreated, 0 ppm, 1 ppm, or 10 ppm) was the second factor. Positioning within the greenhouse was included as a blocking factor and the experiment was replicated three times. All seedlings were watered for 8 weeks and then half had water withheld for 7 days to impose drought while the other half continued to be watered. Ascorbic acid preconditioning increased shoot dry mass, root dry mass, water use efficiency, and photosynthesis in all seedlings while also increasing chlorophyll, relative water content, and leaf area in droughted seedlings. Ascorbic acid preconditioning also decreased membrane injury in droughted seedlings to the point that membrane injury was not significantly different than the watered control. There was strong evidence to support ascorbic acid as a successful seed preconditioning agent in watered and droughted broccoli.

Keywords: antioxidant; *Brassica oleracea*; chlorophyll index; membrane injury; stomatal conductance; water deficit; water use efficiency

Citation: MacDonald, M.T.; Kannan, R.; Jayaseelan, R. Ascorbic Acid Preconditioning Effect on Broccoli Seedling Growth and Photosynthesis under Drought Stress. *Plants* **2022**, *11*, 1324. https://doi.org/10.3390/plants11101324

Academic Editor: Fermin Morales

Received: 20 April 2022
Accepted: 16 May 2022
Published: 17 May 2022

Publisher's Note: MDPI stays neutral with regard to jurisdictional claims in published maps and institutional affiliations.

1. Introduction

Broccoli is a member of the Brassicaceae family that grows best at 13–20 °C [1]. Broccoli grows 60–90 cm tall, forms a branching green stalk, and has dense flower buds that represent the commonly consumed organ [2]. The market for broccoli has expanded significantly in the latter half of the 20th century, with a yield that has increased to approximately 26 million tons globally [3]. The increased popularity of broccoli is largely due to its many desirable nutrients, antioxidants, and other bioactive compounds [2,4]. Broccoli is considered a cool season crop and, like many crops, is vulnerable to several abiotic stresses that might limit productivity [5].

Drought is one abiotic stress critical to photosynthesis and crop yields [6]. Drought may occur because of insufficient rainfall, increased evapotranspiration, or increased salinity in soils [7,8]. Stomata close rapidly after the onset of drought to conserve water, but this is accompanied by reduced CO_2 absorption, photosynthesis (Pn), and growth [9,10]. A reduction in CO_2 absorption causes an imbalance between electron excitation and utilization during photosynthesis, which in turn increases the production of reactive oxygen species (ROS) [11]. ROS serve, in part, as a physiological signal indicative of abiotic stress, but also cause significant damage to cellular membranes and biochemicals [12].

Drought stress can affect a variety of agricultural species. Drought decreased Pn and increased ROS production in several wheat cultivars [13], decreased Pn in apple trees [14], decreased Pn in several forestry species [15], and decreased Pn and yields in several Brassica species [16,17]. Even mild drought can adversely affect broccoli yields where a 20% and 40% decrease in irrigation resulted in 29% and 41% lower broccoli yields, respectively [18]. Closely related crops, such as cauliflower, also had reduced germination, shoot length, root length, curd growth, and dry mass (DM) due to drought [5]. The cumulative effects of drought have reduced global crop yields by 10% from 1964 to 2007 [19]. Drought-induced crop losses of only maize, rice, soy, and wheat from 1983 to 2009 were estimated at USD 166 billion [20].

Seed preconditioning or priming is one of the many technologies available to help mitigate drought stress. Seed preconditioning involves incubating seeds with a seed preconditioning agent (SPA) to induce benefits in developing seedlings and mature plants [21]. Many compounds have been used successfully as SPAs. Salicylic acid, acetylsalicylic acid, and glycinebetaine promoted germination in carrots [22], 5-hydroxybenzimidazole increased yields and stress resistance in tomatoes [23], and pyroligneous acid increased yields in rice [24]. Several naturally occurring antioxidants, such as lycopene, β-carotene, and ascorbic acid (AsA), also increased tomato dry mass and photosynthesis when used as SPAs [25].

Seed preconditioning with AsA is of particular interest because it is inexpensive, readily accessible, and has benefitted several crops. Seed preconditioning with 1 ppm and 10 ppm AsA increased tomato shoot dry mass by 40%, leaf area by 50%, and increased Pn by 223% compared to a control [25]. Preconditioning with 50 ppm AsA increased wheat grain yields by approximately 26% [26]. Finally, preconditioning with 10 ppm and 20 ppm AsA increased rice germination and dry mass [27]. However, there is no literature on the effect of AsA preconditioning on broccoli.

We hypothesize that AsA will promote drought stress tolerance in broccoli. In other species, varieties with increased drought tolerance typically presented with delayed decreases in Pn, higher biomass and leaf area, and decreased membrane injury [13–15,25]. Thus, it is predicted that AsA-preconditioned broccoli would have similar responses. The specific objectives of this research are to investigate the effect of AsA on growth, photosynthesis, and related parameters in watered and drought-stressed broccoli seedlings.

2. Results

2.1. Soil Moisture

Soil moisture was maintained at 38.5 ± 1.5% in all treatments throughout the first 8 weeks of the experiment and was not significantly different between drought and watered treatments until the imposition of drought. However, average soil moisture decreased below 20% after withholding water for 1 day and then gradually decreased to 4.1% after water was withheld for 1 week (Figure 1). Though there was a significant difference in soil moisture between watered and droughted treatments at the end of the experiment (Figure 1), there were no differences in soil moisture between any preconditioning treatments.

2.2. Biomass

Drought stress significantly ($p < 0.001$) decreased shoot DM by approximately 39% compared to watered plants and the second was that preconditioning seeds with 10 ppm significantly ($p < 0.001$) increased shoot DM when compared to a control (Figure 2). The 10 ppm-preconditioned watered seedlings had 35% and 52% higher shoot DM than non-preconditioned or water-preconditioned seedlings, respectively. Further, the 10 ppm-preconditioned droughted seedlings had 42% and 36% higher shoot DM than non-preconditioned or water-preconditioned seedlings, respectively. Differences in shoot size were visually apparent (Figure 3).

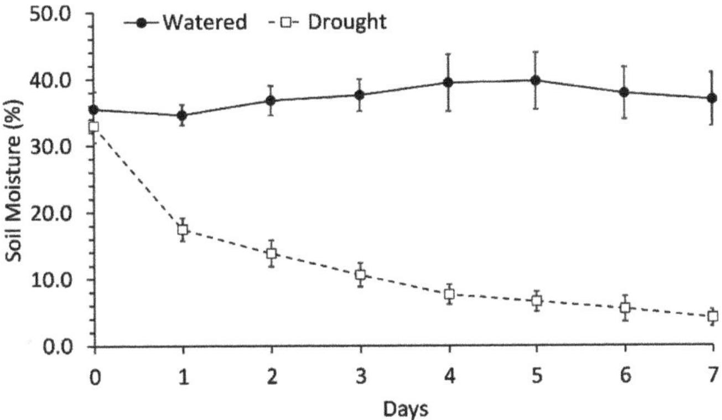

Figure 1. Soil moisture over time in pots during the last week of the experiment. Watered pots continued to receive water each day while water was completely withheld from droughted pots. Each data point represented the mean of 36 pots in each treatment. Error bars indicate standard error.

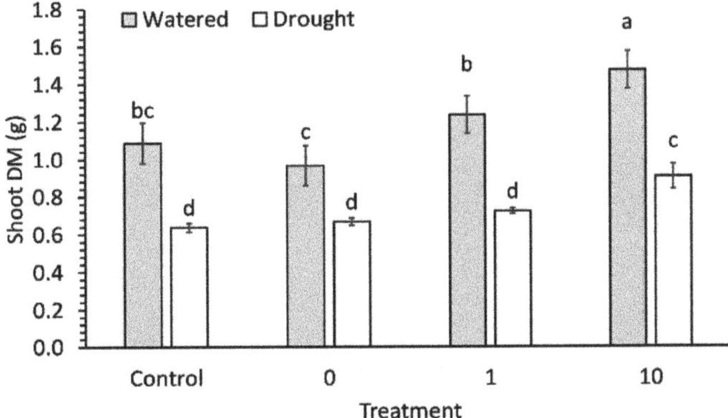

Figure 2. Shoot biomass in 9-week-old broccoli seedlings that were well watered or droughted for 1 week. Each bar represents a mean and error bars represent standard error calculated from 9 replicates. Means with different letters are significantly different ($p < 0.05$) as determined by Tukey's honestly significant difference.

AsA preconditioning had a significant ($p = 0.007$) effect on root DM. Both 1 ppm and 10 ppm preconditioning significantly ($p = 0.007$) increased root DM by 49% and 149% compared to the control, respectively. Only the 10 ppm AsA preconditioning increased root DM in watered seedlings compared to controls (Figure 4a). The increase in DM due to preconditioning was more pronounced in roots than shoots in AsA-preconditioned seedlings, which resulted in a significant decrease in shoot:root ratio of droughted seedlings compared to the control (Figure 4b). There were no differences in the shoot:root ratios of watered seedlings due to preconditioning.

2.3. Photosynthetic Measurements

Pn was significantly ($p < 0.001$) decreased by 68% due to drought compared to watered seedlings (Figure 5a). Ascorbic acid preconditioning significantly ($p < 0.001$) increased Pn in both watered and droughted seedlings. Preconditioning with 10 ppm ascorbic acid increased Pn by 83% compared to a control in watered seedlings. The droughted control had virtually ceased Pn, but droughted seedlings preconditioned with 10 ppm ascorbic acid maintained photosynthetic rates that were not significantly different than watered controls (Figure 5a).

(**a**) (**b**)

Figure 3. Overhead view of 9-week-old broccoli seedlings that were (**a**) watered and (**b**) deprived of water for 1 week. Control seedlings (C) were provided no seed preconditioning while others were provided 0 ppm, 1 ppm, or 10 ppm of ascorbic acid preconditioning.

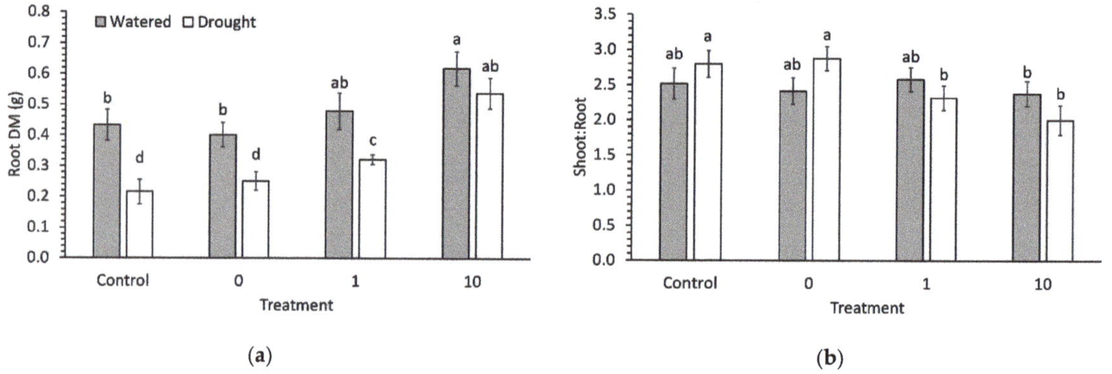

(**a**) (**b**)

Figure 4. (**a**) Root biomass and (**b**) shoot:root ratio in 9-week-old broccoli seedlings which were watered or deprived of water for 1 week. Each bar represents a mean and error bars represent standard error calculated from 9 replicates. Means with different letters are significantly different ($p < 0.05$) as determined by Tukey's honestly significant difference.

Transpiration (E) and stomatal conductance (Gs) were significantly ($p < 0.001$) decreased by 51% and 96%, respectively, due to drought compared to watered seedlings (Figure 5b,c). AsA preconditioning had no significant impact on E (Figure 5b), though Gs was significantly higher in both AsA treatments in watered seedlings (Figure 5c). Consequently, there was a significant ($p = 0.03$) interaction effect between stress and preconditioning on WUE. Preconditioning with 10 ppm ascorbic acid increased WUE by almost 2-fold in watered seedlings, but by 26-fold in droughted seedlings compared to their respective controls (Figure 5d).

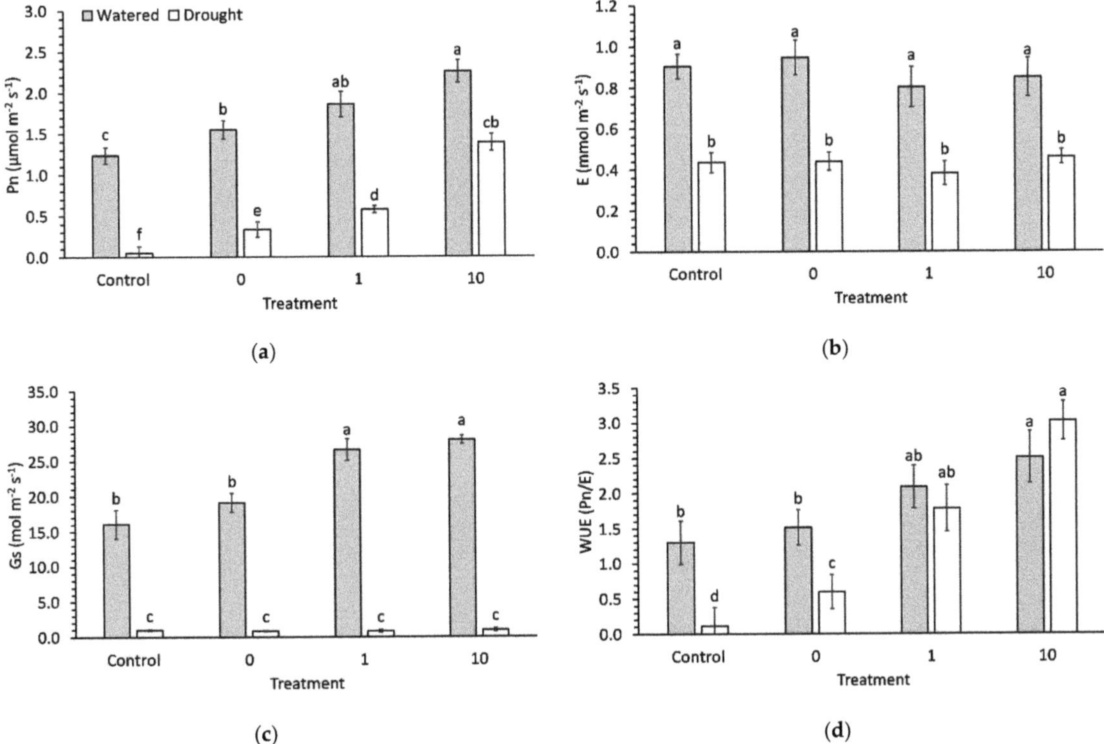

Figure 5. (**a**) Photosynthesis, (**b**) transpiration, (**c**) stomatal conductance, and (**d**) water use efficiency in 9-week-old broccoli seedlings that were well watered or droughted for 1 week. Each bar represents a mean and error bars represent standard error calculated from 9 replicates. Means with different letters are significantly different ($p < 0.05$) as determined by Tukey's honestly significant difference.

2.4. Leaf Measurements

There was a significant ($p < 0.001$) interaction between stress and preconditioning on chlorophyll index (Figure 6a). Drought decreased chlorophyll index by 47% and 31% in seedlings that were not preconditioned or preconditioned with water only, respectively, compared to watered seedlings. However, drought did not affect chlorophyll index in seedlings preconditioned with 1 ppm or 10 ppm ascorbic acid.

There was a significant ($p = 0.028$) interaction between stress and preconditioning on relative water content (RWC) (Figure 6b). Drought decreased RWC by 42% and 29% in seedlings that were not preconditioned or preconditioned with water only, respectively, compared to watered seedlings. Drought also decreased RWC in preconditioned seedlings, but only by 13% and 19% in seedlings preconditioned with 1 ppm and 10 ppm ascorbic acid, respectively, compared to watered seedlings.

Both stress and preconditioning had a significant ($p < 0.001$ and $p = 0.012$) effect on leaf area (Figure 6c). Leaf area was significantly higher than the control due to water or 1 ppm ascorbic acid preconditioning in watered seedlings but was significantly higher than the control due to only 10 ppm ascorbic acid preconditioning in droughted seedlings. Drought generally decreased leaf area by an average of 56% in all seedlings.

There was a significant ($p < 0.001$) interaction between stress and preconditioning on membrane injury index (MII) (Figure 6d). There was no difference in MII between preconditioning treatments in watered seedlings. However, MII was 36% and 40% lower in 1 ppm- and 10 ppm-preconditioned droughted seedlings, respectively, compared to a control. MII in droughted seedlings that had been preconditioned with 10 ppm ascorbic acid was not significantly lower than any watered seedlings.

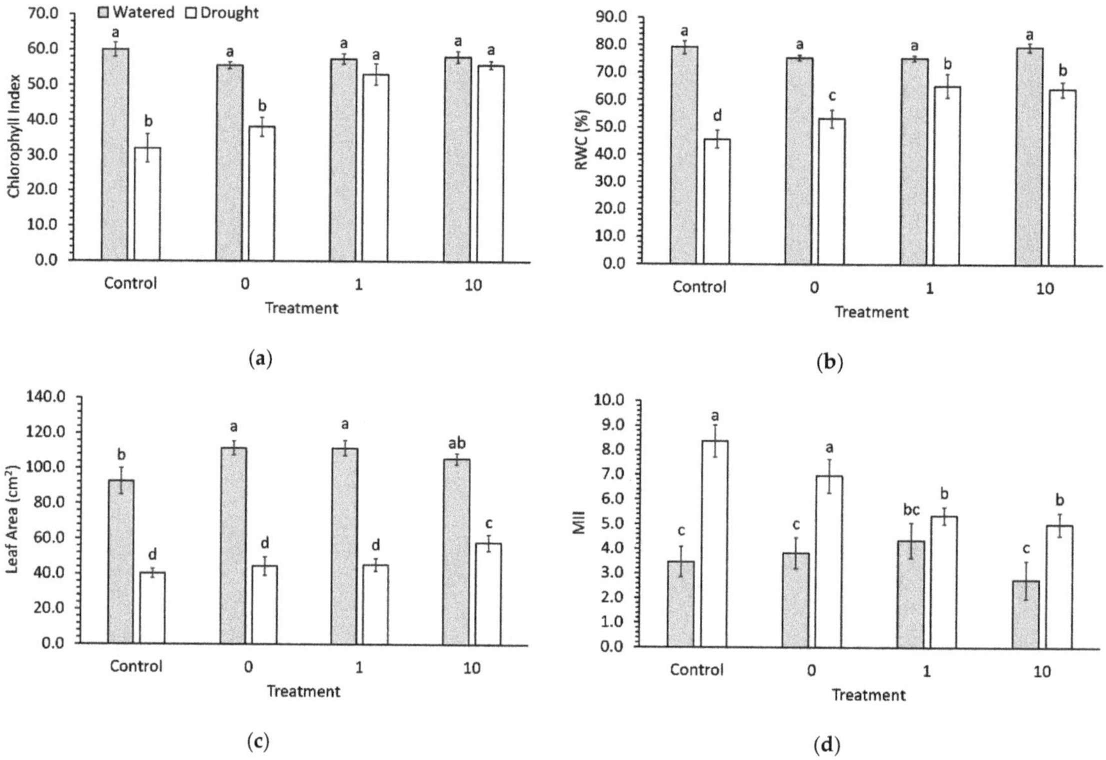

(a)

(b)

(c)

(d)

Figure 6. (**a**) Chlorophyll index, (**b**) relative water content, (**c**) leaf area, and (**d**) membrane injury index in 9-week-old broccoli seedlings that were well watered or droughted for 1 week. Each bar represents a mean and error bars represent standard error calculated from 9 replicates. Means with different letters are significantly different ($p < 0.05$) as determined by Tukey's honestly significant difference.

3. Discussion

AsA increased Pn, WUE, and shoot DM in watered broccoli seedlings, which was comparable to AsA effects on tomatoes [25]. Pn appears to be a key response since Pn would influence both shoot DM and WUE; increased Pn would result in increased CO_2 fixation and an increase in Pn without any change in E would directly correspond to WUE. Although this experiment did not grow plants to maturity, it seems plausible that AsA preconditioning could promote broccoli yields in a similar manner to previous SPAs [23].

AsA preconditioning affected more response variables in drought-stressed seedlings than watered seedlings. Pn, WUE, shoot DM, root DM, leaf area, chlorophyll, and RWC content all increased while MII decreased in AsA preconditioned seedlings exposed to drought. As above, this is comparable to previous results with tomato seedlings [25]. Additionally, as above, the increase in Pn likely explains increases in shoot DM, root DM, and WUE.

Since PM, DM, and WUE were higher in all AsA-preconditioned seedlings compared to their respective controls, these responses in droughted seedlings are not specific indicators of drought stress tolerance. However, chlorophyll, RWC, leaf area, and MII were negatively impacted by drought but largely mitigated by AsA preconditioning, which is indicative of increased stress tolerance. Maintenance of chloroplasts due to AsA preconditioning would result in improved light capture and energy conservation, which would help explain why Pn remained so high during drought [28]. Further, AsA-preconditioned seedlings also had higher RWC during drought, which would improve biosynthesis and chlorophyll proteins and, consequently, Pn [14,29].

The specific mechanism through which AsA confers benefits to developing seedlings remains unclear. One explanation is that since watered seedlings were larger, perhaps seedlings were more apt to survive drought. Several studies report either larger unstressed plants or accelerated germination/emergence due to AsA [22,23,26,30]. A second explanation is that the increased root DM or decreased shoot:root ratio allowed for AsA-preconditioned plants to access more water during drought [25]. Root access of water seems unlikely though because all seedlings were grown in pots and soil moisture never differed among the preconditioned treatments. A third explanation is that, as an antioxidant, AsA is protecting cellular membranes from stress-induced oxidative damage. Certainly, a significant decrease in MII supports the concept of membrane protection, which would include chloroplast membranes and help explain why Pn was maintained during drought. However, a direct antioxidant effect seem unlikely since seeds were treated several weeks before imposed stress, Pn was increased in watered seedlings, and there are instances where other potent antioxidants have had no effect on MII when used as SPAs [25].

For an SPA to provide any benefit to developing or mature plants, it seems likely there must be some change directly within seeds that persists as plants develop. Several SPAs induced new proteins that were not present in untreated or water preconditioned seeds [25]. Further, some SPAs seem to induce an epigenetic effect since next generation seedlings grown from preconditioned parents had similar benefits as their parents [23]. Two previous studies support the concept of AsA-induced proteins; there was an increase in soluble proteins and protease in AsA-preconditioned rice [27] and increased protease, peroxidase, and superoxide dismutase in AsA-preconditioned wheat [26]. Further work is needed to identify which induced proteins are increasing photosynthesis, growth, and/or drought tolerance.

Changes in osmoregulation of droughted seedlings must be considered as a possible mechanism for AsA preconditioning-induced drought tolerance. It is well established that osmoregulation through the accumulation of sugars, ions, amino acids, or other metabolites can promote drought tolerance [13,28,31]. Exogenously applied AsA increased the uptake of several cations and osmolytes proline and glycinebetaine, which increased stress tolerance [32]. Seed priming with other SPAs increased the concentration of several osmolytes that helped mitigate stress [33,34] and it stands to reason that AsA would have a similar effect as an SPA.

4. Materials and Methods

4.1. Experimental Design

This experiment was designed with two factors of interest and one blocking factor. The first factor of interest was seed preconditioning with 4 levels, where seeds were either not preconditioned (control), preconditioned with water only (0 ppm), preconditioned with 1 ppm AsA, or preconditioned with 10 ppm AsA. The second factor was stress level which had 2 levels. All broccoli seedlings were grown and watered for 8 weeks and then half continued to be watered while the other half were deprived of water for 1 week. The blocking factor was based on location within a greenhouse with seedlings placed in 3 different positions. The experiment was replicated 3 times, which required 72 broccoli seedlings overall with 1 seedling per pot.

4.2. Seed Preconditioning

A 1000 ppm stock solution of AsA (Sigma-Aldrich, Oakville, ON, Canada) was made in deionized water. The stock solution was diluted to 10 ppm and 1 ppm while deionized water only was used as the 0 ppm AsA treatment. Each of the three preconditioning treatments was transferred to a 250 mL flask. Exactly 50 seeds were added to each flask and placed in a G24 Environmental incubator Shaker (NB Scientific Co., Inc., Woodbridge Township, NJ, USA) at 150 rpm and 20 °C. After 24 h, flask contents were poured through a sieve screen and seeds were dabbed dry. Control seeds were not preconditioned at all.

4.3. Growing Conditions

Seeds were grown in a greenhouse starting 11 January 2022 in 10 cm pots using Promix (Halifax Seed, Halifax, NS, Canada) as substrate then moved to a greenhouse. There were three trays in different locations within the greenhouse that served as experimental blocks. Each tray was randomly assigned six pots of each preconditioning treatment, for a total of 24 pots per tray. Pots were randomly arranged on each tray.

All seedlings were grown for 8 weeks under the same conditions. The greenhouse was set at 20 °C/15 °C day/night temperatures. All seedlings were watered daily and provided 200 ppm of 20-20-20 fertilizer, 5 ppm of iron chelate, and 2 ppm of micro-chelate (all from Halifax Seed, Halifax, NS, Canada) once per week for the first 8 weeks. However, after the 8th week, half the pots were deprived of water for 1 week to simulate drought stress. Seedlings in the watered treatment were not provided fertilizer for this final week.

All response variables were measured at the end of the 9th week except soil moisture. Soil moisture was measured daily throughout the experiment using an HH2 Moisture Meter (Delta-T Devices, Cambridge, UK) to ensure all pots were being maintained at similar soil moisture and to evaluate the difference in soil moisture during drought.

4.4. Photosynthetic Measurements

Pn, E, and Gs were measured using an LCA-4 Gas Exchange System (ADC Bioscientific, Hoddesdon, UK). The procedure was modified from [25] so that each measurement was taken three times, once after 30 s, 60 s, and 90 s. Pn, E, and Gs were each reported as the average of those three measurements. WUE was calculated as a ratio of Pn:E. Gas exchange measurements were taken on a cloudy day where light intensity in the greenhouse was 238 ± 5.7 μmol m^{-2} s^{-1} provided via ambient light as determined by 3 measurements per block. The temperature was 20 °C with a relative humidity 60% (vapor pressure deficit = 0.94 kPa).

4.5. Intact Leaf Measurements

Chlorophyll index was measured with a Minolta SPAD 504 meter (Konica Minolta, Ramsey, NJ, USA). The newest fully expanded leaf was placed in the sensor and the instrument measured the transmittance at wavelengths of 650 nm and 940 nm [35]. Results were averaged from five readings from each replicate.

Leaf area was measured using ImageJ processing and analysis software (National Institutes of Health, Bethesda, MD, USA). A scaled white paper backboard was placed around seedling stems at soil level and overhead photos were taken of each seedling. Each photo had a scale set before conversion into a binary image. The entire overhead leaf canopy was selected using the flood fill tool to determine area.

4.6. Detached Leaf Measurements

Relative water content (RWC) was measured in all seedlings at the end of the 9th week. One leaf (approximately 0.6 g) was cut from each seedling, weighed for fresh mass (M_f), and then placed in deionized water to reach full turgidity. After 24 h, the leaves were removed from water, surface moisture was removed by dabbing with tissue paper, and leaves were weighed for turgid mass (M_t). Leaves were then dried at 90 °C for 24 h and then weighed again for dry mass (M_d). The following calculation from [36] was used:

$$\mathrm{RWC} = \frac{\mathrm{M_f} - \mathrm{M_d}}{\mathrm{M_t} - \mathrm{M_d}} \times 100 \quad (1)$$

MII uses the percentage of electrolyte leaked into solution to quantify membrane integrity [37]. Centrifuge tubes were filled with 30 mL of deionized water and were allowed to adjust to room temperature (25 °C). The electrical conductivity of the deionized water (EC_W) alone was measured using a CDM 2e Conductivity Meter (Bach-Simpson, London, ON, Canada). Afterward, 1 leaf (approximately 0.4 g) was removed from each seedling and completely submerged in a centrifuge tube. The tubes were sealed and left at room temperature for 24 h. Initial conductivity (EC_0) was measured to quantify the electrolytes leached into solution. Sealed tubes were then placed in a forced-air oven for 4 h at 90 °C to kill tissues and then cooled to room temperature. Final conductivity measurements (EC_f) were taken after equilibrating to 25 °C to determine maximum leakage. MII was then calculated using the following formula:

$$\frac{EC_0 - EC_W}{EC_f - EC_W} \times 100 \quad (2)$$

4.7. Biomass Measurements

Entire seedlings were removed from their pots and then separated into shoot and root. Roots were cleaned of soil then roots and shoots were each dried in a hot air over at 80°C for 48 h. Shoots and roots were weighed to determine DM. The leaf DMs from Section 4.6 were added to their respective shoot DM, so that final values were indicative of the entire shoot.

4.8. Statistical Analysis

Data were submitted to Minitab 19 (Minitab LLC., Centre County, PA, USA) to analyze as a general linear model. The model included main and interactive effects of stress level and preconditioning treatments as well as the experimental block. Statistical assumptions of homogeneity, normality, and independence were verified. Multiple means comparison was completed using Tukey's honestly significant difference at 5% significance.

5. Conclusions

These results support our original hypothesis that AsA preconditioning promotes growth and drought stress tolerance in broccoli seedlings. Pn, root DM, shoot DM, and WUE efficiency were increased in watered and drought-stressed seedlings. Root DM, chlorophyll, MII, RWC, and leaf area were all also improved in drought-stressed seedlings. Improved membrane protection association with chlorophyll index and MII combined with maintained RWC are indicative of drought stress tolerance, which contributed to positive effects of Pn and growth.

From a practical perspective, this research represents a viable SPA technology that could be important for industry. AsA is a relatively inexpensive SPA (CAD~0.20/g) and the required concentration of 10 ppm is relatively low. The cost of preconditioning 50 seeds with AsA based on our methodology was only US $0.0005. Yet, AsA preconditioning improved shoot DM and Pn by up to 52% and 83%, respectively, which would have tremendous value for commercial operations. Further research would be required to determine the effect on yield.

From a physiological perspective, this research presents further evidence of additional roles of AsA in plant physiology. Exogenous application of AsA to seeds has induced several positive effects in developing plants. It seems likely that specific proteins are triggered within the developing broccoli seedlings that protect chloroplasts, membranes, and photosynthesis. AsA represents the impetus for these physiological changes, but there are several other steps that remain unknown to understand the entire physiological mechanism.

Author Contributions: Conceptualization, M.T.M., R.K. and R.J.; methodology, M.T.M., R.K. and R.J.; software, M.T.M., R.K. and R.J.; validation, M.T.M.; formal analysis, M.T.M.; investigation, M.T.M., R.K. and R.J.; resources, M.T.M.; data curation, M.T.M.; writing—original draft preparation, M.T.M.; writing—review and editing, M.T.M., R.K. and R.J.; visualization, M.T.M.; supervision, M.T.M.; project administration, M.T.M. All authors have read and agreed to the published version of the manuscript.

Funding: This project received no external funding.

Data Availability Statement: Not applicable.

Acknowledgments: The authors gratefully acknowledge Krista MacLeod and the greenhouse staff in the Department of Plant, Food, and Environmental Sciences for use of greenhouse space and regular maintenance of plants. The authors also thank Paul Manning for his kind review of this manuscript.

Conflicts of Interest: The authors declare no conflict of interest.

References

1. Diaz-Perez, J.C. Root Zone Temperature, Plant Growth and Yield of Broccoli [*Brassica oleracea* (Plenck) var. *italica* as Affected by Film Mulches. *Sci. Hortic.* **2009**, *123*, 156–163. [CrossRef]
2. Ilahy, R.; Tlili, I.; Pék, Z.; Montefusco, A.; Siddiqui, M.W.; Homa, F.; Hdider, C.; R'Him, T.; Lajos, H.; Lenucci, M.S. Pre- and Post-Harvest Factors Affecting Glucosinolate Content in Broccoli. *Front. Nutr.* **2020**, *7*, 147. [CrossRef] [PubMed]
3. Fahey, J.W. Brassica: Characteristics and Properties. In *Encyclopedia of Food and Health*; Elsevier Inc.: Amsterdam, The Netherlands, 2015; pp. 469–477.
4. Osman, A.S.; Wahed, M.H.A.; Rady, M.M. Ascorbic Acid Improves Productivity, Physio Biochemical Attributes and Antioxidant Activity of Deficit Irrigated Broccoli Plants. *Biomed. J. Sci. Tech. Res.* **2018**, *11*, 8196–8205. [CrossRef]
5. Beacham, A.M.; Hand, P.; Pink, D.A.; Monaghan, J.M. Analysis of Brassica Oleracea Early Stage Abiotic Stress Responses Reveals Tolerance in Multiple Crop Types and for Multiple Sources of Stress. *J. Sci. Food Agric.* **2017**, *97*, 5271–5277. [CrossRef]
6. Awasthi, R.; Kaushal, N.; Vadez, V.; Turner, N.C.; Berger, J.; Siddique, K.H.M.; Nayyar, H. Individual and Combined Effects of Transient Drought and Heat Stress on Carbon Assimilation and Seed Filling in Chickpea. *Funct. Plant Biol. FPB* **2014**, *41*, 1148–1167. [CrossRef]
7. Seki, M.; Umezawa, T.; Urano, K.; Shinozaki, K. Regulatory Metabolic Networks in Drought Stress Responses. *Curr. Opin. Plant Biol.* **2007**, *10*, 296–302. [CrossRef]
8. Vadez, V.; Berger, J.D.; Warkentin, T.; Asseng, S.; Ratnakumar, P.; Rao, K.P.C.; Gaur, P.M.; Munier-Jolain, N.; Larmure, A.; Voisin, A.-S.; et al. Adaptation of Grain Legumes to Climate Change: A Review. *Agron. Sustain. Dev.* **2012**, *32*, 31–44. [CrossRef]
9. Galmés, J.; Medrano, H.; Flexas, J. Photosynthetic Limitations in Response to Water Stress and Recovery in Mediterranean Plants with Different Growth Forms. *New Phytol.* **2007**, *175*, 81–93. [CrossRef]
10. Seminario, A.; Song, L.; Zulet, A.; Nguyen, H.T.; González, E.M.; Larrainzar, E. Drought Stress Causes a Reduction in the Biosynthesis of Ascorbic Acid in Soybean Plants. *Front. Plant Sci.* **2017**, *8*, 1042. [CrossRef]
11. Reddy, A.R.; Chaitanya, K.V.; Vivekanandan, M. Drought-Induced Responses of Photosynthesis and Antioxidant Metabolism in Higher Plants. *J. Plant Physiol.* **2004**, *161*, 1189–1202. [CrossRef]
12. Liu, H.; Sultan, M.A.R.F.; Liu, X.L.; Zhang, J.; Yu, F.; Zhao, H.X. Physiological and Comparative Proteomic Analysis Reveals Different Drought Responses in Roots and Leaves of Drought-Tolerant Wild Wheat (*Triticum boeoticum*). *PLoS ONE* **2015**, *10*, e0121852. [CrossRef] [PubMed]
13. Abid, M.; Ali, S.; Qi, L.K.; Zahoor, R.; Tian, Z.; Jiang, D.; Snider, J.L.; Dai, T. Physiological and Biochemical Changes during Drought and Recovery Periods at Tillering and Jointing Stages in Wheat (*Triticum aestivum* L.). *Sci. Rep.* **2018**, *8*, 4615. [CrossRef]
14. Bhusal, N.; Han, S.-G.; Yoon, T.-M. Impact of Drought Stress on Photosynthetic Response, Leaf Water Potential, and Stem Sap Flow in Two Cultivars of Bi-Leader Apple Trees (*Malus domestica* Borkh.). *Sci. Hortic.* **2019**, *246*, 535–543. [CrossRef]
15. Bhusal, N.; Lee, M.; Lee, H.; Adhikari, A.; Han, A.R.; Han, A.; Kim, H.S. Evaluation of Morphological, Physiological, and Biochemical Traits for Assessing Drought Resistance in Eleven Tree Species. *Sci. Total Environ.* **2021**, *779*, 146466. [CrossRef] [PubMed]
16. Chauhan, J.S.; Tyagi, M.K.; Kumar, A.; Nashaat, N.I.; Singh, M.; Singh, N.B.; Jakhar, M.L.; Welham, S.J. Drought Effects on Yield and Its Components in Indian Mustard (*Brassica juncea* L.). *Plant Breed.* **2007**, *126*, 399–402. [CrossRef]
17. Sabagh, A.E.; Hossain, A.; Barutcular, C.; Islam, M.S.; Ratnasekera, D.; Kumar, N.; Meena, R.S.; Gharib, H.S.; Saneoka, H.; da Silva, J.A.T. Drought and Salinity Stress Management for Higher and Sustainable Canola (*'Brassica Napus'* L.) Production: A Critical Review. *Aust. J. Crop Sci.* **2019**, *13*, 88–96. [CrossRef]
18. IBU Repository. Effects of Water Stress on Yield and Some Quality Parameters of Broccoli. Available online: https://eprints.ibu.edu.ba/items/show/3026 (accessed on 3 May 2022).
19. Lesk, C.; Rowhani, P.; Ramankutty, N. Influence of Extreme Weather Disasters on Global Crop Production. *Nature* **2016**, *529*, 84–87. [CrossRef]

20. Kim, W.; Iizumi, T.; Nishimori, M. Global Patterns of Crop Production Losses Associated with Droughts from 1983 to 2009. *J. Appl. Meteorol. Climatol.* **2019**, *58*, 1233–1244. [CrossRef]
21. Rajasekaran, B. New Plant Growth Regulators Protect Photosynthesis and Enhance Growth Under Drought of Jack Pine Seedlings. *J. Plant Growth Regul.* **1999**, *18*, 175–181. [CrossRef]
22. Lada, R.R.; Stiles, A.; Blake, T.J. The Effects of Natural and Synthetic Seed Preconditioning Agents (SPAs) in Hastening Seedling Emergence and Enhancing Yield and Quality of Processing Carrots. *Sci. Hortic.* **2005**, *106*, 25–37. [CrossRef]
23. MacDonald, M.T.; Lada, R.R.; Robinson, A.R.; Hoyle, J. The Benefits of Ambiol® in Promoting Germination, Growth, and Drought Tolerance Can Be Passed on to Next-Generation Tomato Seedlings. *J. Plant Growth Regul.* **2010**, *29*, 357–365. [CrossRef]
24. Simma, B.; Polthanee, A.; Goggi, A.S.; Siri, B.; Promkhambut, A.; Caragea, P.C. Wood Vinegar Seed Priming Improves Yield and Suppresses Weeds in Dryland Direct-Seeding Rice under Rainfed Production. *Agron. Sustain. Dev.* **2017**, *37*, 56. [CrossRef]
25. MacDonald, M.T.; Lada, R.R.; Robinson, A.R.; Hoyle, J. Seed Preconditioning with Natural and Synthetic Antioxidants Induces Drought Tolerance in Tomato Seedlings. *HortScience* **2009**, *44*, 1323–1329. [CrossRef]
26. Shah, T.; Latif, S.; Khan, H.; Munsif, F.; Nie, L. Ascorbic Acid Priming Enhances Seed Germination and Seedling Growth of Winter Wheat under Low Temperature Due to Late Sowing in Pakistan. *Agronomy* **2019**, *9*, 757. [CrossRef]
27. Anwar, S.; Iqbal, M.; Raza, S.H.; Iqbal, N. Efficacy of Seed Preconditioning with Salicylic and Ascorbic Acid in Increasing Vigor of Ruce (*Oryza sativa* L.) Seedling. *Pak. J. Bot.* **2013**, *45*, 157–162.
28. Bhusal, N.; Lee, M.; Reum Han, A.; Han, A.; Kim, H.S. Responses to Drought Stress in Prunus Sargentii and Larix Kaempferi Seedlings Using Morphological and Physiological Parameters. *For. Ecol. Manag.* **2020**, *465*, 118099. [CrossRef]
29. Jo, H.; Chang, H.; An, J.; Cho, M.S.; Son, Y. Species Specific Physiological Responses of Pinus Densiflora and Larix Kaempferi Seedlings to Open-Field Experimental Warming and Precipitation Manipulation. *For. Sci. Technol.* **2019**, *15*, 44–50. [CrossRef]
30. Chen, Z.; Cao, X.; Niu, J. Effects of Exogenous Ascorbic Acid on Seed Germination and Seedling Salt-Tolerance of Alfalfa. *PLoS ONE* **2021**, *16*, e0250926. [CrossRef] [PubMed]
31. MacDonald, M.T.; Lada, R.R.; Veitch, R.S.; Thiagarajan, A.; Adams, A.D. Postharvest Needle Abscission Resistance of Balsam Fir (*Abies balsamea*) is Modified by Harvest Date. *Can. J. For. Res.* **2014**, *44*, 1394–1401. [CrossRef]
32. Jamil, S.; Ali, Q.; Iqbal, M.; Javed, M.T.; Iftikhar, W.; Shahzad, F.; Perveen, R. Modulations in Plant Water Relations and Tissue-Specific Osmoregulation by Foliar-Applied Ascorbic Acid and the Induction of Salt Tolerance in Maize Plants. *Braz. J. Bot.* **2015**, *38*, 527–538. [CrossRef]
33. Habib, N.; Ali, Q.; Ali, S.; Javed, M.T.; Zulqurnain Haider, M.; Perveen, R.; Shahid, M.R.; Rizwan, M.; Abdel-Daim, M.M.; Elkelish, A.; et al. Use of Nitric Oxide and Hydrogen Peroxide for Better Yield of Wheat (*Triticum aestivum* L.) under Water Deficit Conditions: Growth, Osmoregulation, and Antioxidative Defense Mechanism. *Plants* **2020**, *9*, 285. [CrossRef]
34. Nephali, L.; Moodley, V.; Piater, L.; Steenkamp, P.; Buthelezi, N.; Dubery, I.; Burgess, K.; Huyser, J.; Tugizimana, F. A Metabolomic Landscape of Maize Plants Treated with a Microbial Biostimulant Under Well-Watered and Drought Conditions. *Front. Plant Sci.* **2021**, *12*, 676632. [CrossRef] [PubMed]
35. Martínez, D.E.; Guiamet, J.J. Distortion of the SPAD 502 Chlorophyll Meter Readings by Changes in Irradiance and Leaf Water Status. *Agronomie* **2004**, *24*, 41–46. [CrossRef]
36. MacDonald, M.T.; Lada, R.R. Biophysical and Hormonal Changes Linked to Postharvest Needle Abscission in Balsam Fir. *J. Plant Growth Regul.* **2014**, *33*, 602–611. [CrossRef]
37. Odlum, K.D.; Blake, T.J. A Comparison of Analytical Approaches for Assessing Freezing Damage in Black Spruce Using Electrolyte Leakage Methods. *Can. J. Bot.* **1996**, *74*, 952–958. [CrossRef]

MDPI
St. Alban-Anlage 66
4052 Basel
Switzerland
www.mdpi.com

Plants Editorial Office
E-mail: plants@mdpi.com
www.mdpi.com/journal/plants

www.ingramcontent.com/pod-product-compliance
Lightning Source LLC
LaVergne TN
LVHW080502180726
843515LV00016B/893

* 9 7 8 3 0 3 6 5 9 4 7 1 2 *